INDUSTRIAL ROBOTS

VOLUME I / FUNDAMENTALS

SECOND EDITION

William R. Tanner Editor

Published by:

Robotics International of SME
Society of Manufacturing Engineers
Marketing Services Department
One SME Drive
P.O. Box 930
Dearborn, Michigan 48128

INDUSTRIAL ROBOTS

VOLUME I / FUNDAMENTALS

Copyright 1981 by the
Society of Manufacturing Engineers
Dearborn, Michigan 48121

Second Edition

Third Printing: November, 1983

Library of Congress Catalog Card Number: 81-51986

International Standard Book Number: 0-87263-070-6

Manufactured in the United States of America

SME wishes to express its acknowledgement and appreciation to the following publications for supplying the various articles reprinted within the contents of this book.

Atlanta Economic Review
School of Business Administration
Georgia State University
University Plaza
Atlanta, Georgia 30303

Automation
(Production Engineering)
A Penton/IPC Publication
Penton Plaza
Cleveland, Ohio 44114

Canadian Machinery and Metalworking
Maclean-Hunter Ltd.
Business Publications Circulation Department
481 University Avenue
Toronto, Ontario M5W 1A7
Canada

Foundry Management and Technology
A Penton/IPC Publication
Penton Plaza
Cleveland, Ohio 44114

IEEE Computer
IEEE
345 East 47th Street
New York, New York 10017

Machine and Tool Blue Book
Hitchcock Publishing Company
Hitchcock Building
Wheaton, Illinois 60187

Management Review
American Management Associations
135 West 50th Street
New York, New York 10020

Manufacturing Engineering
Society of Manufacturing Engineers
One SME Drive
P.O. Box 930
Dearborn, Michigan 48128

National Safety News
National Safety Council
444 North Michigan Avenue
Chicago, Illinois 60611

Robotics Today
Society of Manufacturing Engineers
One SME Drive
P.O. Box 930
Dearborn, Michigan 48128

Tooling & Production
Huebner Publications, Inc.
5821 Harper Road
Solon, Ohio 44139

Grateful appreciation is also expressed to:

International Fluidics Services Ltd.
35-39 High Street
Kempston, Bedford MK42 7BT
England

PREFACE

Developed in the early 1960's, industrial robots experienced limited application during their first decade.

Within the last five years, however, industrial robot applications have shown a steady growth in numbers and in diversity of uses.

Industrial robots are available in a wide range of configurations and capabilities from simple "pick-and-place" devices to computer-directed, servo-controlled, point-to-point and continuous path units. Programmable controls, memory systems and up to seven articulations provide industrial robots with a high degree of flexibility and adaptability to many tasks involving manipulation of objects and tools.

Their growth pattern will prevail for the foreseeable future, particularly with the increased capabilities of the robots themselves.

For example, robots are now being used effectively in such mass production applications as painting and welding. They're also being used in applications involving environmental conditions too hazardous for human beings.

And they're getting smarter. At MIT and Stanford, scientists are successfully developing robots capable of making decisions. This is the new science of Artificial Intelligence—a science that promises to revolutionize industry and make the dream of the automatic factory a reality.

With such a range of capabilities, the intelligent applications of robots in industrial operations requires careful evaluation of both available equipment and intended usage. This book provides a beginning for that evaluation.

Since there is so much to be said about industrial robots, the Society of Manufacturing Engineers is presenting a series of two books on industrial robots. Volume II is devoted to specific applications and includes many actual examples of robot usage in a number of industrial areas. Volume I is concerned with fundamentals.

Chapter One, Basics, presents general characteristics and classifications of robots and discusses robotic trends. The book's second chapter, Implementation, includes a user's guide to planning robot applications. The chapter also presents two papers discussing safety.

Chapter Three, Justification, outlines some of the numerous factors cited for justification of the use of robots and details an economic analysis of robot applications. Human Factors is the title of Chapter Four. The chapter discusses resistance to industrial robots and includes an industrial robot public relations checklist. The chapter also includes a paper presenting the view of the United Automobile, Aerospace and Agricultural Implement Workers of America (UAW) on robots.

In the fifth chapter, Capabilities, an overview of robot applications and programmable automation is explored. Chapter Six, Application, briefly explains some of the many applications of today's industrial robot.

The book's final chapter, Advanced Technology, outlines many of the rapidly developing areas of robotics including sensory feedback and adaptive control.

Each installment is written by an expert in the field. The directory of manufacturers, distributors and robotics organizations has been updated and will be helpful sources of additional information.

I wish to express my gratitude to the authors whose works appear in this publication. I also extend my gratitude to the publications which were very generous in supplying some of the material you will find on the pages of this volume. These publications include: *Atlanta Economic Review, Automation, Canadian Machinery and Metalworking, Foundry Management and Technology, IEEE Computer, Machine and Tool Blue Book, Management Review, Manufacturing Engineering, National Safety News, Robotics Today* and *Tooling & Production.* My appreciaiton is also extended to the International Fluidics Services.

Finally, my thanks to Bob King and Judy Stranahan of the SME Marketing Services Department for their help in preparing this book.

William R. Tanner
President
Productivity Systems Inc.
Editor

The informative volumes of the Manufacturing Update Series are part of the Society of Manufacturing Engineers' effort to keep its Members better informed on the latest trends and developments in engineering.

With 50,000 members, SME provides a common ground for engineers and managers to share ideas, information and accomplishments.

An overwhelming mass of available information requires engineers to be concerned about keeping up-to-date. In other words, continuing education. SME Members can take advantage of numerous opportunities, in addition to the books of the Manufacturing Update Series, to fulfill their continuing educational goals. These opportunities include:

• Chapter programs through the over 200 chapters which provide SME Members with a foundation for involvement and participation.

• Educational programs including seminars, clinics, programmed learning courses and videotapes.

• Conferences and expositions which enable engineers to see, compare, and consider the newest manufacturing equipment and technology.

• Publications including Manufacturing Engineering, the SME Newsletter, Technical Digest and a wide variety of books including the Tool and Manufacturing Engineers Handbook.

• SME's Manufacturing Engineering Certification Institute formally recognizes manufacturing engineers and technologists for their technical expertise and knowledge acquired through years of experience.

In addition, the Society works continuously with the American National Standards Institute, the International Standards Organization and other organizations to establish the highest possible standards in the field.

SME Members have discovered that their membership broadens their knowledge throughout their career.

In a very real sense, it makes SME the leader in disseminating and publishing technical information for the manufacturing engineer.

SME

TABLE OF CONTENTS

CHAPTERS

TABLE OF CONTENTS (Cont.)

3 JUSTIFICATION

4 HUMAN FACTORS

THREE LAWS FOR ROBOTCISTS: AN APPROACH TO OVERCOMING WORKER AND MANAGEMENT

5 CAPABILITIES

TABLE OF CONTENTS (Cont.)

6 APPLICATION

7 ADVANCED TECHNOLOGY

CHAPTER 1

BASICS

Commentary

The Robot Institute of America (RIA) defines a robot as "a reprogrammable multi-functional manipulator designed to move material, parts, tools, or specialized devices through variable programmed motions for the performance of a variety of tasks."

This chapter expands upon that definition and classifies robots by mechanical configuration and type of control. Configurations, or coordinate systems, include cartesian or rectilinear, cylindrical, spherical or polar and jointed-arm. Controls include non-servo, point-to-point servo and continuous-path servo. The capabilities and limitations, advantages and disadvantages of the configurations and controls are explored.

The significant features of currently available robots are summarized and some applications are exemplified.

Basics Of Robotics

By William R. Tanner
Tanner Associates

Industrial robots exist in a wide range of capabilities and configurations. However, they commonly consist of several similar major components: the manipulator, the controller and the power supply. These components are described in basic terms. The most common robot configurations are related to the coordinate systems in which they function: cylindrical, spher- ical or jointed-spherical. Significant features and common characteristics of non-servo and servo-controlled robots are explored in depth. Interfacing of robots and associated equip- ment for coordinated interaction is described. Some future trends in robot development are discussed.

GENERAL CHARACTERISTICS

Industrial robots are available in a wide range of capabilities and configurations. Basically, however, they consist of several major components: the manipulator or "mechanical unit" which actually performs the manipulative functions; the controller or "brain" which stores data and directs the movement of the manipulator and the power supply which provides energy to the manipulator.

The manipulator is a series of mechanical linkages and joints capable of movement in various directions to perform the work of the robot. These mechanisms are driven by actuators which may be pneumatic or hydraulic cylinders, hydraulic rotary actu- ators or electric motors. The actuators may be coupled directly to the mechanical links or joints or may drive indirectly through gears, chains or ball screws. In the case of pneumatic or hydraulic drives, the flow of air or oil to the actuators is controlled by valves mounted on the manipulator.

Feedback devices are installed to sense the positions of the various links and joints and transmit this information to the controller. These feedback devices may be simply limit switches actuated by the robot's arm or position measuring devices such as encoders, potentiometers or resolvers and/or tachometers to measure speed. Depending on the devices used, the feedback data is either digital or analog.

The controller has a three-fold function: first, to initiate and terminate motions of the manipulator in a desired sequence and at desired points; second, to store position and sequence data in memory; and third, to interface with the "outside world."

Robot controllers run the gamut from simple step sequencers through pneumatic logic systems, diode matrix boards, electronic sequencers and microprocessors to minicomputers. The controller may be either an integral part of the manipulator or housed in a separate cabinet.

The complexity of the controller both determines and is determined by the capabilities of the robot. Simple non-servo devices usually employ some form of step sequencer. Servo-controlled robots use a combination of sequencer and data storage (memory). This may be as simple as an electronic counter, patch board or diode matrix and series of potentiometers or as sophisticated as a minicomputer with core memory. Other memory devices employed include magnetic tape, magnetic disc, plated wire and semiconductor (solid state RAM). Processor or computer based controller operating systems may be hard wired, stored in core memory or programmed in ROM (read only memory).

The controller initiates and terminates the motions of the manipulator through interfaces with the manipulator's control valves and feedback devices and may also perform complex arithmetic functions to control path, speed and position. Another interface with the outside world provides two-way communications between the controller and ancillary devices. This interface allows the manipulator to interact with whatever other equipment is associated with the robot's task.

The function of the power supply is to provide energy to the manipulator's actuators. In the case of electrically driven robots, the power supply functions basically to regulate the incoming electrical energy. Power for pneumatically actuated robots is usually supplied by a remote compressor which may also service other equipment.

Hydraulically actuated robots normally include a hydraulic power supply as either an integral part of the manipulator or as a separate unit. The hydraulic system generally follows straightforward industrial practice and consists of an electric motor driven pump, filter, reservoir and, usually, a heat exchanger (either air or water). These robots normally operate on petroleum based hydraulic fluid; however, most are available with special seals for operation on fire retardant fluid.

Mechanical arrangements of the manipulator are widely varied among the robots available today, as is the terminology used to describe these mechanical components and motions. The most common configurations are best described in relation to their coordinate systems: cylindrical, spherical and jointed-spherical.

The cylindrical coordinate robots include the Pacer, Versatran and Auto-Place. Their configuration consists of a horizontal arm mounted on a vertical column which, in turn, is mounted on a rotating base. The horizontal arm moves in and out; its carriage moves up and down on a vertical column and these two members rotate as a unit on the base. Thus, the working area or envelope is a portion of a cylinder.

The Unimate and Prab are typical of robots having a spherical coordinate system. Their configuration is similar to the turret of a tank. An arm moves in and out, pivots in a vertical plane and rotates in a horizontal plane about the base. The work envelope is a portion of a sphere.

The third coordinate system, jointed-spherical or jointed-arm, is used by the ASEA and Cincinnati Milacron robots. This configuration consists of a base or trunk and an upper arm and forearm which move in a vertical plane through the trunk. An "elbow" joint is located between the forearm and upper arm and a "shoulder" joint is located between the upper arm and the trunk. Rotary motion in a horizontal plane is also provided at the shoulder joint. The work envelope approximates a portion of a sphere.

These members comprise the major axes or "degrees of freedom" of the robot. As many as three additional degrees of freedom are provided at the extremity of the robot arm in a unit commonly called a "wrist." Wrist axes include "roll" (rotation in a plane perpendicular to the end of the arm), "pitch" (rotation in a vertical plane through the arm)and "yaw" (rotation in a horizontal plane through the arm.)

Additional motions may be provided by mounting the robot on a two-axis (X-Y) table or on a track on the floor or overhead. Many of the robots available are "modular" in design. That is, the user may select as few as two or as many as seven or eight degrees of freedom, depending upon his needs.

A mounting surface is provided on the last axis of the wrist for installation of the tool or gripper with which the robot performs its intended task. These devices are usually unique to the robot application and are thus provided by the user. However, several robot manufacturers offer a selection of devices for grasping parts which may be directly applicable or adaptable to the particular task to be performed.

CLASSIFICATION

Robots may be generally classified as non-servo or servo-controlled devices. For purposes of discussion, this is convenient, since each classification has uniquely common characteristics. The servo-controlled class can be further separated into point-to-point and continuous path devices, each with unique characteristics and applications.

NON-SERVO ROBOTS

Non-servo robots are often referred to as "end point," "pick and place," "bang-bang" or "limited sequence" robots. However, these terms imply limited capability and restricted applicability, which is not necessarily the case. The term "non-servo" is more descriptive and less restrictive than the others used.

A typical operating sequence of a hydraulic or pneumatic non-servo robot is as follows:

Upon start of program execution, the sequencer/controller initiates signals to control valves on the manipulator's actuators.

The valves open, admitting air or oil to the actuators and the members begin to move.

The valves remain open and the members continue to move until physically restrained by contact with end stops.

Limit switches signal the end of travel to the controller which then commands the control valves to close.

The sequencer then indexes to the next step and the controller again outputs signals. These may again be to the control valves on the actuators or to an external device such as a gripper.

The process is repeated until the entire sequence of steps has been executed.

The significant features of a non-servo robot are:

The manipulator's various members move until the limits of travel (end stops) are reached. Thus there are usually only two positions for each axis to assume.

The sequencer provides the capability for many motions in a program, but only to the end points of each axis.

Deceleration at the approach to the stops may be provided by valving or shock absorbers.

It is feasible to activate intermediate stops on some axes to provide more than two positions; however, there is a practical limit to the number of such stops which can be installed.

Although this mode of operation is commonly used on the smaller robots, it is applicable to larger units also.

The programmed sequence can be conditionally modified through appropriate external sensors; however, this class of robots usually is restricted to the performance of single programs.

Programming is done by setting up the desired sequence of moves and by adjusting the end stops for each axis.

Common characteristics of non-servo robots include:

Relatively high speed is possible, due to the generally smaller size of the manipulator and full flow of air or oil through the control valves.

Repeatability to within 0.25 mm (0.010 in.) is attainable on the smaller units.

These robots are relatively low in cost; simple to operate, program and maintain; and are highly reliable.

These robots have limited flexibility in terms of program capacity and positioning capability.

Typical non-servo robots manufactured or marketed in the United States include Auto-Place, Kelate, PickOMatic and Prab.

SERVO-CONTROLLED ROBOTS

The second major class of robots are servo-controlled. The typical operating sequence of a servo-controlled robot is as follows:

Upon start of program execution, the controller addresses the memory location of the first command position and also reads the actual position of the various axes as measured by the position feedback system.

These two sets of data are compared and their differences, commonly called "error signals", are amplified and transmitted as "command signals" to servo valves for the actuator of each axis.

The servo valves, operating at constant pressure, control flow to the manipulator's actuators; the flow being proportional to the electrical current level of the command signals.

As the actuators move the manipulator's axes, feedback devices such as encoders, potentiometers, resolvers and tachometers send position data (and, in some cases, velocity) data back to the controller. These "feedback signals" are compared with the desired position data and new error signals are generated, amplified and sent as command signals to the servo valves.

This process continues until the error signals are effectively reduced to zero, whereupon the servo valves reach null, flow to the actuators is blocked and the axes come to rest at the desired position.

The controller then addresses the next memory location and responds appropriately to the data stored there. This may be another positioning sequence for the manipulator or a signal to an external device.

The process is repeated sequentially until the entire set of data, or "program", has been executed.

The significant features of a servo-controlled robot are:

The manipulator's various members can be commanded to move and stop anywhere within their limits of travel, rather than only at the extremes.

Since the servo valves modulate flow, it is feasible to control the velocity, acceleration and deceleration of the various axes as they move between programmed points.

Generally, the memory capacity is large enough to store many more positions than a non-servo robot.

Both continuous path and point-to-point capabilities are possible.

Accuracy can be varied, if desired, by changing the magnitude of the error signal which is considered zero. This can be useful in "rounding the corners" of high-speed contiguous motions.

Drives are usually hydraulic or electric and use state-of-the-art servo control technology.

Programming is accomplished by manually initiating signals to the servo valves to move the various axes into a desired position and then recording the output of the feedback devices into the memory of the controller. This process is repeated for the entire sequence of desired positions in space.

Common characteristics of servo-controlled robots include:

Smooth motions are executed, with control of speed and, in some cases, acceleration and deceleration. This permits the controlled movement of heavy loads.

Maximum flexibility is provided by the ability to program the axes of the manipulator to any position within the limits of their travel.

Most controllers and memory systems permit the storage and execution of more than one program, with random selection of programs from memory via externally generated signals.

With microprocessor or minicomputer based controllers, subroutining and branching capabilities may be available. These capabilities permit the robot to take alternative actions within a program, when commanded.

End-of-arm positioning accuracy of 1.5 mm (.060 in.) and repeatability of ± 1.5 mm (± .060 in.) are generally achieved. Accuracy and repeatability are functions of not only the mechanisms, but also the resolution of the feedback devices, servo valve characteristics, controller accuracy, etc.

Due to their complexity, servo-controlled robots are more expensive and more involved to maintain than non-servo robots and tend to be somewhat less reliable.

POINT-TO-POINT SERVO-CONTROLLED ROBOTS

One subset of the servo-controlled robot class is the point-to-point robot. This is the typical servo-controlled robot which is used in a wide variety of industrial applications for both parts handling and tool handling tasks.

Significant features of the point-to-point servo-controlled robot are:

For those robots employing the "record-playback" method of teaching and operation, initial programming is rela-tively fast and easy; however, modification of pro-grammed positions cannot be readily accomplished during program execution.

Those robots employing sequencer/potentiometer controls tend to be more tedious to program; however, programmed positions can be modified easily during program execu-tion by adjustment of potentiometers.

The path through which the various members of the manipulator move when traveling from point to point is not programmed or directly controlled in some cases and may be different from the path followed during teaching.

Common characteristics of point-to-point servo-controlled robots include:

High capability control systems with random access to multiple programs, subroutines, branches, etc., provide great flexibility to the user.

These robots tend to lie at the upper end of the scale in terms of load capacity and working range

Hydraulic drives are most common, although some robots are available with electric drives.

Typical point-to-point servo-controlled robots manufactured or marketed in the United States include ASEA, Cincinnati Milacron, Pacer, Unimate and Versatran.

CONTINUOUS PATH SERVO-CONTROLLED ROBOTS

The second subset of the servo-controlled robot class is the continuous path robot. Typically, the positioning and feedback principles are as described previously. There are, however, some major differences in control systems and some unique physical features.

The significant features of the continuous path servo-controlled robot are:

During programming and playback, data is sampled on a time base, rather than as discretely determined points in space. The sampling frequency is typically in the range of 60 to 80 Hz.

Due to the high rate of sampling of position data, many spatial positions must be stored in memory. A mass storage system, such as magnetic tape or magnetic disc is generally employed.

During playback, due to the hysteresis of the servo valves and inertia of the manipulator, there is no detectable change in speed from point to point. The result is a smooth continuous motion over a controlled path.

Depending upon the controller and data storage system employed, more than one program may be stored in memory and randomly accessed.

The usual programming method involves physically moving the end of the manipulator's arm through the desired path, with position data automatically sampled and recorded.

Speed of the manipulator during program execution can be varied from the speed at which it was moved during programming by playing back the data at a different rate than that used when recording.

Continuous path servo-controlled robots share the following characteristics:

These robots generally are of smaller size and lighter weight than point-to-point robots.

Higher end-of-arm speeds are possible than with point-to-point robots; however, load capacities are usually less than 10 kg (22 lbs.).

Their common applications are to spray painting and similar spraying operations, polishing, grinding and arc welding.

Typical continuous path robots manufactured or marketed in the United States include ASEA, Binks, Retab, Trallfa and Versatran.

INTERFACING

Every application requires that the robot interact with some-thing in the execution of its programmed task. Even a simple part transfer operation cannot be successfully accomplished without a part available for the robot to handle or until the robot has been signaled that the part is present. Interfacing the robot and related equipment involves the transmission of information in two directions. A common robot application, spotwelding of an automobile body, will serve to illustrate the extent to which interfacing may be required.

As the automobile enters the robot work station, its body style is determined and this information is sent to the robot for selection of the proper program to be executed. When the body is in place in the work station, this condition is sent to the robot so that it can begin to work. When the robot has manip-ulated the welding gun into its first programmed position, it sends a signal to the weld control to initiate the spotwelding process. When the spotweld is made, the weld control sends a signal back to the robot. The robot then manipulates the spot-weld gun to the next programmed position and again signals the weld control. This process continues until the welding has been completed. When the robot has then moved clear of the body, it sends a signal to the conveyor and the automobile is moved out of the work station.

Similarly, branching can be initiated during the execution of a program by transmission of the appropriate signal to the robot. In a transfer operation, for example, if the location into which the robot is to place a part is already occupied, a signal may be used to interrupt the normal program and branch to a program which directs the robot to an alternative location to dispose of the part.

Moving line operations, either those involving multiple axis line tracking or pacing of the line with the robot on a transporting device, require the interfacing of the robot and the line. In this case, a resolver/tachometer or encoder feedback device driven by the line provides a continuous flow of position and velocity information to the robot. By this means, the robot maintains synchronization with the work while it moves along the line.

Another area of interfacing is the provision of sensory feedback to the robot. One common application is in destacking material. Here, a tactile sensor is mounted on the tool with which the robot handles the parts. The robot is programmed to advance its arm in the direction of the stack, with the programmed stopping point beyond the last piece in the stack. When the sensor contacts a part, its signal interrupts the program, stops the advance of the arm and activates the tool. The program then continues on the next step, removing the part from the stack. This type of sensory feedback is applicable to both non-servo and servo-controlled robots. Other types of sensors, such as proximity detectors, force feedback and vision systems may also be applied to robots in a similar manner.

CONCLUSION

As described here, all robots share some common characteristics. However, they vary greatly in complexity and capability. The complexity of a robot bears a direct relationship to its total ability; the more universal its applicability, the more complex a robot becomes. The trend of robotics is toward greater capability and flexibility and, as a result, greater complexity.

Centrallized control of groups of robots by large computers, perhaps integrated into total manufacturing systems, is under investigation. Greater application of sensory feedback, particularly through the use of low cost vision systems is a reality today. Off-line programming by means of computers which may include interactive graphics systems is being explored.

While all of these developments will enhance its capabilities and apparent complexity, the robot will remain a basically simple device. Mechanical designs will follow conventional practices; servo control likewise will be conventional and controllers will utilize state of the art technology. Thus, the basics of robotics, once mastered, will not become obsolete.

Reprinted from Canadian Machinery and Metalworking, January, 1980

Omens pointing to a boom?

The time may soon arrive to put a robot in *your* shop

By Jake Koekebakker, Associate Editor

Industrial robot salesmen have pursued potential customers for nearly 15 years with nothing less than missionary zeal. Robots, so they proclaimed, were the ideal answer to the problems of a manufacturing industry beset by rising labor costs, challenges by environmental and occupational health and safety authorities, and pressures for improved productivity.

Even though a modest but steadily growing number of robot applications appeared to substantiate this claim (there were some unmitigated disasters too, to be sure), the robot market boom, which the robot makers in their enthusiasm fully expected, failed to materialize. Industry, it seems, just wasn't buying the concept.

Until recently. For the omens appear to signal a turn of the tide. Nobody seems as yet quite certain of the reasons for or the extent of the marked upswing in robot sales that has been observed, but that it is there, of that there is no longer any doubt. The robot's time may have finally come.

Omen: On October 10, 1979, Unimation Inc., Danbury, Conn., until now the largest of the robot manufacturers and one of the first in the field, announced a record order for 100 Unimate robots by Chrysler Corp. It was, said Unimation, the largest sale of robots in U.S. automotive history. Although in its announcement Unimation didn't specify the dollar value, it was believed to be of the order of $6 million.

At about the same time, Cincinnati Milacron Inc., a much more recent entrant in the robot field, but coming up fast, announced an even larger, $8-million order for its T3 "Tomorrow Tool" industrial robot by Volvo Car Corp. of Sweden.

Between them, the two deals alone almost beat total automotive industry robot sales predicted for 1979 only eight months earlier by Frost & Sullivan Inc., New York-based international market research organization. Its figure was $15 million.

Omen: The same Frost & Sullivan market study projected total industrial robot sales for 1979 at close to $80 million. Even if it didn't underestimate sales in other industry sectors as it almost certainly did in the automotive field, the probable outcome would still be pretty respectable for an industry that did, according to F&S, no more than $26 million of robot business in 1977.

And looking further ahead, F&S may be seeing pies in the sky, but by 1985, it says, robot makers should be booking $438 million. Solid growth, Frost & Sullivan researchers call it. The robot industry, used to years of scraping by, wouldn't argue with that.

Omen: Not just one, but at least two prestigious market research organizations felt confident enough about what's happening to stick their neck out. A market study by International Resource Development Inc., Norwalk, Conn., makes far more conservative predictions than F&S, but they, too, would indicate an end to the slow and painful growth of the robot market throughout the late Sixties and early Seventies. In 1979, IRD says, robot sales should total $40 million, doubling to $82 million by 1984 and again nearly doubling to $140 million by 1989 (these, like Frost & Sullivan's, are 1979 current dollars).

Omen: In the summer of 1979, the Society of Manufacturing Engineers (if any organization in North America has been a hotbed of robot buffs, it has to be SME) published a list of robot manufacturers containing 27 names. Only five months later, five new names showed up at the RO-BOTS IV Conference and Exhibition in Detroit. They may have been missed before; if so, they sure weren't hiding their light any longer. They were selling robots. No telling who else might be getting in at any time from here on.

One of the reasons why industry now seems more willing to open its doors to robots than before may be that robots have become better— much better. That fact could be slowly sinking in and superseding a lingering bad taste left by the earliest

About this article

CM&M associate editor Jake Koekebakker first became interested in industrial robots when doing the research for an article on materials handling, Quick response in materials handling. Quick realization that robot technology encompasses much more than only handling, and could, in fact, be one of the most important production tools of the future, prompted Koekebakker to start a permanent file on the subject. Last summer, he attended a three-day seminar on "Robots in Manufacturing," organized by the Society of Manufacturing Engineers and led by William R. Tanner, one of America's leading authorities on robotics. CM&M's associate editor came away with an SME "Certificate of Completion" and much new learning to add to previously gained insights. This feature is the result. Says Koekebakker: "Once you get as deeply involved in a subject as I did in robots, it becomes difficult to know what sources to credit with respect to what you write. I can only say that none of it is of my own invention. It certainly would not have been possible writing this article without the input from Tanner, SME, and Ken Adams of A. F. Mundy Associates, as well as many others I talked to. As for any shortcomings, they are mine."

robot generations, which, most industry observers agree, just didn't work all that well.

In addition, technological capabilities are now available that were still unheard of 10 years ago—particularly in control technology and programmability. Robot makers have discovered electronic logic and computer-type software. They have given some robots unprecedented adaptability to a wide variety of increasingly complex tasks.

Another reason for the increasing popularity of robots is that while labor and other production costs have continued to soar, robot manufacturers have, by and large, held the line on prices.

Whatever the reasons, the 1980s may well be the Age of the Robot. More and more plant managers will almost certainly have to be seriously considering robots as a means of improving productivity. And in doing so, they may find themselves largely on their own. For, even though robots are already being used in numerous applications, each individual situation is, in some respects, unique. Being a relatively new technology, robotics can as yet claim little in-depth experience. What there is of it is often jealously guarded by those that have it in the belief it gives them a competitive edge. They don't want to share that with anybody else.

It's not unlikely, therefore, that in robotics applications the wheel is being reinvented many times over.

The other thing the potential or neophyte robot user has to contend with is a briskly competitive atmosphere in the robot industry itself. Fact is, few robots can do all things, and even less do all things equally as well. Robot salesmen, in their understandable fervour, have been known to conveniently ignore this occasionally (it's nothing new—it happens in any line of business). There are definitely optimum choices to be made in a field that may now include over 30 makes of machines. Many of these are aimed at specific types of jobs. It pays to know which is which.

The word is, beware, learn all you can, and keep your feet on the ground. Which is what this Survey is all about.

It is based on the premise that many plant managers today no longer need to be convinced of the weird and wonderful miracles robots can perform for them. What they need to know is: Can I solve *my* problem by means of a robot and, if so, which one?

This survey is intended to help you find some of the initial answers and identify some possible solutions.

Let's assume you have a specific problem. For example, you want to load a part into a machine tool and, following the machining cycle, take it out again. Whatever the reason, you want to automate that operation.

You have heard robots can do that sort of thing. Yes, but almost every machine loading and unloading operation is different from any other. In fact, for every machine loading and unloading operation robots *can* do, there probably is one they can *not* do.

So, before getting completely stuck on the idea of getting a robot for the job, ask yourself: Are there any other ways to automate it? Keep your options open, and start from square one.

Square one is studying the job. What is it *really* that goes on there? How is it done? Why is it done that way? Could it be done any other way? And where are the potential glitches? Don't just find out what the likely problems are; analyze all that could *possibly* go wrong.

You've got to know that job better

than the operator himself. The more you study it the better. Sit in a corner and watch for a shift or two, if you can get away with it without driving the operator crazy.

At this point look for all the reasons why you should *not* get a robot.

Here's one of the most common reasons why you shouldn't. Your operator, like thousands, nay, hundreds of thousands like him, has got a tote box of parts on one side, and he starts the cycle by picking one part out of the box, and . . . hold it! He did what? Pick one out of a box-full of parts?

A robot can't do that—at least none you can buy today. Robots have no way of telling where the next part is unless it is in the exact same spot as the previous one, or another, precisely predefined spot. Nor can they pick up parts unless these are in exactly the same orientation. You don't want the robot to put the part into the machine backwards, do you? The robot wouldn't know where to reach or how to orientate its gripper, unless you want to install a lot of fancy electro-optical gear, and even if you could afford that, the results would be dubious with today's off-the-shelf technology. They are working on it, but as off today it's not on.

The only way to get around the problem is to present the parts in an orderly way—say, neatly stacked in rows on a pallet (not even all robots can be programmed to handle that). But how are you going to accomplish it, looking back up the line? It could be a real headache. Experience in the field, in fact, indicates that 50% of all the cases where it was thought a robot could be applied turn out to be unworkable due to lack of parts orientation.

All this goes to show that robots are no simple answer. And it demon-

The robot market into the '80s

$ thousands

	1977	1979	1981	1983	1985
Electrical Machinery	3,406	15,840	41,108	58,400	163,812
Automotive	1,924	14,880	21,156	32,120	53,874
Fabricated Metals ..	11,622	16,480	26,144	56,648	67,014
Electronics	78	1,600	11,696	12,264	70,080
Heavy Machinery ...	5,512	12,240	18,576	23,944	12,702
Others	3,458	18,960	53,320	108,624	70,518
Total	26,000	80,000	17,200	292,000	438,000

Source: Frost & Sullivan Inc., New York

The more you are in charge of your own engineering, the better

strates something else, and that is that you've got to think systems. Robots never work in isolation. What you are looking for is a system component, and it's got to be compatible with the rest of the shooting match.

OK, suppose you've figured all the angles, racked your brain, cursed the boss (and, heaven forbid, your wife), and there's nothing for it, a robot seems to be the answer. Now what do you do?

For one thing, get rid of the idea of getting "a robot." You can get 40 robots, maybe more, all of them different. Some, in fact, so different you wonder why they still call them robots.

What you have to do is to start two or more or less concurrent processes. One is to study robots, and the other to engineer your application. The former you can start doing by referring to the list of robots in this Survey, and the latter by reading this: If you are not sure you can handle your own engineering, by all means get outside help, preferably from a qualified consultant (robot systems

Parts must be presented to a robot in an orderly way. Here, ASEA robot picks workpieces off index-ing table for deburring operation

consultants are still a rare breed, but there are some). But *don't* leave it to a robot manufacturer. By all means, treat the manufacturer's salesman with respect, and listen to what he has to say, for he knows his robot, but listen with critical reservation. For, after all, he's there to sell robots—his. The more you are in charge of your own engineering, the better. No-body knows your plant and the job in question as well as you do—if you've done the homework you should have done anyway.

Don't forget, there may be other robot users that could give you valuable tips. Ask the manufacturer's rep about them. He *may* tell you. If not, wonder why not.

What happens next is that you start to match your engineering requirements against a matrix of comparative robot specifications, such as the one presented in this Survey. It will give you an idea of what's available that is compatible with your needs and with your financial resources.

Any generalized description of the major functional parameters of industrial robots tends, as with any equipment, to be limiting when applied to the individual machine. Various robot makers have their own preferred ways of describing their products. The following should only be seen as an attempt to provide the potential user with a set of descriptive terms on the basis of which a comparative evaluation can be made over the broadest possible range of capabilities. It is, in other words, a means for making the initial assessment of what's in the ballpark and what's not.

Axes of motion—To a great extent, the robot's axes of motion, or degrees of freedom, determine the complexity of the aggregate movements it is capable of. The term, in essence, expresses the number of moving joints.

Arm movement—This defines the robot's "reach", or "work envelope". The work envelope usually has one of three shapes, cylindrical, spherical and spheroidal, depending on the basic configuration of the arm and on

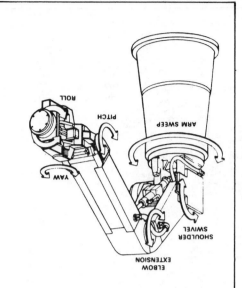

Axes of motion represent complexity of aggregate movements of which robot is capable. Both models shown here have six axes of motion

the major axes of motion. For practical purposes, the description of the work envelope can be simplified by citing only its three major parameters: Degrees of rotation about the centre axis (horizontal arm sweep); vertical motion at both minimum and maximum arm extension; and radial arm extension, measured from the centre axis. It is a simplification, but it gives a rough idea.

Wrist movement—Although wrist movement can make a minor contribution to the shape and size of the work envelope, its main significance is the ability to orientate the gripper or any other end-of-arm tooling. *Pitch* refers to wrist movement in the vertical plane; *yaw* represents movement in the horizontal plane (swing); and the ability to rotate is denoted by *roll*.

End of arm speed—You can get more arguments about a robot's end of arm speed than you can shake a stick at. It varies depending on the axes about which the arm is moving, its position in the work envelope, and the load being carried, to name a few factors. Keeping that in mind, a reasonable, though somewhat simplistic, question to ask still is: How fast can the gripper get from an arbitrary point A to an arbitrary point B in the envelope, empty, and how fast can it move back from B to A fully loaded? The best answer you can expect is a ball-park figure, unless you are willing to accept your answer in the form

The tool may be brought to the work, as a Cincinnati Milacron robot spot welding autobodies is doing (left), or the work may be brought to the tool, as demonstrated by ASEA robot drilling and deburring contactor housings

of a differential equation, or something like that.

Weight carrying capacity—You can get almost as many arguments about weight carrying capacity as about end-of-arm speed, primarily because the two are so closely related. You are still in torque/inertia/momentum/acceleration territory here. Still, it is relevant to ask: What is the weight the robot can practically move from A to B? It is defined as the maximum load that can be carried, at low speed (this is given as a percentage of the maximum speed), and as the load that can be carried at normal operating speed. Again, the answer is a ball park figure, and you can take it with a grain of salt. (Some robot makers circumvent the issue by stating the robot's static load capacity, but what good is that? You want to get the stuff moved!)

Control—Some people like to talk about low-, medium-, and high-technology robots. The inferiority/superiority inference it carries, well, it just is not fair, and what does it tell you anyway? Much more revealing—and technically precise—is to classify robots according to the type of control employed. *Non-servo* robots are the ones sometimes referred to as "end-point" or "limited-sequence" machines. They move between end-points on each axis only (although some intermediate stops can in some cases be provided). They provide relatively high speed; a high degree of repeatability, to within 0.25 mm; and are low in cost, simple to operate, program and maintain, and highly reliable. But they are limited in programming versatility and positioning capability. *Servo-controlled* robots, typically, provide the capability to execute smooth motions with controlled speeds and sometimes even control of acceleration and deceleration, resulting in controlled movement of heavy loads. You can't just throw a 1,000-lb. load through space without servo control. They can be programmed to any position on each axis anywhere in the envelope; more than one program can be stored. End of arm positioning accuracy is of the order of 1.5 mm. (0.050 in.). These machines are more complex, and therefore involve more sophisticated maintenance, are less reliable, and more expensive. An important distinction in servo-controlled robots is between *point-to-point* and *continuous-path* machines. Point-to-point robots are programmed for each end of arm position, without regard for how they get from one point to the next. They are represented primarily by the largest, heaviest-load-capacity machines. In continuous-path machines, on the other hand, the paths that the end of the arm takes through space are defined by the program. Continuous-path robots are smaller and lighter than point-to-point robots, have higher end of arm speeds, and load capacities are generally less than 10 kg. They are used mainly for applications such as spraying paint, arc welding, and grinding and polishing. Some machines, incidentally, act as if they are in the continuous-path category, but really are point-to-point devices; it's just that they provide the capability for defining one heck of a lot of points; each point, nevertheless, must be programmed in.

Memory—The robot memory is part of the controller. It stores the commands that have been programmed in and, through the controller, tells the robot what to do and when. Since the type of memory determines how commands are stored, it gives an indication of the sophistication of the program it is possible to executive, and the degree of flexibility in programming that is possible. *Mechanical step sequencers* include devices such as rotating drums. Other memory devices are pneumatic systems such as *air logic* controllers, electrical memories such as *patch boards* and diode matrix boards and step switches and, finally, electronic memories, which include *magnetic tape* cassettes, floppy *discs*, and *microprocessor*-type devices (ROM, RAM, PROM). It should be pointed out that a combination of memory devices is often employed, such as

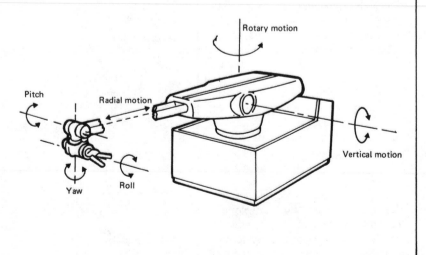

Rotary motion

Pitch

Radial motion

Yaw

Roll

Vertical motion

tape cassettes in conjunction with a microprocessor-type memory.

Programming method—There are three basic approaches to telling a robot what you want it to do: Off-line—by presetting the cams on a rotating stepping drum, for instance, or connecting up air logic tubing (for non servo robots, programming also includes the setting of limit switches is on each axis). This type of programming is usually referred to as *manual*, and is associated with mechanical, pneumatic and electrical memories. As soon as you get into electronic memories, with the potential for more complex programs, you would be all day if you are lucky, or all week if you are not, just wiring up your robot. Instead, what you do is either push buttons, or, and this is almost literally true, you take your robot by the hand and show it what to do. The push-button approach, referred to as *lead-through*, consists of manoeuvring the robot arm from one desired position to the next by means of the control console; all you have to do is push the "record" button at each point, and, once the program is complete, the robot will repeat this sequence from point to point, *ad infinitum*. If you are not so much interested in the exact positions through which the robot arm will sweep, but more in the actual trajectory, you could still program it simply by defining a lot of points. A lot easier would be, though, to use *walk-through* programming; you physically guide the robot arm (or a special teaching arm) through the desired motions, and the robot will repeat *exactly* what you taught it, including any goofs.

Memory capacity—There are several ways memory capacity can be expressed. Simplest is to indicate the number of distinct motions and functions the robot can perform in a single program.

Positioning accuracy—How closely can you program a robot hand to go where you want it to go? Or to put it in another way, how closely should you position a part for the hand to be able to pick it up? Unless equipped with a special sensing device, a robot cannot see what it is doing, so this is an important quantity.

Repeatability—Once you have got a robot hand to go where you want it to go the first time, repeatability will determine how close to that same spot it will return—again and again.

Power—What makes the robot move? It is not a moot question. For one thing, the working environment has a lot to do with the choice of power system. Many of the same criteria apply here as with other types of industrial machinery.

Cooling—Some robots need cooling. The main thing is to make sure you know which type of cooling, and you better talk to your plant engineer, too.

Maximum ambient operating temperature—They say robots can stand the heat of any shop and keep going when a man would long since have dropped in his tracks. Well, there *are* limits. To be sure, end-of-arm tooling may take a lot more before burning its fingers, than the arm and base, and the control console a lot less, particularly if it has a mess of computer logic inside. All of this has got to do also with how much cooling may be required.

End of arm tooling—This is what goes on the end of the arm to do the job. Mostly, it will be a gripper, often referred to as the hand. But it doesn't have to be a hand. Actually, you can put anything you like on the end of the arm. Besides grippers, vacuum pads are the most common off-the-shelf tooling, but put on anything you desire (if you want to pay for custom design)—put on a foot if you like, and you can make it kick the garbage can.

Interfacing—Except in a few applications, most robots need to communicate and interact with the outside world. This can take the form of simply on/off signals by means of electrical, or pneumatic, contacts, or consist of more complex electronic signals. *Inputs* are the number of lines over which the robot will accept signals from the outside world, and *outputs* are the lines over which it will send signals to external equipment.

It would be easy if you could just sit down and say, I need so and so much work envelope and such and such a control scheme, and then just select the set of numbers that most closely fit the picture. But, remember, you are dealing with a system. There are a lot of other factors that have to be taken into account.

Here are some considerations that may help doing that:

☐ There is a rule of thumb in the industry that says that if you can replace two men with a robot, you can probably afford to look at a medium-priced, servo-controlled robot; if you can replace one man, you are probably looking at a non-servo or bang-bang machine; if it is only half a man (half his time, that is) you may have a problem justifying any robot.

☐ Don't simply try and emulate a human operator by a robot. On the one hand, there are things a man can do that a robot can't do—many things, come to think of it. But on the other hand, some robots can do tricks no union would stand for, like working all day long without getting tired (in some plants they mount them on an overhead crane). It's great for saving floor space. Nor does a robot really care whether it is taking the work to the tool or the tool to the work; there are places where they have a robot manipulate heavy, bulky workpieces around a pedestal-mounted welder, i.e., no part fixturing, and getting it in and out of the fixture all day long.

☐ Again, try and think of all the screw-ups you can, and imagine the dire consequences as picturesquely as possible. Such as when a robot puts transmission casings on overhead conveyor hooks, and the hook isn't there when it should be (or it may be swinging). What does the robot do? Ouch! And there is more, much more.

☐ How do you index parts for pick-up when they've got to be in the right spot within, say, 50 thou?

☐ Robots are reputed to be capable of 98% uptime. But they can, and do, break down for the other 2%. And if those 2% come in one or two chunks, it could be a sort of major catastrophe. It might not kill you, but it could cost you a bundle. So, what do you figure in the way of back-up? Can you put a man back into the operation? Yes, *can* you? Suppose you did hang your robot from the ceiling to save floor space. Can the man still squeeze in there? And so forth.

Moral: Take these things into account in your trial layout.

☐ If you can, buy a little extra. Moving a 50-lb load? Figure on 60 lb, or 75 (don't forget end-of-arm tooling—it can weigh too). You don't want to overspecify more than you have to, but what's the price of peace of mind? And after all, you might want to change your mind after you've ordered the machine, or somebody else might do it for you.

▢ Try and think of all these things *before* you get anywhere near deciding which robot you are going to install.

▢ Try and think like a robot—no kidding.

There are a lot of other considerations to keep in mind but they are no different from those related to installing any piece of machinery. Such as: reckon with utility requirements; figure out safety-related consequences of the installation (barriers around work envelope, safety interlocks, etc.); and how will the rest of the plant lay-out be affected. And so forth, and so forth.

One vital item worth worrying over from day one is end-of-arm tooling. It could be a bit of a problem. If you have a simple transfer operation, you can in most cases get off-the-shelf tooling from the manufacturer—in fact, some of them have quite an assortment. But the number of operations where standard tooling does not fit in is far from negligible.

If the manufacturer doesn't offer custom tooling—and many do—you can always go to a tooling shop. Or you can make your own. People have been known to make perfectly acceptable end-of-arm tooling out of a few pieces of dirty old scrap iron. In some cases all you need is a bracket to, say, mount a welding gun. It isn't necessarily complicated, but it *can* be. You better be prepared to experiment, starting early.

Another consideration that merits premeditation is maintenance. Chances are, your maintenance people's idea of a robot is strictly Star Wars style.

Remember the way it was with NC? Well, a lot of maintenance crews did get used to that. All it took was some extra, specialized knowledge.

Robots may be odd beasts, but underneath the cover plates there is really nothing much that would phase an averagely competent plant mechanic or electrician. Certainly on the mechanical side, robots are straightforward and easy to fix.

Of course, you can plan to rely on the manufacturer for maintenance. But robots have a way of quitting mid-morning on Monday or any other time as inconvenient as possible. Don't count on getting a company maintenance man flown down less than 24 hours later. By then, it doesn't make much difference anymore whether he fixes the thing in five minutes, which has been known to occur, or three hours, or six, or twelve. It's all equally embarrassing.

Generally, it's much better to plan having the manufacturer train your own people. Most of them will, if they know what's good for them.

Several skills may be involved—separately or in combination. Number one, probably, is the electrician; some basic training in solid-state electronics would be helpful but not necessary. Your pipefitter may have to look after the hydraulics if your robot has any, which is likely, but there is nothing magic about robot hydraulics. Just a lot of it, in some And, finally, the plant mechanic may have to stand by. No telling what he might have to get into, but those fellows are used to that.

The point is, when you are doing the preliminary evaluation, take into account what maintenance skills you would need and, regardless of who does the maintenance, the location of the manufacturer's service depot(s). A rule of thumb used in the industry, incidentally, is that for every 10 to 12 robots you need one maintenance man per shift.

While you are doing all this planning, you may ask yourself from time to time: Is it all worth it? That's a good question. The fact is, one of the major reasons why the robot industry hasn't up till now seen as much growth as it expected is that all too often ROIs (return on investment) proved to be inadequate.

It is being said that one of the main justifications for robots is their ability to relieve the human worker of undesirable jobs. In practice, that just doesn't turn out to be the key factor—not in North America, anyway. While there are some robot installations that were put there specifically to meet occupational health and safety regulations, most robots are bought for the hard-nosed reason of reducing production costs or improving productivity. The main criterion, in other words, is: What does the robot mean to the bottom line?

There are several ways of measuring this, of course, and they are no different from those followed in assessing any capital equipment. You'll probably want to go with a straight ROI calculation, or look at pay-back time.

There are, of course, some "non-economic" justifications too, although you can hang a dollar sign on almost anything you want to. They should enter the picture. Health and safety certainly is one of them. And the fact that the robot doesn't get tired—therefore: more consistent productivity; but also note the potential for better quality, and reduction of scrap and rework. And there will be less worry about absenteeism, goofing off and messing up.

One robot bonus some people are still fond of mentioning is its flexibility, namely that it could be moved somewhere else when the job it is bought to do expires for some reason or other. That *may* be true for *some* robots, for *some* tasks. In practice, it rarely turns out that way. Industry experts say very few users have ever moved their robots. They usually stay put, wherever they were placed originally, doing what they were bought to do there.

Who knows? Not that many robots have been anywhere all that long—long enough, in any case, to outlast their original job, and who is to say that day may not come? But in terms of the more or less immediate requirements, reckoning with the average pay-back times, which are no less than seven years, it's a tenuous argument at best. Don't use it.

The objective of this article is to assist the potential robot user in doing some of the preliminary planning. Hopefully, what it will lead to is the possibility to identify two or three serious candidates. That will only be the beginning of the real nuts and bolts evaluation, talking brass tacks with the manufacturers.

One hazard in compiling a listing such as the one that follows is missing someone. From the outset, the objective was to include *all* North American sources of industrial robots and all of the products they offer. In all,

Pick and place

Several among the manufacturers of mechanical transfer devices objected that their products are not really robots. Given the definition of a robot as a *programmable* device, which pick-and-place machines are not, this objection is justified. But in some cases, this may be all you need. The manufacturers of mechanical transfer devices, therefore, are listed separately, although their products are not included in the Survey listings.

SPECIFICATIONS	Source Model	ASEA IRb-6	ASEA IRb-60	Auto-Place 10	Auto-Place 50	Binks MK II-90
Axes of motion		3, 5 or 6	3, 5 or 6	6	6	6
Movement						
arm—horizontal (degrees of arc)		340	330	200 (360)	200 (360)	85
vertical (min. mm, max. mm)		520-1510	170-1900	0-508	0-127	2057
radial (mm)		950	2000	940 (1093)	940 (1093)	914
wrist—pitch (degrees of arc)		180	195	(270)	(270)	190
yaw (degrees of arc)			300	(270)	(270)	190
roll (degrees of arc)		360	360	270	270	190
End of arm speed						
average max. unloaded (m/s)				2.5	1.5	1.7
normal operating, loaded (m/s)				1.2	.74	var.
Weight carrying capacity						
max. at low speed (kg @ %)		10 @ 60	100 @ 60	4.5 @ 25	14 @ 25	27.2 @ 50
at operating speed (kg)		6	60	1	4	13.6
Control						
non-servo				✔	✔	
servo—point to point		✔	✔			
continuous path		✔	✔			✔
Memory						
mechanical step sequencer						
air logic				✔	✔	
patch board (electrical)						
magnetic tape/disc						
microprocessor		✔	✔		✔	✔
other (specify)						
Programming method						
manual				✔	✔	
lead-through		✔	✔			
walk-through						✔
Memory capacity (no. of steps)		250-15,000	250-15,000	24 (100+)	24 (100+)	2x300 sec.
Positioning accuracy (± mm)		0.1	0.2	1.3	1.3	4
Repeatability (± mm)		0.05	0.1	0.6	0.6	4
Power—hydraulic						
pneumatic				✔	✔	✔
electric		✔	✔			
Cooling—air						✔
water						✔
none		✔	✔	✔	✔	
Max. ambient operating temperature (°C)		50	50	50	50	45
End of arm tooling						
off the shelf—gripper		✔	✔	✔	✔	
vacuum pad				✔	✔	
other (specify)				Magnets	Magnetic	
custom design		✔	✔	✔	✔	✔
Interfacing						
Inputs (number)		16+	16+	Any number	Any number	16
Outputs (number)		14+	14+	Any number	Any number	16
Potential application(s)		1 3 10 17* 20	1 3 10 17* 20	2 7 8 9 11 14 15 17 18*	2 4 7 9 11 12 14 15 17 18*	Paint spraying
Number installed—U.S.		not spec'd	not spec'd	300	400+	7
Canada		not spec'd	not spec'd	6		
Overseas		not spec'd	not spec'd	70	10	40+
Price of basic unit F.O.B. manufacturer (U.S.$)		60,000	90,000	70,000	12,000	75,000
Name of Canadian representative		ASEA Ltd	A. F. Mundy Associates Canada Ltd.	A. F. Mundy Associates Canada Ltd.	Binks Mfg. Co. of Canada Ltd.	
Location(s) of service depot						
parts		White Plains, NY. Montreal Toronto	White Plains, NY. Montreal Toronto	Troy, Mich. Rexdale Ont.	Troy, Mich. Rexdale, Ont.	Franklin Park, Ill.
service		same	same	same	same	same

*1—arc welding 2—assembly 3—deburring 4—die casting 5—drilling 6—flame cutting 7—forging 8—foundry 9—gauging 10—grinding 11—heat treating 12—injection moulding 13—investment casting 14—machine loading & unloading 15—metal stamping 16—palletizing 17—parts handling 18—press feeding 19—routing 20—spot welding 21—spray painting 22—tracking.

SPECIFICATIONS	Source / Model	Cincinnati Milacron T³	Cincinnati Milacron HT³	DeVilbiss TR-3000	General Numeric Fanuc 1	Industrial Automates 9500
Axes of motion		6	6	6	5	
Movement						
arm—horizontal (degrees of arc)		240	240	135	210	90 (120)
vertical (min. mm, max. mm)		0-3960	0-3960	2044	500	178
radial (mm)		2465	2615	2185	800	609
wrist—pitch (degrees of arc)		180	180	176		180
yaw (degrees of arc)		180	180	176		180
roll (degrees of arc)		240	240	210	90/80	180
End of arm speed						
average max. unloaded (m/s)		1.3	0.9	1.7	1	0.76
normal operating, loaded (m/s)		1.3	0.9	var.	1	0.76
Weight carrying capacity						
max. at low speed (kg @ %)		45	100		20	4.5
at operating speed (kg)		45	100	4.53	20	4.5
Control						
non-servo						✔
servo—point to point					✔	
continuous path		✔	✔	✔		
Memory						
mechanical step sequencer						
air logic						
patch board (electrical)						
magnetic tape/disc					✔	
microprocessor		✔	✔	✔	✔	✔
other (specify)						
Programming method						
manual					✔	
lead-through		✔	✔			✔
walk-through				✔		
Memory capacity (no. of steps)		700+	700+	2 hr.	80	1000
Positioning accuracy (± mm)		1.3	<1.3	0.35	1	0.3
Repeatability (± mm)		1.3 (0.6)	1.3	2	1	0.3
Power—hydraulic		✔	✔	✔		
pneumatic						
electric					✔	✔
Cooling—air				✔		
water		✔	✔	✔		
none					✔	✔
Max. ambient operating temperature (°C)		50	5-50	35	45	not spec'd
End of arm tooling						
off the shelf—gripper		✔	✔		✔	✔
vacuum pad					✔	✔
other (specify)				spray gun	✔	magnetic
custom design		✔	✔		✔	✔
Interfacing						
Inputs (number)		8+	8+	13+ TTY	8	16
Outputs (number)		8+	8+	5+ TTY	14	16
Potential application(s)		5 9 16 19 20 22*	16 17 20 22*	Spray painting	not spec'd	12 14 17*
Number installed—U.S.		} >200	} >200	100	2	15
Canada				11		1
Overseas				900	30	5
Price of basic unit F.O.B. manufacturer (U.S.$)		63,000	68,000	68,000	30,000	11,845
Name of Canadian representative		Cincinnati Milacron Cda. Ltd.	Cincinnati Milacron Cda. Ltd.	DeVilbiss (Cda.) Ltd.	Hall Smith Co. Ltd.	
Location(s) of service depot						
parts		Cincinnati Montreal Toronto Windsor	Cincinnati Montreal Toronto Windsor	Toledo, Ohio Barrie, Ont.	Elk Grove, Ill. Burlington, Ont.	West Allis, Wisc.
service		same	same	same	same	same

*1—arc welding 2—assembly 3—deburring 4—die casting 5—drilling 6—flame cutting 7—forging 8—foundry 9—gauging 10—grinding 11—heat treating 12—injection moulding 13—investment casting 14—machine loading & unloading 15—metal stamping 16—palletizing 17—parts handling 18—press feeding 19—routing 20—spot welding 21—spray painting 22—tracking.

SPECIFICATIONS	Source / Model	Manca (modular)	Modular Machine Mobot	Nordson	Prab 4200/5800	Prab Versatran E
Axes of motion		1-7	1-7	6	3-5	3-7
Movement						
arm—horizontal (degrees of arc)		100-360	360	116	270	240
vertical (min. mm, max. mm)		10-1,250	761	2788	826	761
radial (mm)		1,250	20,000	1000	1473	761 (1068)
wrist—pitch (degrees of arc)		to 360	360	240	90	220
yaw (degrees of arc)		to 360	360	240	90	220
roll (degrees of arc)			360	240	180	359
End of arm speed						
average max. unloaded (m/s)		1.1	0.5	2.5	1.27	1.27
normal operating, loaded (m/s)		to 1	0.3	var.	1.27	1.27
Weight carrying capacity						
max. at low speed (kg @ %)		to 1,700	200 @ 10	25 @ 30	56.7	56.7
at operating speed (kg)		to 1,700	200	15	45.4	45.4
Control						
non-servo		✔	✔		✔	
servo—point to point			✔			✔
continuous path			✔	✔		✔
Memory						
mechanical step sequencer		✔	✔		✔	
air logic						
patch board (electrical)		✔				(✔)
magnetic tape/disc			✔			
microprocessor			✔	✔		✔
other (specify)		prgr. cntr.			prgr. contr.	
Programming method						
manual		✔	✔		✔	✔
lead-through			✔		✔	✔
walk-through				✔		
Memory capacity (no. of steps)		optional	unlimited	30 min.	24-100	4000
Positioning accuracy (± mm)		0.1	1	2	0.20	0.76
Repeatability (± mm)		to 0.01	1	2	0.05	0.51
Power—hydraulic		✔	✔	✔	✔	✔
pneumatic		✔	✔			
electric		✔	✔			
Cooling—air					✔	✔
water		✔		✔	✔	✔
none			✔			
Max. ambient operating temperature (°C)		not spec'd	65	43	50	50
End of arm tooling						
off the shelf—gripper		✔	✔		✔	✔
vacuum pad		✔	✔		✔	✔
other (specify)		magnetic				
custom design		✔	✔	spray gun	✔	✔
Interfacing						
Inputs (number)		optional	unlimited	8	8	50
Outputs (number)		optional	unlimited	16	8	50
Potential application(s)		14 17 18*	part handling	finishing	4 5 7 8 11 12 14 15 17 18*	All ex'pt spray paint
Number installed—U.S.			50+	10	not spec'd	not spec'd
Canada				2	not spec'd	not spec'd
Overseas		>300		25	not spec'd	not spec'd
Price of basic unit F.O.B. manufacturer (U.S.$)		not spec'd	15,000	85,000	20,000-35,000	35,000-55,000
Name of Canadian representative				Nordson Cda. Ltd.	Can-Eng Mfg. Co.	Can-Eng Mfg. Co.
Location(s) of service depot parts		Rockleigh N.J.	San Diego	Amherst, Oh. Scarborough, Ont.	Kalamazoo, Mich.	Kalamazoo, Mich.
service		Rockleigh NJ Menomonee Falls, Wisc.	San Diego	same	same	same

*1—arc welding 2—assembly 3—deburring 4—die casting 5—drilling 6—flame cutting 7—forging 8—foundry 9—gauging 10—grinding 11—heat treating 12—injection moulding 13—investment casting 14—machine loading & unloading 15—metal stamping 16—palletizing 17—parts handling 18—press feeding 19—routing 20—spot welding 21—spray painting 22—tracking.

SPECIFICATIONS	Source Model	Prab Versatran F	Reis 1215	Rimrock 195	Seiko 100	Seiko 200
Axes of motion		3-7	5(6)	5	4	3
Movement						
arm—horizontal (degrees of arc)		300	270	190	—	90 (120)
vertical (min. mm, max. mm)		to 1520	1200	—	50	20
radial (mm)		to 1520	1500	2032	200	—
wrist—pitch (degrees of arc)		270	360	—	—	—
yaw (degrees of arc)		180	±180	—	—	—
roll (degrees of arc)		300	360	90	90-180°	—
End of arm speed						
average max. unloaded (m/s)		1.27	1.2	1.016	0.2	0.5
normal operating, loaded (m/s)		1.27	1.2	∼1.016	0.2	0.5
Weight carrying capacity						
max. at low speed (kg @ %)		to 906	25	27.21 @ 57	1.5	0.3
at operating speed (kg)		to 680	5-25	27.21	1.5	0.15
Control						
non-servo				✔	✔	✔
servo—point to point		✔	✔			
continuous path		✔				
Memory						
mechanical step sequencer					✔	✔
air logic					✔	✔
patch board (electrical)		(✔)			✔	✔
magnetic tape/disc						
microprocessor		✔	✔		✔	
other (specify)				prgr. contr.	prgr. contr.	prgr. contr.
Programming method						
manual		✔		✔	✔	✔
lead-through		✔	✔		✔	—
walk-through			✔	✔		—
Memory capacity (no. of steps)		4000	682+	23	256	256
Positioning accuracy (± mm)		1.3	0.6	0.25	0.01	0.01
Repeatability (± mm)		1.3	0.6	0.25	0.01	0.01
Power—hydraulic		✔		✔		✔
pneumatic			✔		✔	✔
electric			✔	✔	✔	
Cooling—air		✔				
water		✔				
none			✔	✔	✔	✔
Max. ambient operating temperature (°C)		50	50	80	60	60
End of arm tooling						
off the shelf—gripper		✔	✔	✔	✔	✔
vacuum pad		✔			✔	✔
other (specify)						
custom design		✔		✔	✔	✔
Interfacing						
Inputs (number)		50	8	6	4+	4+
Outputs (number)		50	8	2	4+	4+
Potential application(s)		All ex'pt spray paint.	not spec'd	Die casting	Press Feeder	not spec'd
Number installed—U.S.		not spec'd	—	—	not spec'd	not spec'd
Canada		not spec'd	—	—	not spec'd	not spec'd
Overseas		not spec'd	20	—	not spec'd	not spec'd
Price of basic unit F.O.B. manufacturer (U.S.$)		55,000- 110,000	55,000	20,000- 60,000	4,000	6,000
Name of Canadian representative		Can-Eng. Mfg. Co.	—	Vacumet Ltd.	Hall Smith Co. Ltd.	Hall Smith Co. Ltd.
Location(s) of service depot						
parts		Kalamazoo MI	Elgin IL	Columbus, Ohio	Torrance, CA	Torrance CA Burlington, Ont.
service		same	same	same	same	same

*1—arc welding 2—assembly 3—deburring 4—die casting 5—drilling 6—flame cutting 7—forging 8—foundry 9—gauging 10—grinding 11—heat treating 12—injection moulding 13—investment casting 14—machine loading & unloading 15—metal stamping 16—palletizing 17—parts handling 18—press feeding 19—routing 20—spot welding 21—spray painting 22—tracking.

SPECIFICATIONS	Source / Model	Seiko 400	Seiko 700	Sterling-Detroit ROBOTARM	Thermwood E	Unimation 1000
Axes of motion		4	5	3		3 to 5
Movement						
arm—horizontal (degrees of arc)		—	90 or 120	90°	135	208
vertical (min. mm, max. mm)		100	40	203.2	2134 mm	900-1900
radial (mm)		400 (700)	300	—	762, 2108	1060
wrist—pitch (degrees of arc)		—		optional	180	90-180
yaw (degrees of arc)		—		optional	180	90-180
roll (degrees of arc)		(180)	90/180	120°	270	
End of arm speed						
average max. unloaded (m/s)		0.57	0.6	0.61	1.02	1
normal operating, loaded (m/s)		0.57	0.6	0.61	1.02	var.
Weight carrying capacity						
max. at low speed (kg @ %)		4	1	272	Not spec'd	22.9
at operating speed (kg)		4	0.5	272	8.2	12
Control						
non-servo		✔	✔	✔		
servo—point to point					✔	✔
continuous path					✔	✔
Memory						
mechanical step sequencer		✔	✔			
air logic		✔	✔			
patch board (electrical)		✔	✔			
magnetic tape/disc					(✔)	
microprocessor					✔	
other (specify)		prgr. contr.	prgr. contr.	prgr. contr.		CMOS
Programming method						
manual		✔	✔	✔		
lead-through					✔	✔
walk-through					✔	
Memory capacity (no. of steps)		256	256	1000	3000/5 min	up to 256
Positioning accuracy (± mm)		0.025	0.025	0.1	3.2	1
Repeatability (± mm)		0.025	0.025	0.1	3.2	1.3
Power—hydraulic				✔		✔
pneumatic		✔	✔		✔	✔ (wrist)
electric		✔	✔		✔	
Cooling—air					✔	✔
water						
none		✔	✔	✔		
Max. ambient operating temperature (°C)		60	60	Any	50	50
End of arm tooling						
off the shelf—gripper		✔	✔	✔		✔
vacuum pad		✔	✔	✔		✔
other (specify)		✔	✔	✔	✔	
custom design		✔	✔	✔	✔	✔
Interfacing						
Inputs (number)		4+	4+	16	8	6
Outputs (number)		4+	4+	16	8	6
Potential application(s)		Assembly	not spec'd	4 7 12*	17 21*	7 12 14 17 18*
Number installed—U.S.		not spec'd		500	—	not spec'd
Canada		not spec'd	1500	50	—	not spec'd
Overseas		not spec'd		15	—	not spec'd
Price of basic unit F.O.B. manufacturer (U.S.$)		7,500	8,000	25,000	45,000	23,950-24,950
Name of Canadian representative		Hall Smith Co. Ltd.	Hall Smith Co. Ltd.	J. S. Bulmer	—	CAE Morse Ltd.
Location(s) of service depot						
parts		Torrance CA Burlington, Ont.	Torrance CA Burlington, Ont.	Detroit, MI	Dale IN	Danbury, Ct. Farmington Hills, Mississauga,
service		same	same	same	same	same

*1—arc welding 2—assembly 3—deburring 4—die casting 5—drilling 6—flame cutting 7—forging 8—foundry 9—gauging 10—grinding 11—heat treating 12—injection moulding 13—investment casting 14—machine loading & unloading 15—metal stamping 16—palletizing 17—parts handling 18—press feeding 19—routing 20—spot welding 21—spray painting 22—tracking.

SPECIFICATIONS	Source Model	Unimation 2000/2100	Unimation 4000	Unimation Puma 250	Unimation Puma 500/600	Unimation Arc Welding
Axes of motion		3 to 6	3 to 6	3 to 6	5	5
Movement						
arm—horizontal (degrees of arc)		208 max	200	330 waist	320	90
vertical (min: mm, max. mm)		1890/2480	1672-2830	not spec'd	not spec'd	1000
radial (mm)		1041/1365	1321	not spec'd	not spec'd	not appl.
wrist—pitch (degrees of arc)		211	226	240	200	50
yaw (degrees of arc)		200	320	630	300	180
roll (degrees of arc)		360	370	630	270	165
End of arm speed						
average max. unloaded (m/s)		1	1	1.5	0.3	0.01
normal operating, loaded (m/s)		var.	not spec'd	not spec'd	not spec'd	0.01
Weight carrying capacity						
max. at low speed (kg @ %)		136	205	not spec'd	2.27	5 at 100
at operating speed (kg)		12-50	90	3.5	not spec'd	5 at 100
Control						
non-servo						
servo—point to point		✔	✔	✔	✔	
continuous path		✔	✔	✔	✔	
Memory						
mechanical step sequencer						
air logic						
patch board (electrical)						
magnetic tape/disc						
microprocessor				✔	✔	
other (specify)		Plated wire	Plated wire			semi-cond
Programming method						
manual						
lead-through		✔	✔	✔	✔	✔
walk-through						
Memory capacity (no. of steps)		2048	up to 2048		not spec'd	not spec'd
Positioning accuracy (± mm)		1	not spec'd	not spec'd		1
Repeatability (± mm)		1.27	2.03	.05	.1	1
Power—hydraulic		✔	✔			
pneumatic						
electric					✔	✔
Cooling—air		✔	✔			✔
water						
none						
Max. ambient operating temperature (°C)		50	50	45	45	50
End of arm tooling						
off the shelf—gripper		✔	✔	✔	✔	
vacuum pad		✔	✔			
other (specify)						torch holder
custom design		✔	✔	✔	✔	
Interfacing						
Inputs (number)		9	9	8	8	1
Outputs (number)		9	9	8	8	4
Potential application(s)		1 4 6 7 8 11 12 13 14 16 17 18 19 20	1 4 6 7 8 11 12 13 14 16 17 18 19 20*	2 14 17*	1 2 14 17 20*	Arc welding
Number installed—U.S.		not spec'd	not spec'd	not spec'd	not spec'd	not spec'd
Canada		not spec'd	not spec'd	not spec'd	not spec'd	not spec'd
Overseas		not spec'd	not spec'd	not spec'd	not spec'd	not spec'd
Price of basic unit F.O.B. manufacturer (U.S.$)		33,000- 56,000	48,000- 70,000	35,000	35,000	35,000
Name of Canadian representative		CAE Morse Ltd.	CAE Morse Ltd.		none	
Location(s) of service depot						
parts		Danbury, Farmington Hills, Mississauga,	Danbury, Farmington Hills, Mississauga,	Danbury, Conn. Farmington Hills, Mich.	Danbury, Conn. Farmington Hills, Mich.	Danbury, Conn. Detroit, Mi.
service		same	same	same	same	same

*1—arc welding 2—assembly 3—deburring 4—die casting 5—drilling 6—flame cutting 7—forging 8—foundry 9—gauging 10—grinding 11—heat treating 12—injection moulding 13—investment casting 14—machine loading & unloading 15—metal stamping 16—palletizing 17—parts handling 18—press feeding 19—routing 20—spot welding 21—spray painting 22—tracking.

Robot suppliers

ASEA Inc.,
Four New King Street, White Plains, N.Y. 10604 (914) 428-6000.

ASEA Ltd.
10300 Henri-Bourassa Blvd. W., Saint Laurent, Que., Box 700, Post. Sta. Cartierville, Montreal, Que. H4K 2J8 (514) 332-5350

AutoPlace, Inc.,
1401 East 14 Mile Road, Troy, MI 48084 (313) 585-5972; Cdn. rep.: **A. F. Mundy Associates Canada Ltd.,** 66 Racine Rd., Rexdale, Ont. M9W 2Z7 (416) 745-6976

Binks Manufacturing Co.,
9201 West Belmont Ave., Franklin Park, IL 60131; (312) 671-3000.

Binks Manufacturing Co. of Canada Ltd.
17 Vansco Rd., Toronto, Ont. M8Z 5J5, (416) 252-5181

Cincinnati Milacron Inc.
Industrial Robot Division, 4701 Marburg Ave., Cincinnati OH 45209; (513) 841-8386.

Cincinnati-Milacron, Canada Ltd.
122 N Queen St., Toronto, Ont. M8Z 2E4; (416) 233-3216

The DeVilbiss Co.
300 Phillips Ave., Toledo, OH 43692; (419) 470-2169.

DeVilbiss (Canada) Ltd.
P.O. Box 3000, Barrie, Ont. L4M 4V6 (705) 728-5501

General Numeric Corp.
390 Kent Ave., Elk Grove, IL 60007; (312) 640-1595; Cdn. rep: **Hall Smith Co. Ltd.,** 3505 Mainway, Burlington, Ont. L7M 1A9; (416) 335-6008.

Industrial Automates, Inc.
6123 W. Mitchell St., West Allis, WI 53214 (414) 327-5656.

Manca, Inc.
Leitz Building, Rockleigh, NJ 07647 (201) 767-7227.

Modular Machine Co.
2785 Kurtz St., San Diego, CA 92110 (714) 224-3494.

Nordson Corp.
555 Jackson St., Amherst, OH 44001 (216) 988-9411.

Nordson Canada Ltd.
849 Progress Ave., Scarborough, Ont. (416) 438-6730

Prab Conveyors, Inc.
5944 E. Kilgore Rd., Kalamazoo, MI 49003 (616) 349-8761; Cdn rep: **Can-Eng Manufacturing Co.,** 6800 Montrose Rd., Niagara Falls, Ont. L2E 6V5 (416) 356-1327.

Production Automation Corp.
12956 Farmington Rd., Livonia, MI 48150 (313) 261-0890.

Reis Machines,
1450 Davis Rd., Elgin, IL 60120 (312) 741-9500

The Rimrock Corp.
1700 Rimrock Rd., P.O. Box 19127, Columbus, OH 43219 (614) 471-5926.

Seiko Instruments, Inc.
2990 West Lomita Blvd., Torrance, CA 90505 (213) 530-3400; Cdn rep: **Hall Smith Co. Ltd.,** 3505 Mainway, Burlington, Ont. L7M 1A9 (416) 335-6008.

Sterling-Detroit Co.
261 E. Goldengate, Detroit, MI 48203 (313) 366-3500; Cdn rep: **J. S. Bulmer,** 725 Fernhill Blvd., Oshawa, Ont. (416) 576-0612

Thermwood Co.
P.O. Box 436, Dale, IN 47523, (812) 937-4476

Unimation, Inc.
Shelter Rock Lane, Danbury, CT 06810 (203) 744-1800; Cdn rep; **CAE-Morse Ltd.,** 4500 Dixie Rd., Mississauga, Ont. L4W 1Z6 (416) 625-5161.

Mechanical Transfer Devices

Automation Devices, Inc.
Automation Park, P.O. Box AD, Fairview, PA 16415; (814) 474-5561

Livernois Automation Co.
25200 Trowbridge, Dearborn, MI 48124 (313) 278-0200

Pickomatic System, Inc.
37950 Commerce St., Sterling Heights, MI 48077 (313) 939-9320

R & I Manufacturing Co. Inc.
91 Prospect St., Thomaston, CT 06787 (203) 283-0127.

The Van Epps Design & Development Co. Inc.
4760 Van Epps Rd., Cleveland, OH 44131 (216) 661-1337.

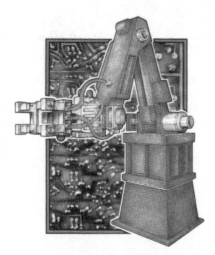

Robots are Molding a New Image–Part I

Reprinted from Foundry Management and Technology, May, 1979

No longer filled with the razzle-dazzle market optimism of the 60s, robot builders today view the U.S. market cautiously and confidently. For foundrymen, industrial robots offer new economies and production clout. Nonetheless, utilization of robotics technology and equipment in the foundry is evolutionary, not revolutionary. This report describes current robots, their applications, and the reasons why they have become a credible, proliferating reality in manufacturing.

By NORBERT G. SEMAN / Associate Editor

USE OF AUTOMATION to perform industrial work depends as much on society's need to change its attitudes and work habits as on its ability to make quantum technical advances. The advent of industrial robots has signalled new opportunities in manufacturing that are due to an on-going process of evolution in robotics technology and to changing economic conditions. Robots have become more credible because of their increasing practicality and low cost.

In part, growth of the robot industry has been the result of a 250% increase in U.S. labor cost over the last 10 years while average robot cost increases held closer to 50%. More importantly, robot builders have tried to create products that are readily adaptable to end-users' needs and incorporate sophisticated automation systems to interface with modern manufacturing operations.

Models in use today perform a limited number of preprogrammed motions consistently and reliably, to a point that is beyond man's physical limitations, with a high degree of maneuverability. They are flexible and easy to train for jobs under fixed conditions. However, robots are nonadaptive in that they are unable to replace human workers who use muscles, senses, and brains to perform industrial work under varying conditions.

Industrial robots are common among diecasters and no longer are foreign to investment casters. In the future, they may join ranks in the cleaning, heat treatment, shakeout, core dipping, core machine loading, mold pouring, and other areas of foundries.

Using the right robots in the right applications can help foundrymen meet the challenges of increasing productivity, cutting costs of material and labor, reducing scrap, attaining quality and safety, and increasing utilization of facilities.

The Robot Defined — The industrial robot has been defined most recently as a programmed *manipulator* designed to perform useful work automatically without human assistance. In a different definition, a robot is a commercially available, mechanical, programmable device that independently performs manipulative job functions.

In this sense, robots fill the gap between special-purpose automation and human endeavor. They have demonstrated an ability to perform work that requires simple, repetitive motions and, therefore, can relieve human operators from fatiguing, hazardous, or monotonous tasks.

Terms like "teachable," "self-contained," and "program-controlled" often are applied to robots. However, robots are best understood in terms of their real capabilities. Essentially, they are off-the-shelf automation. It is the robot's ability to be easily taught or reprogrammed that distinguishes it from other types of automated handling equipment.

The term *robot* derives from the Czech word "robota," meaning servitude or drudgery, and its Old Slavic equivalent "rabota" — work. It gained international currency as the result of a play entitled "R.U.R." (Rossum's Universal Robot) written in 1920 by Karl Capek, a Czech playwright, who conjured up a world mechanized to the extent that all labor was performed by machine-made men.

Capek's criticism targeted on the dissemination of American technological progress in Europe, which he felt contributed to the depersonalization of human workers. Valid though his social criticism may have been in the limited context of the years between the two wars, it sidesteps the positive, supportive role industrial robots play in improving work life and eliminating hazards in the workplace.

The Robot Industry — Robots in 1979 are an accepted and proven method of automating manual operations in hostile environments. They are being produced in the

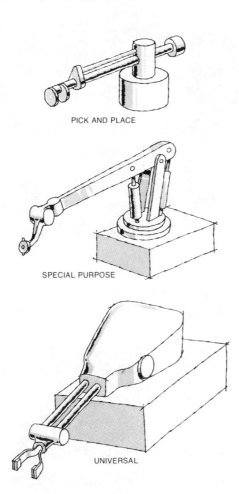

PICK AND PLACE

SPECIAL PURPOSE

UNIVERSAL

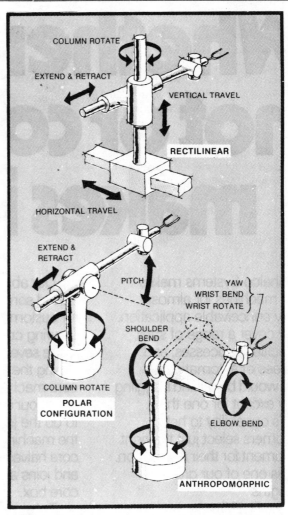

COLUMN ROTATE

EXTEND & RETRACT

VERTICAL TRAVEL

RECTILINEAR

HORIZONTAL TRAVEL

EXTEND & RETRACT

PITCH

YAW
WRIST BEND
WRIST ROTATE

SHOULDER BEND

COLUMN ROTATE

POLAR CONFIGURATION

ELBOW BEND

ANTHROPOMORPHIC

One way of classifying industrial robots is by manipulative functions. These include the pick-and-place models with simple gripper design, robots that perform special jobs such as spraying and coating, and those that can be used to perform any type of manipulative function. Courtesy SME.

Robots move according to three basic design variations pictured above. The design emulates man's physical capabilities by simulating his movements and activities through "degrees of freedom" or freedom of movement similar to man's waist, wrist, elbow, shoulder, and fingers.

U.S. by about ten major manufacturers. A trend developed within the last few years has witnessed automakers conducting direct research and design studies on robots destined for that branch of industry.

This work clearly has been a cooperative effort from the standpoint of independent robot builders who have benefited from the developmental work and the financial resources of the giant automaking firms. Automakers still rely on independents for robot hardware, in the form of components and complete units.

Existing robot applications are estimated at between 3,000-4,000 units worldwide, excluding Japan. Many of the industrial robots produced in the U.S. have been exported to Europe and Japan, where their use has been necessitated and encouraged for several reasons that lack a parallel in the U.S. at this time. Among them are recurring labor shortages, stated productivity goals, government loans, and other incentives for the purchase and implementation of capital equipment.

Prices for U.S. robots range from around $10-12,000 to over $100,000 for high-technology, increased capacity models. Annual sales volume for 1979 is expected to reach $50 million or a total of 1,500 units.

To promote robotics technology and to increase the manufacturing base for robots in the U.S., a trade association, the Robot Institute of America (RIA), was founded by the Society of Manufacturing Engineers in 1974. SME currently serves as the technical society for robot manufacturers and researchers in this country. RIA membership includes corporate users, researchers and consultants, component suppliers, individual engineers, and nine robot manufacturers.

According to John J. Wallace, outgoing president, RIA, industrial robots have gained wide acceptance in manufacturing from both management and labor. The chief motivation for applying robots in the U.S., he says, has been cost effectiveness. In the past, U.S. end users considered hazard prevention an insufficiently compelling reason for buying a robot.

In addition, nobody in the U.S. ever bought a robot — or much else for that matter — because of government tax credits or similar incentives. Wallace believes that robots have been grossly underused in the U.S. because of an economic environment that provides no encouragement for the purchase of robots or any other type of productivity-oriented capital equipment.

The robot industry now believes, however, that heightened concern for safety in the workplace may create more demand for industrial robots. The OSHA "no-hands" ruling, which, until it was revoked in December 1974, restricted where a worker could place his hands relative to the dies in mechanical presses, represented this concern. Regulated worker exposure to noise, heat, and dirt in hostile environments; particulate and toxicological level limitations; and recent policy of OSHA's Directorate of Health Standards Programs to advocate engineering and work practice controls in preference to personal protective equipment could accelerate robot demand and sales. Low-cost medium-technology robots and other

Founded in 1974 as a trade association dedicated to advancing robotics technology, the Robot Institute of America provides knowledgeable spokesmen to discuss the practicality of industrial robots. Above are the outgoing president, Jack J. Wallace, Prab Conveyors Inc., left, and Jerry Kirsch, Auto-Place Inc., incoming president.

manipulators that are making it possible for smaller firms to mechanize and automate operations on a cost-effective basis could be affected particularly.

Capability Package Concept — In terms of cost savings, industrial robots available today, whether high or medium-technology, are an automation bargain. The trend is toward universality and versatility to produce a standard unit that can handle diverse manufacturing problems.

"Time and time again, we find ourselves replacing custom-made automation, which was designed to do only a limited job (and failed to do it successfully), with a standard

This heavy-duty electrically driven robot handles aluminum castings in final machining area. Permissible ambient and handling capacity ranges are 0 to 40 C and 6 to 60 kg.

Robot applications today range from picking up a 100-lb welding gun and putting spot welds on rocker and shroud panels of car bodies to performing slurry dips in investment casting plants.

Industrial robots are the basic tool in turnkey welding systems. Minicomputer-based control is programmed by a simple, hand-held "teach" pendant (inset). The operator needs only a thorough knowledge of the physical job the robot is expected to perform. No computer experience or complex calculations are required. Cathode ray tube displays pertinent data. Design permits complex six-axis motion anywhere throughout a 1,000-cu-ft envelope approximating a 16-ft hemisphere around the base.

robot, which is sold at a fraction of the cost of custom-made automation," Wallace says.

Industrial robots now represent years of development, refinement, and debugging. Technical advances and evolutionary options have increased robot reach, payload capability, memory, and innovations in multi-axis, end-of-arm tooling. Testing procedures, advance programming by manufacturers, and reduced time for installation and startup reportedly are providing significant cost savings over special-purpose automation.

Pioneering the Robot Frontier — Robots are being discussed and investigated by an increasing number of scientists and researchers worldwide. To provide a forum for robotics experts, the 9th International Symposium and Exposition on International Robots convened March 13-15, 1979, in Washington, D.C. Robotics specialists from 13 countries presented 56 papers in 12 sessions devoted to such topics as recent progress with "sighted" robots; new applications in machining, production and assembly; improved control systems, software, and the languages that are making robots more productive; and breakthroughs in proximity and tactile sensing.

To promote technology transfer in an industry that knows no national boundries, an informal, multinational group known as the "Robotics Council" exists, with a closed membership of about 60 management persons from the automotive, agricultural, finishing and coating, assembly, and other industries. A general membership of about 1,000 persons also has been developed.

Several robotic organizations function worldwide to foster initiatives in research and development

Diecasting Update . . . Robots

A substantial shift has occurred in the diecasting industry. An escalation in the price of zinc a few years ago forced a shift to the use of aluminum, which accelerated a trend toward automatic ladling of metal in cold chamber units and an increase in the tonnage capacity of the new machines.

Several things affected robot utilization. The larger tonnage machines require robots of increased strength and linear stroke to manipulate large castings at high speed. Cycle times have increased also, allowing the use of slower, less sophisticated units for part removal without prohibitively affecting total production output.

However, this development also increased the market for the robots with greater capability. The "lead it

by the hand to teach it" methods allow quick setup for trimming or other secondary functions, and die care lubrication programs have been developed to help cool specific problem areas in larger dies.

Long-time users now find value in utilizing the robots for insert loading in multiple-cavity dies and also find that the removal of 65-lb transmission case castings can be performed more quickly automatically than by manpower. Large castings sometimes must be manipulated carefully around obstacles in the die area or the tie bars, thereby requiring four to five axes of motion. Many die care programs also require four to five axes of motion to reach all die projections. — **William E. Uhde,** Unimation Inc.

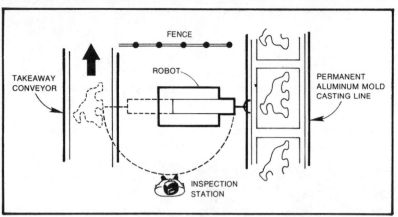

One large foundry uses a robot to unload aluminum exhaust manifolds from permanent mold equipment, to present them to operator for visual inspection, and, after inspection, to place them on conveyor.

Robot builders evaluate equipment from the standpoint of their own product liability. For some models, standard equipment includes hazard guarding such as fences to prevent accidental entrance of a worker into the operating area.

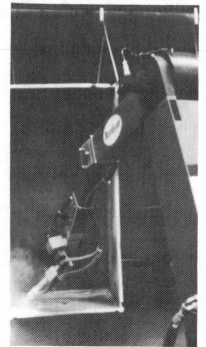

Types of Robots — Varying degrees of complexity in design and control, specific manipulative capabilities, and the unique characteristics of proliferating robot applications differ greatly by individual manufacturer and complicate the classification process.

From the standpoint of control, robots in general may be broadly classified as **non-servo** and **servo-controlled** types.

Non-servo robots often are referred to as "pick and place" or "end point" robots. They are characterized by a few degrees of freedom or limited sequence capability, utilizing hydraulic and pneumatic control valve and actuator systems. Programming involves setting up the desired sequence of moves by adjusting end stops for each axis. Typical applications are simple loading/unloading and transfer operations.

Servo-controlled robots constitute the second major class. Significant features include the ability to command the manipulator's members to move and stop anywhere within their limits of travel rather than only at extreme points. Velocity, acceleration, and deceleration of various axes moving between programmed points can be controlled. Memory capacity exceeds that common to non-servo types. Drives are usually sophisticated hydraulic or electric mechanisms that utilize state-of-the-art servo-control technology.

Servo-controlled manipulators have two subgroups: point-to-point robots and continuous-path robots, often referred to as medium-technology and high-technology machines, respectively.

and to organize seminars, conferences, and exhibitions. Robotics activity worldwide now attracts an ever increasing number of adherents and manufacturing engineers. Leading robotic organizations include the *Association Française de Robotique Industrielle* (AFRI), France; *British Robot Association* (BRA); *Japan Industrial Robot Association* (JIRA); *Robot Institute of America* (RIA); and *Societa Italiana perla Robotica Industriale* (SIRI), Italy.

Special-purpose industrial manipulators satisfy complex-motion requirements in priming, spray-painting, and other coating applications. This model uses continuous-path programming to provide smooth-flowing, highly flexible movement. All six robot axes are moved by completely sealed rotary actuators that prohibit dirt from entering the hydraulic fluid. Such a robot might perform core and mold wash operations in high-production foundries.

Reprinted from Foundry Management and Technology, June, 1979

Robots are Molding a New Image – Part II

The first installment of this article provided an introduction into industrial robotics by highlighting some of the men and the machines that have brought robots to the attention of manufacturing engineers worldwide. In this concluding installment, two major types of equipment are explained in terms of their applicability to the foundry, with emphasis on some important, recent developments.

By NORBERT G. SEMAN / Associate Editor

TWO GENERAL TYPES of industrial manipulators are helping to meet the challenge of the foundry industry. They are programmable robots and teleoperator manipulators.

Programmable Robots — Prior to 1977, two main types of programmable industrial robots were commercially available. They were the "point-to-point" and the "continuous-path" machines. The term "controlled-path" robot also has gained currency as a result of the use of minicomputers for extended controller memory capability and, more recently, microprocessors for cost savings.

Point-to-point (PTP) robots move in discrete steps from one point or location in space to another. During the "teaching" phase of the operation, each of these points has to be recorded in sequence. At any one point in space the robot's program may be interrupted by input signals from interlocking equipment, or it may provide output signals to operate external equipment.

Continuous-path (CP) robots operate, in theory, in an infinite number of points in space that, when joined, describe a smooth compound curve. This curve usually is developed during the programming or "teaching" phase, which is carried out by an operator. The controlled-path type of machine is less common and utilizes a computer control system with the computational ability to describe a desired path between any preprogrammed points. Each axis or degree of freedom can be controlled and actuated simultaneously to move to those points. The computer calculates both the desired path and the acceleration, deceleration, and velocity of the robot arm along the path.

Load capacities of PTP and CP robots are essentially a function of width, motion, inertia, and other factors. Although they differ by individual manufacturers, as do sizes, a present-day capacity range of between 300-500 lb on a fully extended medium-technology robot arm with ± .008 in. accuracy on repeated positioning is reported by an RIA spokesman.

Man and Machine — Although not classified as a robot in strict terms,

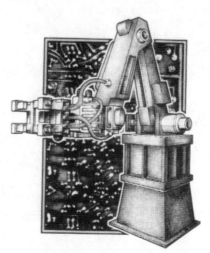

but worthy of mention here because of its related design, application history, and functional attributes, the teleoperator unit covers industrial manipulators that are man-operated, with the operator in the control-loop on a real-time basis.

Teleoperator manipulators use master-slave operator's controls with force feedback. Balanced use of man and machine, which permits the operator to transmit his inherent intelligence and dexterity through the machine to the task, is based on cybernetic anthropomorphous machine systems (CAMS) technology.

Teleoperator manipulators are *cybernetic* in that man is retained in the system; they are controlled by a human operator with responsive, decision-making capability. They are *anthropomorphous* in that they resemble man in form and duplicate his manipulative powers. Essentially, the machine system serves as a physical extension of man's strength, endurance, and task-performing functions. For this reason, such machines are said to have "instinctive" control, which has proved of interest in satisfying the relatively unstructured applications of the foundry.

The first application of a teleoperator manipulator in the foundry industry reportedly was for a shakeout application in an automotive foundry in the early 1970s. Other applications have included furnace slagging, shotblasting, ladle skimming, palletizing, and metal pouring installations. A load capacity range of 125-7,000 lb is presently available for manipulators of this type. Six degrees of movement include horizontal extension, hoist, azimuth rotation, yaw, pitch, and roll. A servo-control system transmits a small proportion of the load force to the operator's hand, thus giving him "instinctive control" of the job.

Back to Basics — Industrial robots exist today in a wide range of capabilities and design configurations (see May, p. 31). In general, however, they share three main components: the manipulator, the controller, and the power supply.

The manipulator consists of a series of mechanical linkages and joints capable of motion in various directions to perform the desired type of work. These linkages commonly are actuated by pneumatic or hydraulic cylinders, hydraulic rotary devices, or electric motors. In dirty, hazardous areas, completely sealed rotary actuators offer critical protection for internal parts. Their use probably will become more common on future industrial

robots, according to one manufacturer.

Sensing the positions of the various linkages and joints and transmitting this data to the controller is accomplished by feedback devices. These include limit switches, encoders, potentiometers or resolvers, tachometers to measure speed, and, more recently, visual and tactile sensors.

The controller initiates and terminates manipulator motion in a predetermined sequence, stores position and sequence data in memory, and interfaces with external devices. Controllers range from simple step sequencers, pneumatic logic systems, and diode matrix boards to minicomputers and microprocessors for cost advantages over computers. Two-way communication between the controller and auxiliary devices can be provided to allow the manipulator arm to interact with other equipment related to the robot's assigned task.

Power supplies function to provide energy to the actuating devices of the manipulator. They include input power regulators, compressors, and hydraulic power packages.

Union Concern — In the past, labor unions such as the UAW have voiced no objection to robot utilization *per se* in manufacturing operations. In fact, the reaction of the work force often has been initially hesitant, but generally receptive and favorable if it means a job that is less hazardous and more desirable.

Recently, however, the UAW Skilled Trades Department approved resolutions for a contractual ban on layoffs that would result from a reduction in the workforce due to the introduction of a technological advance or change. Both robot adherents and end-users so far have countered this with assurances that displaced workers will be redeployed to other job functions.

In this year of major contract negotiations, the UAW has taken a historic stand on the utilization of industrial robots. That position will affect collective bargaining in the months ahead and may echo widely throughout other manufacturing

Operator-controlled manipulator with load capacity up to 7,000 lb isolates operator from noise and heat in a controlled environment cab. Inset shows a 3,000-lb capacity manipulator with arm and hand load control, 110 deg of four-way vertical and horizontal "wrist" action and force feedback. Photos courtesy G-E and Action Machinery Co. (insert).

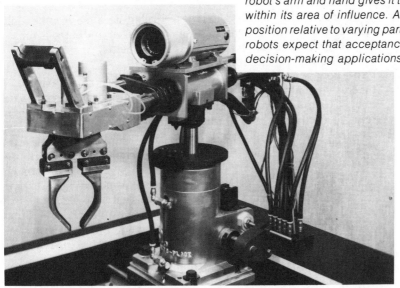

Video-adaptive robot with seven axes of motion represents important step in the progression of robot intelligence. Video-feedback positioning of the robot's arm and hand gives it the ability to search for a randomly placed part within its area of influence. After finding the part, the robot adjusts hand position relative to varying part-presentation conditions. Builders of "sighted" robots expect that acceptance of their product will lead to inspection and decision-making applications. Courtesy Auto-Place Inc.

industries in the near future. For details see box below.

Advantages — According to Walter Gray, formerly with General Electric Co., the prevailing trend in foundry production, beleaguered as it is by energy, materials, environmental, and safety and health considerations, "is toward man becoming the controller, not the doer." To alleviate or at least manage those problems, foundries have opted for automation in the form of industrial robots and manipulators that offer a viable solution to cost control, scrap reduction, pollution control, safety and health standards, attaining greater productivity, improving quality control, and reducing labor costs.

Industrial robots and other manipulators may be considered in foundries in any work function that involves the handling of parts, tooling, flasks, or process materials requiring movement, positioning, sorting, or transporting over limited distances. As noted earlier, such potential applications could include charging or tapping furnaces; handling of sand, flasks, and cores; handling of cast parts in shakeout or cleaning rooms; manipulating grinders, cutoffs, and other process tools; pouring molds, and others.

Another important advantage of manipulator utilization, whether of the programmed or teleoperator type, has been the reduction of exposure of human workers to undesirable environments and the subsequent benefit of reduced turnover in the labor force. Motivation for considering robot manipulators in such applications also has included improved economics through reduced labor, increased productivity, and improved worker health and safety by reducing exposure to noise, heat, dirt, and machine hazards.

High levels of reliability and availability achieved by industrial robots today allow them to function up to 98% of the time, according to several sources. Absenteeism in areas of worker discontent has and will continue to exert motivational pressures on foundry managers to acquire robots and other industrial manipulators. That trend has involved three basic factors that remain valid today: 1. Part orientation, including loading/unloading clearance. 2. Required cycle times. 3. Load capacities.

Conclusion — The approach to a foundry robot application project should be based on sound principles of methods and work simplification. To become enamoured of robotry initially is to risk losing sight of goals and needs. Caution must prevail in weighing additional options against the value of keeping the robot simple to insure its success.

Recent developments have enhanced the capabilities of industrial robots, but the basics of robotics, as pointed out earlier, remain simple and are not doomed to obsolescence. Providing foundrymen with an effective solution to specific problems, industrial robots are molding a new image.

Acknowledgment — The following organizations generously supplied photographs, commentary, and technical data for this article: Action Machinery Co.; ASEA Inc.; Auto-Place Inc.; British Robot Association; Gebr. Boehringer GmbH, FRG; Gebr. Bühler AG, Switzerland; Central Foundry Div., GMC; Cincinnati Milacron Inc.; DeVilbiss Co.; Eaton-Kenway, sub. of Eaton Corp.; Manigley & Cie, Switzerland; Nordson Corp.; OSHA, Dept. of Labor; Re-entry & Environmental Systems Div., General Electric Co.; Robot Div., Prab Conveyors Inc.; Robot Institute of America; Seiko Instruments Inc.; Society of Manufacturing Engineers; SRI International; Unimation Inc.; United Auto Workers, Skilled Trades Dept.

Off-The-Shelf Automation

Forget the name "robot." Think instead of a piece of automatic handling equipment that can be programmed to do not just one, but a large number of production jobs. Think also of a machine that can replace human labor where the job is dull, simple or hazardous. Then think "robot."

The robot is saddled with a science-fiction image which unfortunately inhibits a realistic evaluation of its capabilities. Today's robots are not human-like machines, nor are they ever likely to be. The majority perform only a few preprogrammed motions. Only the most advanced have crude TV "eyes" or sensors that provide a primitive sense of touch.

But the industrial robot does offer the engineer a number of advantages worth considering. The robot is essentially a piece of off-the-shelf automated equipment that can be put on line with a minumum of debugging. The chief difference between a robot and any other piece of automated equipment is that the robot can be easily reprogrammed. It is not limited to one job or one sequence of motions.

Robots have had their greatest success in simple transfer operations such as loading and unloading presses, die-casting machines, or molding machines. In part fabrication, the robot has been used for both automatic painting and welding.

Loads from a few pounds to several hundred pounds can be handled. The objects moved can be as diverse as steel billets, glass tubes or explosives.

A robot is most effectively used in medium-length production runs. The volume must be large enough so that manual labor is not practical yet small enough so that total automation with special-purpose machines is not economic.

Why they're needed

Sales figures for robots have never been spectacular. After more than a decade of production and promotion, the estimated robot population in the U.S. is less than 2,000 units. Despite a past strewn with failed companies and over-promoted designs, market sages see a bright future for the robot because it offers solutions to three of industry's most difficult problems:

OSHA regulations: Existing and proposed worker safety rules will require extensive modification or replacement of production line equipment. The rule which restricts where a worker can place his hands within a machine and the proposed lower sound levels are just two examples. Where a large number of machines are involved, it may be less costly to convert to robot operation than to change equipment.

Productivity: Robots do not belong to unions, get no paid vacations, and don't have hangovers on Monday morning. They also do not balk at changes in work rate or machine assignment. The robot therefore offers the potential for less down time, more consistant, faster production, and less waste.

Labor cost: At the present time, the robot has, in most cases, taken over only those jobs where humans can not or are unwilling to work. These include those jobs that are hazardous, strenuous, or boring. But, as labor costs continue to rise, the robot will become a stronger candidate for a wider spectrum of blue-collar jobs.

Unfortunately, replacing men with robots is not a simple matter of firing the man and plugging in the machine. To use the robot most effectively, the entire production operation must be analyzed and possibly reorganized to take advantage of the robot's capabilities.

Rarely is it cost-effective to bring in just one robot. The exceptions are the less sophisticated robots which may perform a simple transfer function that influences only one machine, or a single robot to work in a very hazardous environment.

Generally, several robots are

How many axes?

Robots are available with two to six axes of motion. Most frequently, three to five axes are used in industrial applications. This Unimate sketch shows a six-axes robot. Normally, the robot is supplied with five axes. The sixth, wrist swivel, is added as an option. The Cincinnati Milacron 6CH Arm has six axes and can lift a 175-lb load and move it at 50 ips.

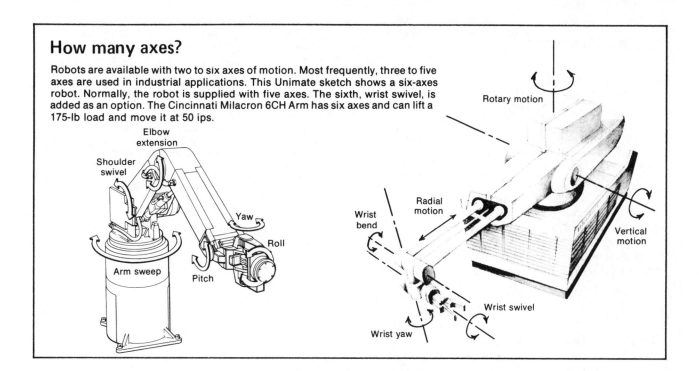

Three robots work arrangements

Work surrounds robot: Robot remains stationary and removes part from die-casting machine.

Robot moves to work: New Series F Versatran robot from AMF moves along a track at speeds up to 24 ips. By using a track for added mobility, one robot can serve several machines or work stations.

Work moves to robot: Famous Lordstown, Ohio assembly line uses a row of Unimate robots to spot weld car bodies as they move by on an assembly line.

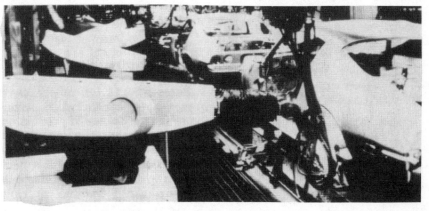

In the U.S., the three "high-technology" robots presently available are made by Unimate, Versatran, and Cincinnati Milacron.

Unimate, to date the most successful robot manufacturer in terms of units sold, offers two series of robots, the 2,000 and the 4,000. Both are in a turret-on-a-box configuration with up to 1,024 steps in their solid-state memories. A maximum of six axes of motion are available, although special units may be built to order. Unimate robots are hydraulically actuated, except for the hand, which may be pneumatic. Because the robots "understand" computer language, an entire bank of robots may be under the command of a single computer.

Versatran, a division of AMF, has been in the robot field since 1963. Recently introduced Series F robots are hydraulically actuated and servo-controlled with analog feedback. Three versions are available with payloads of 250, 350, and 550 lb. Microprocessor control can provide 4,100 positions with a 3-axis unit. Positioning repeatability is ±0.050 in.

Cincinnati Milacron, a new entrant into the robot field, offers the *6CH Arm.* It is said to be the only robot to have a jointed arm. This design reportedly provides a larger operating envelope than other arms. The memory is a minicomputer which can be linked directly with a master computer. Hydraulic actuation is used on the six axes.

At the other end of the technical-sophistication scale are the limited-sequence or pick-and-place robots. One unit of this type is the *Auto-Place,* a pneumatically actuated unit with a 24-step, air-logic programmer. Two versions are made, one rated for a 10-lb load, the other for 30 lb. The unit may be mounted directly on the unit it is serving.

Another pneumatically actuated design is the *Robotarm* from Stirling-Detroit. This company specializes in pick-and-place operations on die-casting machines. The arm is integrated with the design of the machine and not floor mounted.

The *Auto-Mate,* another unit in the limited-sequence category, is turret mounted and pneumatically actuated. The program is solid state, but a pneumatic system can be used if the robot is to operate in an explosive environment. With solid-state programming, the steps are first calculated on a special chart, then punched in with buttons. Load capacity is 10 lb and base price is around $10,000.

A "mid-technology" robot is the *Prab,* which has the load capacity of more complex machines (70 lb) but is controlled by a rather simple program. To establish the program on the hydraulically actuated machine, tabs are set on a rotating drum. The tabs trip switches to activate the robot's four axes as the drum turns. Base price for the Prab is around $15,000.

Some robot manufacturers key their designs to a specific process or type of industrial machine. The Prab, for example, specializes in serving various automatic metal-working machines.

A British-designed robot calls *Binks,* now available in the U.S., handles only painting. The unit can be programmed to paint several objects with different shapes. Sensors on the assembly line tell the robot which object is in front of it and the robot selects the appropriate painting sequence.

The Norwegian-built *Trallfa* robot has been teamed with U.S.-made DeVilbiss painting equipment and ASEA welding components for automated welding. The 64-step robot has six axes of motion.

needed for the cost savings in production to exceed the costs of the robots plus their maintenance. For the more sophisticated robots, four or five units are a minimum.

What can they do?

Each robot has its own unique characteristics, such as number of axes, type of arm motion, type of programming, load capacity, and speed. Selection requires a careful pairing of these capabilities and the job the robot is to fill.

Cost: Base price can range from around $3,000 for a limited-sequence robot to $85,000 for a computer-controlled unit. But this is only the beginning, in some cases. Additional equipment might be needed, such as special-purpose hands or manipulators. If the robot must interface with a complex production process, additional sensors and interconnections will be needed. Cost of rearranging plant equipment or production lines to best utilize the robot, plus maintenance equipment and personnel, must also be considered.

Arm: All robots operate with some type of arm and hand configuration. Arm configuration determines where the robot can reach. Straight arms, for example, can not reach over objects. Articulated arms can. Arm positioning is based on rectangular, spherical, or cylindrical coordinate systems.

Axes: The complexity of the motions a robot can perform is determined by the number of axes the robot has. Simple transfer operations may need only two motions, such as rotation and grip. A robot that can weld some recessed point on a product may require six axes (arm rotation, elevation, reach, wrist bend, swivel, and yaw).

Load: Company specifications often give the robot's load capacity under static conditions. This is not necessarily the load the arm can take at full reach and maximum speed. Maximum load depends not only on the strength of the arm but the speed at which the load is to move and how fast the load is accelerated and decelerated.

Hand design: Most robots have a simple grip hand as standard. Optional hand or gripper designs with multiple fingers or suction cups are usually available as additional-cost options.

Actuation: Robot arms are moved by hydraulic or pneumatic actuators, electric motors, or, in simpler designs, mechanical linkages. Pneumatics are preferred for high speed motions with fairly light loads. Hydraulics, the most common actuation method, offers the ability to hold heavy loads for sustained periods without slip. Electric drives are said to be the simplest to control. Damping or accelerating and decelerating characteristics must be considered with evaluating actuators if heavy loads are anticipated.

Speed: Cycle time is critical when evaluating the robot's ability to increase productivity. The robot must be able to meet, or exceed the cycle time of the machine or conveyor it serves.

Space: Robots can be set up in three ways:

Robot moves to the work, the least used method, has the robot ride on a rail or other guide system. With this arrangement, one

Programming your robot

Manual: In the simple robots, programs are created and changed by physically fixing stops, setting limit switches, inserting punched cards, arranging wires, or, in the case of air-logic units, fitting air tubes. Most of the robots using this type of programming are the so-called limited-sequence machines which have two to four axes of motion and a relatively fews steps in their program.

Walkthrough: More complex robots, in which magnetic tape or discs or a computer memory generates the robot commands, the program can be "taught" by walking the robot through the operating sequence. The operator manually moves the hand and arm, then signals the memory to record the motion or location.

Leadthrough: The memory is similar to that needed for walk-through programmed. But, in this case, the operator drives the robot through the sequence. With the control console or a separate control box, the robot is slowly moved through the program. At the end of each motion, the action is entered into the robot's memory.

Programming for all robots may be point-to-point or continuous path. In the more complex robots, point-to-point is about the same as continuous path because the individual points may be only a fraction of an inch apart.

Simple point-to-point programs are used where only the end points of the cycle are important. For example, the robot might pick up a part from a conveyor and place it on a machine.

Continuous path is needed where the entire motion is significant, as when painting, or when the hand must grasp a moving object.

Manual programming of the Auto-Place robot requires interconnecting a series of pneumatic valves in the 24-step air-logic box. The program can be enlarged by interconnecting two or more logic boxes. Stops on the Auto-Place arms must also be set to establish reach length and degree of turn.

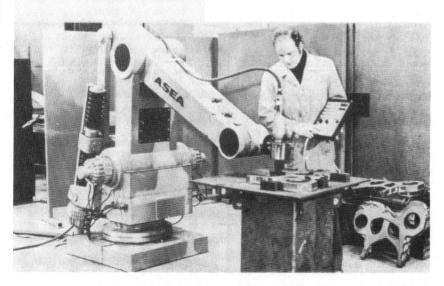

Control box allows the programmer to drive the robot through the operating sequence on the Swedish-built ASEA unit. When each motion is properly completed it is recorded in the memory.

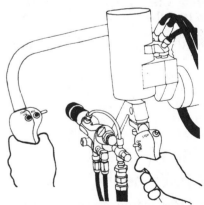

Control handles allow the programmer to walk the robot through its program. This robot, which is designed specifically for painting, has one control to establish arm motion and another for the painting sequence. The Norwegian-built Trallfa robot is fitted with painting equipment and marketed in the U.S. by DeVilbiss Co., Toledo, O.

robot can serve several widely separated work stations or machines.

Work moves to the robot has the robot stationary and the work moving by. Removing parts from an assembly line or painting moving parts are typical examples.

Work surrounds the robot, the most commonly used configuration, has the robot stationary amid the machines it serves. Where floor space is tight, this can be a problem. Required floor area is greatest for robots that have the arm and control unit as a single package. Less or no floor space is needed by those robots which have an arm that can be mounted directly on the tool or machine it serves with the control in a remote location. The remote control location can be an asset when the robot must operate in a hostile environment.

Survivability: Robots require special protection and shielding when operating in high temperatures, where vibration is severe, or in corrosive environments. Special protection may also be needed if there are sharp temperature shifts or electromagnetic radiation is present.

Programming: Most robots operate with an open-loop program in that they repeat the same set of motions continuously without modification. Robots are programmed in one of three ways: manually inserting the program, physically leading the robot through the program, or leading the robot through the program using machine controls.

What's next?

Robot manufacturers are reluctant to talk of a new generation of "smarter" robots because industry has only begun to utilize the capabilities of existing robots. But a number of research projects are underway and, if industry's interest in robots grows as predicted,

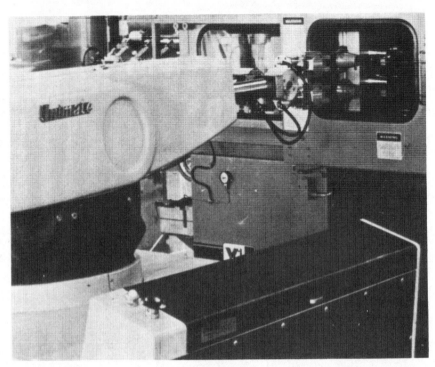

Unimate 2000 series industrial robot has dual hand to load and unload an injection-molding machine. After removing the two plastic parts, the robot sequence includes fitting the parts together and feeding them to an ultrasonic welding machine to the left of the robot.

these smarter robots will eventually appear on the plant floor. It is one of the robot's peculiar advantages that a more sophisticated design does not make the earlier design obsolete. A new design will take on more demanding jobs while the older robot can continue in its assigned task.

The next generation of robots will probably free the robot from its part-orientation limitation. Present robots must operate with parts that are of the same general size and orientation. These parts must also be in the same location. The robot can not "look" for a part. New robots may be able to find specific parts with TV eyes or orient parts as required. A memory linked to the eyes may be able to tell the arm which part to select or be capable of sorting out or removing wrong or broken parts.

More sophisticated sensors are also being developed which can "feel" the difference between various part sizes or part orientations.

The next stage of development may be robots that can understand spoken commands or convert printed language into operating commands. Elementary intelligence in the robot's programs may offer the ability to change programs on their own or modify programs to meet new situations.

Although these gee-whiz advances promise much for the future, the greatest advance in robot utilization will may be brought about by more effective plant management techniques. For example, "group technology" would be a boon to robots. In this technique, the parts to be manufactured are classified into families and the machine tools to create the parts are assigned to these families. Parts are never placed into bins for storage or transfer to other areas but maintain their orientation throughout the manufacturing process.

Computer aided manufacture (CAM), automatic warehousing, and automatic inspection may also be carried out more effectively if the robot is incorporated into the plan.

CHAPTER 2
IMPLEMENTATION

Commentary

This chapter addresses the activities involved in implementing robot installations, from initial surveys through system start-up and beyond.

A systematic approach to identifying and engineering robot applications is presented. Related topics cover the issues of determining if a robot is appropriate and, if so, which one to choose.

The integration of the robot into a total system is explored in depth, from planning, through interfacing, to launching. Guidelines are provided for planning a safe installation of the robot in the workplace. Robot reliability and maintenance expectations and requirements are reviewed and related to installation and training.

The information contained in this chapter represents the collective experience of many of the major robot users, most of whom were pioneers in some aspects of robot application. Failures as well as successes are described and analyzed.

Presented at the First North American Industrial Robot Conference, October, 1976

A User's Guide To Robot Applications

By William R. Tanner
Ford Motor Company

The development and implementation of industrial robot applications can best be approached through a logical sequence of
steps. Due to the inherent limitations of robots, it is important to the success of any robot application that these
limitations be carefully considered.

Alternatives to the usual fixed position index, short cycle
approach to robot applications include parallel operations with
longer, more complex cycles, moving line operations with tracking systems, unusual attitudes and positions of robots and part
handling instead of tool handling.

Maintenance and programming should be handled by training of
in-house personnel, rather than by service contracts with the
manufacturers. The necessary expertise is within the capabilities of most plant maintenance personnel and their respon-
sibilities should include both programming and maintenance of
the robots.

INTRODUCTION

The use of industrial robots in production operations is a
relatively new aspect of manufacturing engineering. The development and implementation of robot applications generally
follows the same basic sequence as any other manufacturing
process. However, the robot's unique combination of flexibility and limitations requires some special consideration for
successful application.

Through ten years experience with robots and their use in a
variety of tasks within the universe of automotive assembly
operations, a systematic, sequential approach to robot applications has been developed and is presented here. Several
alternatives to "traditional" robot applications are also
presented and their potential advantages are explored. Maintenance and programming requirements are reviewed and the development of in-house capabilities in these areas are discussed.

A list of "do's and don'ts" representing significant considerations based upon actual experience is included as a check list
for the development of robot applications.

APPLICATIONS DEVELOPMENT

The first step in robot applications development is to become
familiar with the basic capabilities and limitations of the
equipment available. Today's robots range from simple pick-
and-place devices to multi-axis, computer controlled machines,
with handling capabilities from a few ounces to half a ton or
more and operating in fixed stop, point-to-point or continuous

path modes. To simplify this task, first determine what kinds of operations you want to perform with robots and then concentrate your research on the machines which have the capabilities to do those jobs. For example, if you are going to apply adhesives, paint or similar materials to an object (which generally involves a spraying operation) you are probably going to need a continuous path control machine, which narrows the field of search considerably.

There is a great deal of specification information available from the robot manufacturers which can serve as the primary source of data. In order to make the analysis easier, develop a matrix so that a direct comparison of such features as load capacity, number of axes, work envelope, memory capacity, control system, speed, price, options and so forth can be quickly made between the various robots. This matrix will be a valuable reference tool during the evolution of specific applications. Also, note any unique features of the available robots for future reference. Make note of specific questions about the capabilities and limitations of various robots and review these with the manufacturers. In essence, the more familiar you become with the equipment available, the easier it will be to first determine whether any robot is capable of doing a specific task and to secondly narrow the field down to the best choice or choices.

Armed with all of this knowledge, you are now equipped to make an initial survey for potential applications. At this point, look for tasks which meet certain criteria: The operation must be within a robot's capabilities in terms of load, work envelope and complexity; the operation must not require that judgement be applied except in simple terms; and, the operation must be one which justifies the use of a robot.

The first of these criteria, not exceeding the robot's capabilities, is quite straightforward in terms of load and work envelope. The complexity of the task will determine the memory capacity and/or control complexity required, as well as the number of axes of motion necessary.

In terms of judgement requirements, the robot is totally dependent upon outside information, except for its programmed positions. Presence or absence of a part is one bit of information that is usually necessary and providing this information is easily handled; however, determining if the part is good or bad can be significantly more difficult and if this determination is necessary, you should probably consider robotizing some other task. Bear in mind that a robot is a device which is programmed to move through and stop in various positions in space. It performs this task with a fair degree of accuracy and repeatability; however, if the object it is to handle or work on is not consistently oriented or located with at least a similar degree of accuracy, the robot will fail. Likewise, if the tool a robot is handling malfunctions, the robot will still go through its motions even though it will not accomplish anything.

The last criterion, justification, may involve a number of factors such as economics, health, safety, job enrichment or product quality. Here, you probably will have some direction or policy to guide you, at least in the initial survey stages.

Your initial survey will yield a list of potential applications which will require more detailed study to implement. Before going on to that step, consult with the various robot manufacturers whose equipment could do the jobs. Here's where your capabilities matrix comes in handy -- in deciding which manufacturers to consult. Ask their applications engineers to look at the jobs on your list and to evaluate each of them. In the process, they may also see some other potentials which you have overlooked. It is strongly recommended that you have more than one manufacturer look at the jobs. Get at least two, and if possible, more opinions. If only one out of three or so opinions is positive, you should probably pass on that job, at least until you have more experience. Another source of help in this screening process might be other people with similar robot applications (if you know of any) or other major users of robots. Often the manufacturers can give you some leads in this area.

Even though you have solicited their advice, it is wise to consider the manufacturer's opinions and recommendations with some caution. Remember that their business is to sell robots and although it is unlikely that any of them would deliberately mislead you, they tend to be somewhat "optimistic" at times.

After the first broad survey for potential robot applications, you are ready to zero in on the details. For the first installation, select the simplest task on your list. The intricate, involved tasks may be more challenging from an engineering standpoint and may even show a greater potential return on investment, but it's best to avoid the temptation to tackle the tough ones until you have established a fair degree of competence and a reasonable level of confidence.

Once the first application is selected, get out and study the job. If it is a manual operation (and it probably will be) make sure that you know everything that the man has to do. Remember, the robot is extremely limited in its decision making capabilities. Be sure that the man doesn't have to use some sense, such as sight or touch, that the robot doesn't have or that he doesn't have to make a value judgement in executing his task. At the same time, be sure that everything you see the man do is necessary. Don't be mislead by some extraneous action or process that the robot couldn't handle, but that the man really doesn't have to do either.

At this point, it is also advisable to consider any alternatives other than robots to do the task at hand. If the job can be done easier or cheaper with special-purpose automation, and if such automation won't become obsolete in the near future, then a robot may not be the best way to go. In other words, don't use robots simply because they are novel or new or different.

Another thing to look at is the possible advantages of mounting a robot in something other than the usual feet-on-the-floor attitude. Many robots can be inverted, suspended over a work station or off of a wall or similar support. An unusual position of a robot might provide an easy solution to an otherwise difficult problem. Of course, you have to select a robot which will operate in the unusual attitude you've planned for it. Also, consider reversing the usual bring-the-tool-to-the-work approach by having the robot carry the work to the tool.

In analyzing the task, try to anticipate all the things that could go wrong with anything associated with the job. This is where many apparently easy applications become failures. For example, if the operation involves transferring parts to a conveyor, does the hook or basket always stop in the same place? Is the hook stable when the conveyor stops or does it swing like a pendulum? Is the incoming part always in the same place and have the same orientation? If the delivery conveyor brings another part before the robot is ready for it, what happens? If the robot misses a hook, what should it do, try again or wait for the next hook? Will it even know that it has missed? If an empty hook arrives at the next operation, what is the effect? Remember that robots perform their tasks in a set sequence and most cannot adapt to changing conditions the way a man does. Take your time in looking at the operation - most of the problems will become evident eventually. When you are looking, try to think like a robot, by visualizing the motions it will go through in performing the task.

The next consideration is backup for the robot. Although robots have become extremely reliable (average long term experience is about 98% uptime) they will still break down sometimes. To assure that a robot breakdown isn't going to cause a loss of production, a backup system must be provided. This could be simply a stockpile of parts to feed subsequent operations or a manual backup system. On multi-robot installations, it may be possible to use the robots themselves to back up each other, but this probably won't provide 100% production capability. Another point to consider in developing the backup plan is this: when the robot goes down, it probably won't be back on line in just a few minutes. It's often a complex device and troubleshooting, even with experienced personnel, can go slowly.

By now, you should be ready to select the robot for your specific task and to engineer the application. Be sure that the robot chosen has sufficient reach, speed, memory, program capacity and load capacity to do the job. Provide some extra capacity if possible. Don't forget that the load on the robot will include the tooling as well as the part and that high speed operations may also impose significant inertia loads on the equipment. Consider the environment. Intrinsic safety or explosion proofing may be required or protection of the robot from contamination by dust, paint overspray, metal particles, excess heat, etc.

Once the robot has been selected, make a layout of the installation to determine its location, possible interferences, facilities changes required (if any) and the like. At this time, also determine what interfaces will be required between the robot and other equipment and plan for the provision of utilities such as electrical power, compressed air, cooling water, etc. Adequate interlocks and guards must be provided to protect personnel in the area (remember the robot can't see) and also to protect the robot from material handling equipment or other possible damage. End-of-arm tooling must also be provided. Here the robot manufacturer may be able to assist you, particularly in the design of part grippers for handling applications. If line tracking is required, the installation and inter-connection of a suitable feedback device must be provided. Also, provide the backup system to protect production when the robot is down, as well as spare parts and test equipment for maintenance.

Assuming that you have justified the installation and ordered the robot, you should do as much preparatory work for the installation as possible ahead of time. Service drops and floor preparation can usually be done in advance, as well as some interfacing and rearrangements and the development of the end-of-arm tooling. Maintenance and programming training, if done at the manufacturer's facility, can also be handled before the robot arrives but don't do it too far in advance and some refresher training should be also planned once it is on site.

When the robot arrives, the installation and inter-connection can be completed, the maintenance training can be performed and the programs developed. Generally, the robot manufacturer will provide assistance during the installation and startup. As with any new equipment installation, you should anticipate some startup problems. Programs will have to be refined, tooling adjusted, timing and interlocks tuned in, etc. However, if you have done a thorough job of engineering the application, these startup problems will be minimized.

Once the operation is in production, monitor the operation pretty closely for a while. Keep track of downtime (a log is a good idea) so that you can identify recurring problems, not only with the robot, but also with the peripheral equipment. At some point you should also make a comparison between estimated and actual costs, savings and performance for future reference. And, of course, continued surveillance of the operation may suggest ways to improve it.

ALTERNATIVE APPROACHES

The more or less traditional approach to robot applications is on relatively short, repetitive cycles. However, memory capacities of many robots are such that longer more complex cycles are feasible and should be considered, particularly in tool handling operations. Instead of shuttling a job past a series of robots in a sequential fashion, consider parallel lines of duplicate operations of longer duration. For one thing, the loss of cycle time for parts transfer becomes less, percentage-wise, and for another, the parallel systems afford some backup capability. Tooling limitations may inhibit total effectivity, however, programming of tool changing can, in some cases, be used to increase the robot's flexibility to the point where parallel operations become feasible.

Also, consider moving line operations, as long as the parts can be accurately located on the conveying device. Some robots are capable of simultaneous line tracking in more than one axis at a time, which enhances moving line operations. Other robots have traversing capability, enabling them to move parallel to a line and in synchronization with it. It may be feasible in other cases to build a conveying system which moves the parts parallel to a single axis of robot motion. One potential advantage of moving line operations is cycle time improvement. On an operation with a 1-minute cycle time, for example, the moving line approach will result in a 5% or less loss for part movement into the work station, as contrasted with up to 25% lost cycle time to index into a stop station. Another advantage is an increased work envelope, since the robot can start working on something as soon as it enters the work area and continue after a portion of the object has left the work area.

In tool handling applications, the traditional approach is for the robot to move the tool, with the part fixed in a location. Consider reversing the process by moving the parts past a stationary tool. In spotwelding smaller parts, for example, the usual approach is to manually load a fixture, then weld the parts with the robot moving the welding gun and manually unload and reload the fixture. With a stationary welder, such as a pedestal welder and the robot's gripper including locating and clamping details, the parts could be manually loaded into a simple positioning fixture, or perhaps preloaded in quantity into parts dispensers. The robot would then pick up the parts, move them through the welding sequence and also dispose of the parts onto a conveyor or in a container, with an obvious reduction in manual intervention required. Of course, this approach is limited by the size of the parts and by tool access.

You should also consider integrated robot systems as part of development of a new product, process or facility, as opposed to an add-on approach. There are several advantages to this.

The overall cost should be lower, since an assembly system which can accommodate robots probably won't cost much more than one which can't and you avoid the rearrangement costs associated with the add-on approach. The system will be more flexible than special-purpose automation and more efficient than one to which robots are applied as an "add-on". The system can often be launched as a manual operation to ease the problems of startup and debugging, with the robots being added later without seriously interrupting production if they were included in the original concept.

MAINTENANCE AND PROGRAMMING

It is both desirable and advantageous to develop in-house programming and maintenance capabilities, rather than relying on a service contract with the robot manufacturer. The in-house capability is certainly going to reduce reaction time, which is significant since an idle robot isn't earning its keep. Your maintenance people will be responsible for the peripheral equipment anyway and what may often first appear to be a robot problem turns out to be actually a problem with some related equipment which is interlocked with the robot. Train your own people, make sure you cover all shifts and make sure also that they have adequate tools, test equipment and spare parts to do their job.

Because of the high reliability of robots, it is difficult, particularly in installations of only a few robots, to keep proficiency in trouble-shooting and maintenance high. One suggestion in regard to keeping up maintenance skills is to provide for regular retraining. This might be done as part of a periodic overhaul by having the manufacturer's service personnel assist in the overhaul and retrain at the same time. Another approach, which has been effective in plants with larger robot populations, is to provide a spare machine. This is used as a replacement or turn-around machine in a regular, periodic, in-house rehab program in which each machine in turn is removed from production for scheduled major maintenance and replaced by the turn-around unit. This system keeps the proficiency of the maintenance personnel high since they are doing maintenance on a continuing basis and also provides fill-in work, with the overhauls performed on a "time available" arrangement when the maintenance people aren't otherwise occupied. It also provides some added backup capability in the event of a major problem with a robot on line by quickly replacing it with the spare machine.

Give your maintenance people the total responsibility for the robot's performance. That means doing the programming as well as keeping it in operation. They are certainly capable of handling the task with adequate training and it avoids a conflict between what the robot is programmed to do and how it actually performs.

OTHER FACTORS

In considering robots for your plant or business, be sure you
consider the total potential for them. If you can only see a
place for one or two, it might be better to forget them entirely. Training, maintenance, parts and equipment inventories can
be a real burden for one or two robot applications. Of course,
in such a situation you can rely on the manufacturer for parts
and service, but you must recognize that his reaction time may
be slow since you aren't his only customer, and he doesn't perform this service for free, either.

Another consideration, if you have a number of robot applications, is to try to use the same kind of robot on all of them.
This approach may result in some "overkill" on certain jobs,
but it simplifies maintenance and keeps down the spares and
tools inventory. Many robot manufacturers build families of
similar machines which often use the same controls and share
many common components. This affords some degree of flexibility of choice without introducing more complexity into the
maintenance task.

DO'S AND DON'TS

DO: - Become thoroughly familiar with an operation before you
put a robot on it so that you avoid unpleasant surprises.
Engineer the application. Try to develop the ability to
visualize the robot going through its motions.

- Start with the simplest applications, even if the payoff
isn't as good as with more complex ones. Remember that
corollary of Murphy's law - "If you have a 50-50 chance
of success, there is a 75% chance of failure".

- Get your management to back you up. Total commitment by
everyone is necessary for success.

- Include "production" people in your efforts. Get their
ideas (they are probably a lot closer to the job than you
are) and make them feel that they are part of the action.
Remember, they have a stake in the success of the robot,
too, and they're not as likely to blame everything on the
robot if they were involved in the development of the
application.

- Be honest in answering questions from the workers in the
plant. They are genuinely interested and are vital to a
successful application.

- Provide comprehensive maintenance training of sufficient
personnel to cover all shifts and give them the tools
necessary to do their job.

- Provide adequate spare parts to keep the robots running.

The manufacturers can recommend what spares should be stocked. Also, set up a system for regular, prompt return of those defective parts which are handled on an exchange-repair basis.

- Provide backup for production, either standby facilities or reserves of parts for subsequent operations.

- Protect people from the robots and vice-versa. Make sure that OSHA and other safety standards are observed.

- Use your imagination. Consider alternatives to the usual floor mounting of robots and some of the other concepts previously described. Don't simply imitate a man with a robot - there may be a better way.

- Make a record of your robot programs. Some have capability to record on cassette or punched tape. At the very least, write up a sequence sheet so that you can recover quickly in case of memory loss or control system failure. And, keep the records up-to-date. Make new tapes or sequence sheets whenever there is a program change.

DON'T: - Try to do something to something that isn't consistently located. Don't forget that robots generally can't adapt to variances in their environment.

- Try to do something that requires using a maximum capability of the robot, such as reach, load capacity or speed. You need to keep something in reserve to take care of contingencies or in case something changes. Don't get into a situation where the robot becomes a bottle neck.

- Believe everything the manufacturer tells you or rely totally on him for engineering the application or making the installation. Don't forget that you know more about your business than he does, or that he isn't usually going to be around when something happens.

- Assume that the robot program, once made, will be good forever. Things change, sometimes gradually and sometimes suddenly and you must be prepared to change the programs accordingly.

- Assume that every problem with an application will be a robot problem. Make sure that the related equipment is maintained and operating properly. Don't forget that the robot doesn't know or care if the tool it is handling is working or not and don't forget that the robot can't tell you when its tool needs attention.

- Put the robot somewhere where a malfunction might go

undetected for any length of time. It will fill a
system with scrap as easily as with good parts.

CONCLUSION

There has been a great deal of progress in the evolution of
robots in the past ten years, in terms of versatility, capability and reliability. Cost effectiveness increases steadily
and there is a wider selection than ever before. The variety
of applications of robots has kept pace with this development.
Robots are still basically simple, but don't let this simplicity mislead you. Applications must still be carefully surveyed and engineered for success. At the same time, don't be
afraid to use them. Common sense and straightforward approaches to their use will yield good results.

Reprinted from Tooling & Production, August, 1979

Vern's rules of thumb when applying robots

by **Vernon E Estes**
Manager, Process Automation & Control Systems
General Electric Co
Schenectady, NY

The Manufacturing Engineering Consulting and Applications Center of General Electric is one of the best equipped in the US. We are an applications oriented service with a total of nine robots used in an ongoing effort to evaluate the latest state-of-the-art equipment. The Process Automation and Control Systems Operation is responsible for applying the robots in real-world applications. Here's a list of do's and don'ts that reflect our practical experiences in robotics. It has been compiled to benefit people thinking about using this form of automation to help their business.

• Implementation should start in hostile areas

Areas considered hostile by you may not be considered hostile by those being replaced. Ideal start-up would occur in an area that has received OSHA citations or where you have experienced Workmen's Compensation cases.

• Consider applications where productivity is lagging

Most robot installations have resulted in significant productivity improvements—20 percent or more. This is especially true on repetitive jobs that can be demeaning to the human spirit.

• Evaluate long-term needs

Try to identify all possible applications from the start and evaluate vendors on how many they can handle. Each brand name of equipment purchased will require spare parts, test equipment and recorders. Unnecessary expense occurs if duplicate parts are purchased for different brands of robots.

• Implementation cost will be indirectly proportional to the cost of the robot

The more flexible, expensive robots have less precise positioning requirements for the robot, parts orienters and equipment being serviced. Computer features usually allow their capability to be expanded by software. Less expensive robots use stops for positioning and are sequentially programmed. They are normally equipped with a fixed number of sequences that are not readily expandable.

• Keep it simple

When first applying a robot, try a simple but economically lucrative application. Many people are fascinated by the technology and feel it applies to only complex tasks. The point here is to avoid failure that could dampen management spirit.

• Assume that if it can happen, it will

This is Murphy's law. If parts do not always eject from a press or a poorly positioned part could damage equipment, or if people are to work in the proximity of the robot, you had better correct these problems before installing a robot. Sensors can eliminate many problems. Estimate sensor necessity by the cost of possible damage without it, compared to the cost of adding the sensor. No cost is too great for sensors required to protect humans.

• Don't expect vendors to furnish turnkey implementations

What you are really interested in is *integrated turnkey*. That means there is an operator or worker just as responsible for the implementation as you expect your vendor to be. Most robot systems are interfaced to something in the shop that you know more about than the robot vendor does. To make the marriage between the robot and your equipment successful, the same effort will have to be made as in a successful human marriage. Such a marriage requires a 60 percent effort on both parts to make it 100 percent successful.

• Don't forget people requirements

Robots require maintenance by people. They consist of some type of electrics or electronics and mechanical devices. The people who maintain these devices must understand that they affect the accuracy as well as the overall functionality of the robot. You will also need programmers who know the hazards of poor programming in relation to damage to both equipment and humans. ■

Reprinted from Robotics Today, Spring, 1980

Can I Use a Robot?

Here are seven rules of thumb which will help you determine whether a robot is the right choice for your particular application

WILLIAM R. TANNER
President
Tanner Associates

Today's industrial robots are available in a wide range of capabilities and price ranges and are being used in a wide variety of manufacturing operations. Choosing the best robot for a particular application can be difficult, but an objective, systematic analysis of the operation and of the robot's capabilities can simplify this task.

Before such an analysis is made, however, the basic question, "Can I use a robot?" must be answered. Implicit in this question is the fact that an industrial robot does not fit every situation. There are a number of factors which should be considered in deciding whether or not to use a robot on a particular job. These factors include:

- Complexity of the operation
- Degree of disorder
- Production rate
- Production volume
- Justification
- Long-term potential
- Acceptance

Let's examine each of these factors in detail and develop some rules of thumb to apply to the basic question. Describing all of the conditions under which you *can* use a robot would take up this entire issue of ROBOTICS TODAY, and more. We will, instead, point out a number of situations in which a robot is not the best answer. This approach might seem rather negative. However, you, the reader, should take a positive approach. Assume that you *can* use a robot unless one or more

of the rules of thumb clearly applies.

Complexity of the Operation

Although simple robots exist and are well suited to simple tasks, there are operations where a cylinder, a valve, and a couple of limit switches are sufficient. In other cases, a gravity roll or chute may suffice to transfer and even reorient a part from one location and attitude to another.

At the other end of the scale, operations which require judgment or qualitative evaluation should be avoided. Checking and accepting or rejecting parts on the basis of a measurable standard can be done with a robot. If, however, the only feasible measuring system is human sight or touch, then a robot is out of the question.

In the same vein, operations which involve a combination of sensory perception and manipulation should also be avoided. An example would be a machine tool loading operation which requires that the part being loaded into the chuck be rotated until engagement of a notch with a key is felt, after which the part is fully inserted. While the development of a "hand" for the robot capable of doing this is technically feasible, the complexity of the hand and its potential unreliability will certainly reduce the overall probability of success. Of even greater complexity are operations requiring visual determination of random spindle orientation and orienting the part to match.

Rule: *Avoid extremes of complexity.*

Degree of Disorder

Robots cannot operate in a disorderly environment. Parts to be handled or worked on must be in a known place and have a known orientation. For a simple robot, this must be always the same position and attitude. For a more complex robot, parts might be presented in an array; however, the overall position and orientation of the array must always be the same. On a conveyor, part position and orientation must be the same and conveyor speed must be known.

Sensor-equipped robots (vision, touch) can tolerate some degree of disorder; however, there are definite limitations to the adaptability of such

robots today. A vision system, for example, enables a robot to locate a part on a conveyor belt and to orient its hand to properly grasp the part. It will not, however, enable a robot to remove a part, correctly oriented, from a bin of parts or from a group of overlapping parts on a conveyor belt.

A touch sensor enables a robot to find the top part on a stack. It does not, however, direct the robot to the same place on each part if the stack is not uniform or is not always in the same position relative to the robot.

Rule: *Disorder is deadly.*

Production Rate

Small nonservo (pick-and-place) robots can operate at relatively high speeds. There is a limit, though, on the capability of even these devices. A typical pick-and-place cycle takes several seconds. A rate requiring pickup, transfer, and placement of a part in less than about three seconds cannot be supported by a robot.

Operations which require more complex manipulation or involve parts weighing pounds rather than ounces require even more time with a robot. The larger servocontrolled robots are able to move at speeds up to 50 ips (1270 mm/sec); however, as speeds are increased, positioning repeatability tends to decrease. For a rough estimate of cycle time, one second per move or major change of part orientation should be allowed. In addition, at least a half-second should be allowed at each end of the path to assure repeatable positioning. In handling a part, allow another half-second each time the gripper or handling device is actuated.

Rule: *Robots are generally no faster than people.*

Production Volume

There are two factors related to production volumes to consider. In batch manufacturing, the typical batch size must be considered. In single-part volume manufacturing, the overall length of the production run is important.

In small-batch manufacturing, changeover time is significant. Recalling that a robot needs an orderly environment, it is possible that part orienting and locating devices may need to be changed or adjusted before each new batch is run. A robot's end-of-arm tooling and program may also have to be changed for each new batch of parts. Generally, people do not require precise part location—their hands are instantly adaptable and their "reprogramming" is intuitive. A robot becomes impractical when its changeover time from batch to batch approaches ten percent of the total time required to manufacture the batch of parts.

If a single part is to be manufactured at high annual volumes for a number of years, special-purpose automation should be considered as an alternative to robots. Per operation, a special-purpose device is probably less costly than the more flexible, programmable automation device, or robot. Single-function, special-purpose devices may also be faster and more accurate than robots. Where flexibility is required or obsolescence is likely, robots should be considered; where these are not factors, special-purpose automation may be more efficient and cost effective.

Rule: *For very short runs, use people. For very long runs, use fixed automation.*

Justification

The application of an industrial robot can represent a significant investment in capital and in effort. Economic justification must therefore be carefully considered. On the balance sheet, increased productivity; reduced scrap losses or rework costs; labor cost reduction; improved quality; improvement in working conditions; avoiding human exposure to hazardous, unhealthy, or unpleasant environments; and reduction of indirect costs are among the plus factors. Offsetting factors include capital investment; facility, tooling, and rearrangement costs; operating expenses and maintenance cost; special tools, test equipment, and spare parts; and cost of downtime or backup expense. Ballpark estimates of the potential costs and savings should be made to indicate whether or not a reasonable return on the investment can be expected.

"Management direction", "following the crowd," and emotion are no substitutes for economic justification and, in the long run, will not support the application of robots. In some cases, safety or working conditions may override economics; however, these are usually exceptional circumstances.

Rule: *If it doesn't make dollars, it doesn't make sense.*

Long-term Potential

Another consideration is the long-term potential for industrial robots in the particular manufacturing facility. Both the number of potential applications and their expected duration must be taken into account.

Because of its flexibility, a robot can usually be used on a new application if the original operation is discontinued. Since the useful life of a robot may be as long as ten years, several such reassignments may be made. Unless the first application of the robot is to be of relatively long duration, its possible reapplication must be considered. In the process of justifying the initial investment, the cost of reapplying the robot should also be included. If the initial application is of significantly shorter duration than the robot's useful life and no follow-on applications can be foreseen, it can seldom be justified.

As with any electromechanical device, an industrial robot requires some special knowledge and skills to program, operate, and maintain. An inventory of spare parts should be kept on hand. Auxiliary equipment for programming and maintenance or repair may also be required. Training of personnel, spare parts inventory, special tools, test equipment and the like may represent a sizeable investment. The difference between the amount invested in these items to support a single robot or to support half a dozen or more robots is insignificant.

Maintenance and programming skills and reaction time in case of problems tend to deteriorate without use. Few opportunities will normally arise to exercise these skills in support of a single robot. Under these conditions, the abilities may eventually be lost and any serious difficulty with the robot may then result in its removal.

Rule: *If I only need one, I'm better off with none.*

Acceptance

Not everyone welcomes robots with open arms. Production workers are concerned with the possible loss of jobs. Factory management is concerned with the possible loss of production. Maintenance personnel are concerned with the new technology. Company management is concerned with effects on costs and profit. Collectively, all of these concerns may be reflected in a general attitude that, "Robots are OK, but not here."

It is important that everyone's attitude toward robots be known. One large company has developed an attitude survey questionnaire which, when completed and scored, indicates the probability of success of robot installations. The survey also identifies the areas where support is lacking and areas where action should be taken to change perceptions and attitudes. If, in spite of such action, the probability of acceptance is still low, the installation of robots is not recommended.

Even if a formal survey is not made, it is essential to know whether a robot will be given a fair chance. Reassignment of workers displaced by a robot can be disruptive. Training of personnel to program and maintain the robot can upset maintenance schedules and personnel assignments, and new skills may even have to be developed. The installation and startup can interrupt production schedules, as can occasional breakdowns of the robot or related equipment. Unless everyone involved is aware of these factors and is willing to accept them, the probability of success is poor.

Rule: *If people don't want it, it won't make it.*

Back to the question, "Can I use a robot?" The approach to take is to assume that you *can*, unless one or more of these rules clearly applies.

Robots are being used effectively in a wide variety of manufacturing operations. The successes are applauded and publicized, the failures are not. Using these few simple rules in making the initial decision on a potential robot application may help make all of your robot installations successful ones. ■

Presented at the Robots V Conference, October, 1980

Robots Are Easy. . .It's Everything Else That's Hard

By David Cousineau
General Electric Company

This technical paper discusses the General Electric Company's Power Transformer Department's initial steps in the field of Robotics that led to the development of an extremely useful industrial robot Application Matrix. Embracing Kepner-Tregoe (1) decision analysis techniques and using appropriate weights and measures, the Application Matrix was employed to evaluate and quantify potential robotic opportunities. The paper then delves into Power Transformer's first industrial robot application and the convoluted process of system design, integration and implementation. The title will be brought into focus as the writer develops the idea, that of the total system, the industrial robot itself was the easiest component to deal with.

INTRODUCTION

The title "Robots Are Easy....It's Everything Else That's Hard" is not a statement of fact but simply a "grabber" designed to get the reader's attention and yet at the same time reveal the main thrust of this paper. Robots are not really easy but the degree to which they are uncomplicated to install, program, operate and maintain is a tribute to their manufacturers' excellence. By and large, industrial robot manufacturers have done a superb job of designing and building robots so that the integrated whole (manipulator, controller and power supply) is relatively easy to apply. From project selection through implementation, when a balanced view is taken of all the elements that go into a successful installation, the robot itself emerges as the easiest system component with which to contend.

What is the "everything else that's hard?" One of the things that's difficult is project selection, or more accurately, the "best" robot application with which to introduce robotics into a large manufacturing organization. Other items that were found to be significantly challenging to effect total system integration were:

> Upstream operational changes
> Parent equipment modifications and retrofits
> Required rearrangement
> Ancillary equipment needs
> Special process development
> Special purpose equipment
> System interface
> Safety considerations
> Warning and alarm systems
> Down stream changes

(1) Kepner-Tregoe and Associates, Inc.
Princeton, New Jersey

This paper is presented in two parts. Part one discusses Project Selection and how to use the Industrial Robot Application Matrix. Part two discusses Power Transformer's first robot installation and the kinds of things that had to be done to effect a totally integrated automated system.

As the scope of these discussions is quite broad, a certain amount of depth is sacrificed in the interest of space.

PART I-PROJECT SELECTION

Background

Initial steps toward project selection in a large manufacturing operation are tentative at best and frought with uncertainty, high level "advice", over-zealous supporters, recalcitrant detractors, and "not in my area" attitudes. The neophyte roboticist is on the one hand apparently overwhelmed with opportunities and on the other has no objective way to sort out the real from the imaginary. Some proposed applications might at first look financially attractive but closer scrutiny could reveal that extensive modifications to existing equipment and/or expensive ancillary equipment would greatly diminish apparent return on investment. Other potential applications may require very few equipment modifications or retrofits but volume may be lacking. Still others may have high volume, a repetitive routine and simple equipment modification but are so highly automated already that they are not labor intensive enough to justify anticipated capital equipment costs.

Struggling with these and other variables, the writer had a growing feeling that there must be some way of putting it all into perspective with the hope of logically and perhaps numerically evaluating potential industrial robot applications before committing resources. Gestation of the foregoing gave birth to what is now called, in Power Transformer, the Industrial Robot Application Matrix.

Application Matrix (See Figure I)

In using this matrix, Proposed Applications A, B, C, etc. are entered in the appropriate columns in the spaces provided. Objectives that apply in varying degrees to the proposed applications are already listed in the left hand column with room for more if it is felt they are needed. Note that the objectives are divided into two categories; MUST and WANT.

MUST Objectives

MUST Objectives are, by definition, GO/NO GO and because of their extreme importance are automatically weighted at 10 so that "GO PROJECTS" can be further evaluated as to their relative desirability. In other words, MUSTS are mandatory and application success depends on them.*

*Note: In a pure Kepner-Tregoe decision analysis MUSTS are not weighted or scored.

Five MUSTS are Listed in the Objectives Column of the Matrix

A. Healthy Business

Large companies sometimes have product lines that are either unhealthy or scheduled for exit. Unless there is a concerted effort to restore a product line or business to health, an industrial robot is of little long range help.

B. Available Funding

Irrespective of apparent project worthiness, if funding is not available further study is fruitless. Thorough and convincing project evaluation, however, could make funds available during the next budgeting period.

C. Large Volume of Parts

As industrial robots are used to fill the void between manual operation and hard automation, a large volume of parts is another application absolute. This can be further quantified by scoring which is described further on.

D. Highly Repetitive Routine

A large volume of parts does not necessarily mean a highly repetitive routine, as in some cases the routine or operation may vary from one part to another or be too complex for robotic application.

E. Labor Intensive

Though some robots are installed in development situations or to remove workers from a hostile environment, it is apparent that the prime mover of industrial robots is productivity. If there is not enough direct and indirect labor savings to justify an industrial robot installation, continued application evaluation is wattless.

WANT Objectives

While MUSTS are mandatory, WANT objectives are desirable and to the degree they are desirable the application under study is enhanced. As WANTS are not black and white, weighting is used in an attempt to quantify the relative impact a given WANT has on the viability of proposed projects. WANTS were subjectively weighted as follows:

        Low impact on proposed applications...... 1 to 3
        Medium impact on proposed applications... 4 to 6
        High impact on proposed applications..... 7 to 9

(Note: Weights can be changed to suit a given user's situation)

<u>Want Objectives (Continued)</u>

Additionally, it was discovered that WANTS fell into four basic, logical sub-categories; Operational, Environmental, Climatic and Financial.

A. <u>Operational (Nuts & Bolts)</u>

1. Operational Simplicity:

The simpler the operational sequence is the better, for as minipulative moves become more complex higher technology robots are required to perform them. High technology robots are more intelligent and have many advantages over low technology robots but are considerably more expensive.

2. Few Parent Equipment Modifications:

Here is an area where in a lot of money can be spent very quickly as modifications and retrofits to accommodate robotic operation and interface do not come cheaply. Equipment that was designed to be operated by humans is generally not the best set up for robotic operation. Modifications and retrofits must be defined and at least scope estimates applied.

3. Minor Rearrangement Needs:

While project impact is generally not as significant as equipment modifications and operational complexity (except for major multi-unit robotic systems) rearrangement is itself labor intensive and hence costly.

4. Inexpensive Ancillary Equipment:

At this writing industrial robots are still pretty much one-arm blind idiots, as visual and tactile sensing is yet in the development stage. Accordingly, parts transfer systems or parts presentation equipment of some kind is required to preposition and orient the parts for the robot to pick up. Generally these systems are quite expensive, often even more expensive than the robot itself.

B. <u>Environmental (Work Place)</u>

1. Diminish Hazardous Situations:

Fortunately, much of the work that robots are suited for humans are not. Die casting machines, forging presses and punch presses have perhaps removed more fingers than surgical amputations from all other causes. While application impact is not viewed as "high", careful consideration of environmental WANT objectives could add significant points to a given proposed project.

B. __Environmental (Work Place) Continued__

2. Improve Hostile Environment:

   Removing operators from a work station environment that is hot, dirty or noisy can only be seen as positive.

3. Eliminate Strenuous Work:

   Another "warm fuzzy" for industrial robots and the degree to which particular application under study eliminates strenuous work, increasing scores should be given.

C. __Climatic (Emotional)__

1. Shop Management Support:

   The inherent productivity advantages of industrial robots may not be as readily apparent to some shop operations managers as others. For a robotic application to succeed, both shop management support and involvement is necessary. Within the same manufacturing organization this can vary widely and should be taken into account when scoring this WANT. Significantly, complete application success cannot be claimed unless or until the project engineer can divest himself of psychological ownership.

2. Low Labor Sensitivity:

   For a variety of emotional, psychological and physical reasons, labor sensitivity can also vary widely from one manager's area to another's. Whether this variance stems from internal pressures or external forces is not the point, but simply the "sensitivity index" as it is perceived. Cognizant of this, the matrix user can award appropriate points.

D. __Financial (Dollars)__

1. High Return on Investment:

   Unless an installation is a development application, acquisition of project funding usually comes down to justification. Though there is a definite interrelationship between many of the listed objectives, the bottom line is unquestionably return on investment. How this is calculated and viewed varies from one company to another and is beyond the scope of this paper. An analysis of savings, however, should take into account not only the obvious direct labor savings but indirect labor, benefits, overtime premium, night shift bonus and direct and indirect material savings as well. Applications can and often do have a favorable impact on both upstream and downstream operations which should also be quantified and included in ROI calculations.

D. Financial (Dollars) Continued

2. Low Funding Requirements

If the MUST objective of Available Funding is a "Go" then
the least expensive application is more desirable from a
purely capital consumption standpoint. Though a simplistic
view if taken by itself, funding requirements do deserve
consideration; even if relative impact on project selection
is low compared to return on investment, it should be measured.

Information

INFO column is provided for brief notes and may be used to illuminate
scoring decisions. Often useful when reviewing decisions at a later
date.

Score

Scoring is used to measure the degree of impact for a particular
objective as it applies to a specific application under consideration.
In order to keep scoring simple it is suggested that the following
system be used:

Unfavorable or low degree of impact...... 1 to 3
Favorable or medium degree of impact..... 4 to 6
Very favorable or high degree of impact.. 7 to 9

Though they are at times awkward, note that objectives are stated so
that elevated scores indicate that they are true to a greater degree.

Weighted Score

This is simply the product of weight times the score.

Point Total & Application Selection

To arrive at point total, weighted scores are merely added together.
Application selection and prioritization can then be made based on these
totals. Unless there is significant point spread it may be prudent to
go back through the matrix to be sure scoring will stand up under
scrutiny. This is particularly important with the heavily weighted
"swingers" such as all the MUSTS and such WANTS as high return on
investment, few parent equipmemt modifications, operational simplicity
and inexpensive ancillary equipment.

SUMMARY

This application matrix was successfully used in the Power Transformer Department to evaluate and score some seven potential robotic installations. In addition, another dozen possible applications were looked at but never made it to the matrix as they could be discarded after only a cursory review. The application that was selected as our "best shot" is a financial and emotional success beyond our expectations and provided an opportunity for the Power Transformer Department to acquire experience and expertise with this flexible form of automation. The industrial robot application matrix is not a panacea for evaluating potential robotic installations but an extremely useful tool—especially if adapted to suit a users needs.

INDUSTRIAL ROBOT APPLICATION MATRIX

OBJECTIVES		PROPOSED APPLICATIONS											
		A				B				C			
MUST	WT	INFO	GO/NO	SC	WT SC	INFO	GO/NO	SC	WT SC	INFO	GO/NO	SC	WT SC
HEALTHY BUSINESS	10												
AVAILABLE FUNDING	10												
LARGE VOLUME OF PARTS	10												
HIGHLY REPETITIVE ROUTINE	10												
LABOR INTENSIVE	10												
WANT	Wt.	INFO		Sc	Wt Sc	INFO		Sc	Wt Sc	INFO		Sc	Wt Sc
OPERATIONAL SIMPLICITY	8												
FEW PARENT EQUIPMENT MODIFICATIONS	9												
MINOR REARRANGEMENT NEEDS	4												
INEXPENSIVE ANCILLARY EQUIPMENT	7												
DIMINISH HAZARDOUS SITUATIONS	3												
IMPROVE HOSTILE ENVIRONMENT	2												
ELIMINATE STRENUOUS WORK	2												
SHOP MANAGEMENT SUPPORT	5												
LOW LABOR SENSITIVITY	6												
HIGH RETURN ON INVESTMENT	9												
LOW FUNDING REQUIREMENTS	1												
Total													

(Left margin vertical labels: FINANCIAL CLIMATIC ENVIRONMENTAL OPERATIONAL)

FIGURE 1

PART II - APPLICATION

System Description (See Figure 2)

The Unimate Industrial Robot System located in the Bushings Operation of the Power Transformer Department in Pittsfield, Massachusetts is used to perform the secondary operations on the Re-X (T) capacitor bushing product family.

Re-X Capacitor bushings are manufactured in a continuous process facility. Powdered raw materials are precisely weighed in batches, thoroughly mixed and then introduced into a melting furnace. The resulting molten glass is then drawn off into steel molds. The cast body -still a transparent glass- is removed from the mold when the temperature has been lowered. The process of converting this glass into a ceramic takes place in a recrystallizing kiln.

The product family consists of three basic Re-X Capacitor bushings rated at 15 KV, 8.7 KV and 1.2 KV shown in Figure 3 - shown left to right, respectively.

The operations performed by the industrial robot system are: Flattening the bottom flange of the bushing and aligning the ceramic portion so that it is perpendicular to the flange; buffing the lateral area of the ceramic portion so as to remove any blemishes that may be on the petticoats; packing the finished product into corrugated shipping containers.

Before an operator can start the robot system he must close the personnel Safety Gate and connect the Safety Interlock. Proceeding to the Remote Control Console, he then starts all the equipment contained inside the perimeter fencing and pushes a reset button that releases the Unimate from remote hold which enables it to be started from its own console. The Sequence Controller is reset and its selector switch placed in "auto". The cycle is initiated at the Unimate's control console by pushing the cycle start button and turning the selector switch from "hold" to "run".

Working in a counter clockwise direction, the Unimate picks up a bushing from the Buffered Parts Transfer System and carries it to the Hydraulic Press. It loads the die, cycles the press then regrasps the bushing and takes it to the Buffing Machine. At this point the Unimate maintains control of the bushing as it rotates it 320º inside the buffing machine to effect petticoat cleaning. It then transfers the completed part to the Box Positioner, Indexer & Escapement and loads the first cell of the corrugated shipping container. In a repetitive fashion the Unimate continues to get bushings from the buffered parts transfer system, loads the press, cycles the press, buffs the petticoats and progressively packs the shipping container. When the box is full, the Unimate cycles the boxing escapement which automatically ejects the filled container and positions an empty one in the loading position.

(T) Trademark of the General Electric Company

Operation Simplification

A. Upstream

During the course of application investigation and evaluation it was discovered that the flattening press die was being used to make a secondary sort in addition to the basic visual sort that was being performed after recrystallization. Press operators were required to "try" each bushing in the die set and cycle the press only on those that fit and handle aside those that did not.

In addition to impinging on systemic output, this operational variable would have been difficult if not impossible to contend with robotically. It was therefore eliminated by simply building appropriate sorting guages and installing them up stream so the bushing sorters could effect a complete and total sort with one handling of the product.

B. Downstream

For many years Re-X capacitor bushing have been packed in two layer corrugated boxes with precut egg-crating assemblies and fillers to protect the product and assure customer receipt in perfect condition. Robotic packing in two layers would have required programming a three dimensional array and very expensive special purpose automation to feed preassembled egg-crating and fillers as needed.

Accordingly, the KISS method ("keep it simple, stupid") was followed and the shipping container was redesigned and simplified so a single layer fillerless packing configuration was created. The new design was easier to handle, increased packing density and was considerably less expensive.

Parent Equipment Modifications & Retrofits (See Figures 4 & 5)

For an industrial robot to attend another piece of equipment it has to be properly interfaced so as to monitor various machine conditions and control parent equipment cycles. In this application several things had to be changed or modified on the Denison Hydraulic Press to accomplish this requirement.

To begin with, the press electrics had to be changed so the Unimate could initiate the cycle after the press die had been succesfully loaded. This modification was taken one step further by incorporating a key operated selector switch so that in the event of a major system breakdown the press could also be operated manually via conventional twin palm buttons.

FIGURE 4
UNIMATE LOADING FLATTENING PRESS

FIGURE 5
DETAILS OF PRESS MODIFICATIONS AND SENSOR ARRANGEMENT

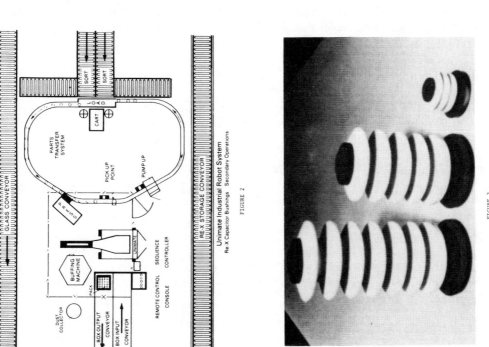

GLASS CONVEYOR

DUST COLLECTOR

BOX OUTPUT CONVEYOR

BOX INPUT CONVEYOR

BUFFING MACHINE

REMOTE CONTROL CONSOLE

SEQUENCE CONTROLLER

PACK

UNIMATE

STORAGE

PARTS TRANSFER SYSTEM

PICK-UP POINT

PUMP UP

CART

LOAD

SORT

SORT

RE-X STORAGE CONVEYOR

Unimate Industrial Robot System
Re-X Capacitor Bushings Secondary Operations

FIGURE 2

RE-X CAPACITOR BUSHING PRODUCT FAMILY

FIGURE 3
RE-X CAPACITOR BUSHING PRODUCT FAMILY

Parent Equipment Modifications & Retrofits (Continued)

The press die was rebuilt and fitted with a pneumatic actuator so that the die cavity, which is mounted on a slide to facilitate loading, could be moved in and out of the press. In addition, the die set and press were retrofitted with a series of photo-electric and proximity switches to monitor important conditions and preclude a smash-up. Conditions monitored were: Die cavity empty and in the load/unload position on the slide (thru-beam pair and proximity switch); press ram up/down (proximity switches); die loaded, bushing properly seated and slide in "press" position (second thru-beam pair and proximity switch); die in load/unload position with part raised and ready for removal (same non-contact sensors as "a" but logic reversed); part successfully removed from die cavity (again same thru-beam pair used in "a" with original logic restored.)

Finally, in order to control this profuse electronic dialogue, a hardwired solid state modular Logic Controller was designed by our in-house "wizard," Mr. C. B. French, built by our own electricians and interfaced with the above sensors, actuators and the Unimate itself. By so doing, this integrated functional subsytem required only one output and two input connections to the Unimate.

Rearrangement Requirements

Most major equipment installations require some degree of rearrangement of existing facilities to accommodate change. Suffice it to say that in this robotic installation a good degree of rearrangement was required and though not "hard" or difficult, it was nonetheless involved and was therefore accomplished in advance of the actual installation so as to minimize start-up time and negative shop impact.

Two points are to be made here:

1. Required rearrangement should be carefully worked out and delineated in advance so that the impending installation will be the best possible "fit" for total line integration.

2. To the degree possible, this rearrangement should be completed in advance of the actual robot installation and its peripherals to compress implementation.

Ancillary Equipment

In retrospect, ancillary equipment needed for this application fell into two broad categories, General Purpose and Special Purpose, and will be discussed separately.

A. General Purpose (See Figure 6)

Until visual or tactile sensing is readily available for general application, parts must be presented in a pre-oriented manner for even high technology robots such as a Unimate to pick up. This is most often accomplished by use of Parts Transfer Systems with as much buffering added as can be afforded to provide system "elasticity". Using escapements, shot bolts, meters and a variety of both contact and non-contact sensors, these systems generally use roller bearing chuting mounted at an appropriate pitch to gravity convey parts to the robots pick-up point.

For the capacitor bushing application, a number of systems were investigated with the hope of uncovering equipment that would feature helical storage towers and not require the use of part pallets. Because of financial constraints the quest was not 100% successful but Automation Service Inc. of Warren, Michigan designed and built a relatively inexpensive and reliable system that did fulfill our needs. Using urethane part pallets (to prevent bushing damage) in roller bearing chuting with a pneumatic pump-up, a closed-loop system was provided with 45 minutes of buffered part storage ahead of the Unimate. In this way, the last manual operation and the automated system did not impinge on one another.

B. Special Purpose

Three special purpose pieces of equipment were needed to complete the automated functional arrangement:

1. An Automatic Buffing Machine
2. Some sort of Box Positioner, Indexer & Escapement
3. A remote Control Console

1. Automatic Buffing Machine (See Figures 7 & 8)

In the manual method of cleaning bushing petticoats a conventional pedestal buffing lathe was used with firm Scotch-Brite (R) wheels. Obviously, this would be less than desirable for robotic operation as a compliant wheel was needed to conform to the bushings petticoats and centering and rotational forces would have been excessive for the required hand tooling.

After carefully working out design parameters with several vendors, Eagle Bridge Machine & Tool Company of Eagle Bridge, New York was selected to design and build a Special Purpose Buffing Machine.

(R) Registered Trademark of the 3M Company

FIGURE 6
PICKING UP BUSHING FROM BUFFERED PARTS TRANSFER SYSTEM

FIGURE 7
LOADING SPECIAL PURPOSE BUFFING MACHINE

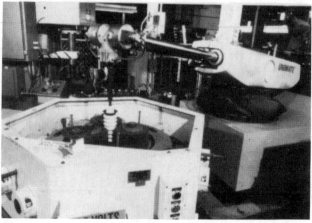

FIGURE 8
DETAILS OF BUFFING OPERATION-LEXAN (T) COVER REMOVED
(T) TRADEMARK OF THE GENERAL ELECTRIC COMPANY

1. Automatic Buffing Machine (See Figures 7 & 8) Continued

Three Scotch-Brite flap wheels of appropriate density and coarseness were located 120° apart and mounted on vertical spindles. Driven by three 3-H.P. motors, the wheels move in on pneumatically actuated slides. The Unimate places the bushing inside the buffing machine and rotates it 320° while proceeding downward so that the complete lateral area of the bushing is cleaned. In addition, three limit switches monitor the slide positions to assure that the buffing wheels are fully retracted during loading and unloading. Wheel engagement is individually and collectively adjustable to compensate for wheel wear and media penetration.

2. Box Positioner, Indexer & Escapement (See Figures 9 & 10)

From the onset of this project it was decided that, to the fullest extent practical, a fully automated integrated system requiring minimal operator interface would be the paramount objective. To that end it was recognized early on that some sort of mechanism to feed and consistently position empty shipping containers was needed. Further, this device would also have to possess the ability to eject filled containers internal to the robot's process time so as not to erode the inherent output capabilities of the balance of the system.

Lacking experience with this very specialized requirement, it was further decided to place this responsibility with the selected robot vendor. As a Unimate #2005B was determined to be the best choice for this application, an order was placed with Unimation Inc. of Danbury, Connecticut for a box positioner, indexer and escapement. Unimation Inc. in turn subcontracted this purchase to Production Equipment Company of Meriden, Connecticut who designed and built an excellent piece of special purpose automation to carry out the needed functions. This equipment was later interfaced with a belt conveyor which holds a two hour queue of empty boxes ahead of the positioner and feeds on demand. Filled containers are simply ejected onto an inclined roller conveyor.

3. Remote Control Console (See Figure 11)

Finally, it was decided that to complete the functional arrangement a remote or central control console would be used to start and stop all the equipment contained inside the perimeter fencing. The console was also designed (another "home-brew") so that all incoming and outgoing signals to and from the Unimate could be monitored via indicating lights. Further, it was fitted

FIGURE 9
UNIMATE PACKING CORRUGATED SHIPPING CONTAINERS

FIGURE 10
DETAILS OF BOX POSITIONER, INDEXER & ESCAPEMENT

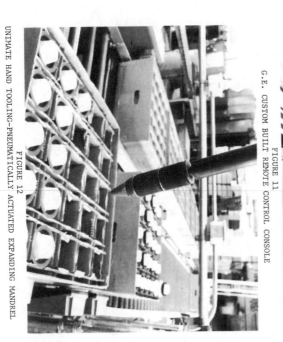

FIGURE 11
G.E. CUSTOM BUILT REMOTE CONTROL CONSOLE

FIGURE 12
UNIMATE HAND TOOLING—PNEUMATICALLY ACTUATED EXPANDING MANDREL

3. <u>Remote Control Console (See Figure 11) Continued</u>

with a key operated switch so the entire system could be shutdown and locked out from a single point. This signal monitoring capability coupled with the "remote hold" and alarm system lights proved to be of great benefit during the start-up and debugging phase as it made the integrated whole essentially self-diagnostic.

<u>MISCELLANEOUS</u>

A few odds and ends were incorporated into the industrial robot system that bear mentioning as they have proved very useful to the point that they could be viewed as "up-time boosters."

Using photoelectric sensors, a Warning System featuring flashing lights and an intermittent beep was installed on both the box input conveyor and the filled carton output conveyor. At any time during operation, if there are fewer than two empty boxes ahead of the packing station or more than six filled shipping containers on the system output conveyor, the appropriate warning device will alert the operator to attend the systems needs.

Similarly, an Alarm System has been incorporated into the arrangement so that if the Unimate is detained for any reason an enunciator is sounded alerting the operator (and foreman) as to malfunction. This is accomplished by programming the robot to initialize a pre-set timer during the first program step of the base routine. In this way, if the robot cannot complete the entire part program in the allotted time and reinitialize the timer, the alarm is activated.

As the internal diameter of the bushings vary widely, considerable concern was given to the possibility of the robot's Hand Tooling failing to successfully release the part in the shipping container. Cost and complexity ruled out visual systems so a simple whisker switch was mounted in the return path of the supposedly empty hand tooling. This way, if a bushing was not released it would trip the switch. The whisker switch was connected to the Unimate's "remote hold" function so if activated the robot would come to an immediate stop (hold). This feature alone prevented more than one smash-up, for after placing a completed part in the shipping container the next operational element is to grasp a new one!

Finally, the entire industrial robot system was enclosed in perimeter fencing. A safety interlock was installed on the personnel gate and connected to the Emergency Stop circuit so that inadvertent opening to gain system entrance would shutdown the entire functional arrangement. However, to facilitate programming and permit entrance into the restricted area for assorted equipment adjustments (i.e. buffing wheels) interlock by-passes were installed by wiring appropriate blocks to the Unimate's "teach/repeat" and "run/hold" selector switches.

Off-The-Shelf Purchases

A number of "off-the-shelf" purchases were made to complete the system, not the least of which was the industrial robot itself. Before acquisition, a thirteen page industrial robot bidding specification (plus six attachments) was prepared and reviewed with several vendors three of whom responded with definitive proposals. Accordingly, a vendor analysis was prepared and the Unimate Model 2005B manufactured by Unimation Inc. of Danbury, Connecticut was selected as the best choice for this particular application and, coincidentally, was the least expensive.

Other off-the-shelf purchases are delineated as follows:

-Unimate Sequence Controller for directing the robot's packing sub-routine.
-Hand Tooling designed by Unimation Inc. which grips the bushings internally via a pneumatically actuated expanding mandrel. Hand tooling also boasts .100 compliance which was needed to accommodate eccentric bushing bores. (See Fig. 12)
-Shipping container input belt conveyor.
-Unimate Cassette Recorder for back up storage of part programs.
-Unimate Equipment Tester for diagnostics and troubleshooting.
-Foreman/Maintenance Training at the Unimate School.
-Start-Up Assistance (well worthwhile, system on line in 5 days).

Analysis of Results

As stated earlier, the success of this industrial robot application far exceeded our expectations both in dollars saved by cost improvements and by establishing a climate of receptivity that permeated virtually all levels of management and labor. System uptime has been excellent and what little downtime that occurred was not attributable to the Unimate. At this writing more than 80,000 bushings have been processed through the system.

Other system benefits are highlighted here:

o A substantial output and productivity improvement in excess of 300% from a cycle reduction of approximately 80%.

o Elimination of the need for hand packing the finished product.

o Elimination of an inprocess inventory bottleneck in the Re-X capacitor bushing line.

o Increased packing density for a 100% improvement in floor space utilization.

o Elimination of a stultifying and boring job in a seasonally hostile environment.

Analysis of Results (Continued)

o Enhanced safety as the operators were no longer required to manually load and unload the flattening press or hand hold bushings against an exposed buffing wheel to effect cleaning.

o Provided an opportunity for the Power Transformer Department to acquire experience and expertise with this flexible form of automation in a geographically isolated area with low labor sensitivity, good shop management support and in a financially attractive situation.

Conclusion

The conclusions of this paper are perhaps best revealed by bringing the title back into focus and discussing its part separately.

"Robots Are Easy"....is an argueable opinion and not a concise statement of truth. It depends upon one's point of view for from a manufacturer's standpoint, industrial robots are incredibly complicated programmable manipulators into which have gone thousands of hours of research and development before release for manufacture. In the writer's opinion, it is because the manufacturers have done their job so well that from a user's viewpoint industrial robots are by themselves relatively easy to install, program, operate and maintain.

"It's Everything Else That's Hard"....refers to all the other components of a totally integrated industrial robot system wherein the challenges lay and were ultimately surmounted. Parent equipment modifications, general and special purpose automation and a host of other "cold uglies" had to be subdued so the robot could be accommodated. Indeed, approximately two thirds of total project expenditures were consumed by these needs while the balance was used for the robot, hand tooling and related accessories.

Finally, the writer is of the opinion that one successful robotic application does not an expert make, but he has learned something of considerable value and would like to share it with his fellow professionals.

Presented at the Robots II Conference, October, 1977

Analysis Of First UTD Installation Failures

By Gennaro C. Macri
Ford Motor Company

The considerable acceptance of UTDs (robots) by the Automotive Industry has generated more interest in other areas of high production manufacturing and assembly. However, with the extensive utilization of machines in 1977 by the Automotive Assembly Division of the Ford Motor Company, we have learned that the "first" machine installed at any location is the most important. The proper initiation of the "first" application cannot be over emphasized. The basic but diverse criteria which must be analyzed and/or implemented before "launch" of the UTD are all too often neglected. Each of the more important criteria are examined. Examples are given of the effect on UTD applications when these criteria are not fully considered.

INTRODUCTION AND BACKGROUND

Industrial UTDs or Universal Transfer Devices (commonly called Robots) are receiving more and more consideration for use in automotive assembly. Our reasons for this have been reduced changeover time and cost, environmental considerations, safety, employe job enrichment and program flexibility. Other segments of manufacturing are also beginning to install more UTDs for much the same reasons, but some of them are finding their first installations not as successful as they had hoped - in some cases "disastrous". We in the automotive industry have experienced the same problems even though we may not openly admit it.

Why do these failures happen? The answer is sometimes simple, but more often than not it is complex. We hope some of your answers will be found in the details of this paper.

We have made no effort to compare or analyze any specific application in this paper, but to relate all the criteria the lack of which contributes to failure of a robot or any other new technology. Failure by definition can mean complete lack of any success or just not completely successful. A success value of 99% or less can be considered a failure. In most of the first applications studied the success value was less than 99% and in two cases complete failures.

In our Division the first successful UTD application took place in 1970 and since then we have installed over one hundred machines and expect to install a great many more over the next few years. But with those already in operation not all of them could be said to be or have been 100% successful. Even in the instance where two specific installations were identical except for the geographic location of the assembly plants involved we had one fully successful and the other about 85% successful. Both were planned with the same criteria or considerations as outlined below, but after analysis we find that in some instances one criterion is more important in a specific plant than in another. The consideration in this case was "Management Education and Familiarization". How are you going to evaluate which consideration is more important in a specific application? We don't know nor do we think there is a fool-proof method for finding out. The only answer we have at this time is that training or education is included as part of two of the listed considerations. Therefore, training could be the most important item.

It should be noted at this point that all of the information presented in this paper represents the collective experience of the Automation Assembly Department, Automotive Assembly Division, Ford Motor Company. Obtaining information regarding the true results of robot applications from other users has not been good nor has obtaining such information from the robot manufacturers been successful. We can understand the reluctance. No one will readily admit to anything other than "smashing" success, particularly the sales people of the equipment manufacturers.

The basic criteria or considerations are as follows and are explained in the body of this paper.

    I.   MANAGEMENT EDUCATION AND FAMILIARIZATION

    II.   THE "RIGHT FIRST" APPLICATION

    III.   WHICH ROBOT WILL BEST SUIT THE APPLICATION

    IV.   PRODUCTION "BACK-UP" CAPABILITY

    V.   SPECIAL TOOLING REQUIREMENTS

    VI.   PLANT LAYOUT AND FACILITIES

    VII.   SAFETY

    VIII.   TECHNICAL TRAINING, SPARE PARTS AND TEST EQUIPMENT

This is not so much a technical paper as one that is conversational in tone. The information that is presented is handled in much the same manner as one would explain to a new engineer just joining a UTD application group. The tone may also indicate a negative position regarding UTD applications. It is only because we are dealing with failure and the reasons why failures come about. We are fully dedicated to more and better applications of robots and that is why we present this analysis.

MANAGEMENT EDUCATION AND FAMILIARIZATION

In a large organization such as the Ford Motor Company where we have many divisions, each with many plants, we find that each division and then each plant must be educated on an individual basis. In our particular division we have twenty-two assembly plants employing about 48,000 people. Our plants are from coast to coast in the United States and we have three in Canada. In no way can we say they are all the same. Each plant has an individual personality which may bause a difference in application of the listed criteria and it is a matter of judgement or past experience that will define the difference. No set rules will apply, if the difference is unknown then apply all criteria equally.

UPPER MANAGEMENT

At any of the divisions, there are two levels of management to be informed about UTD technology. Each level must be treated in a different manner. The upper divisional level must be familiarized with what this type of equipment can and will do for the company. The reasons for implementation must be explained, in detail, along with comparisons of present and other known methods of doing the same type of operations. It is at this upper level of management that basic policy regarding the implementation of any new technology must be made. Lower levels of management will rarely assume any of the risks involved. Also the very important loop of "information feedback" up to this upper level of management regarding the outcome of any installation and objective analysis data does not normally take place. Without such information feedback the basic policy cannot be improved for better implementation of the new technology.

It is at the upper levels of management that we have to avoid getting trapped by the "overzealousness" of the powerful. Back in 1972 a general manager of another division made a statement to his staff regarding the fact that the Assembly Division had some twenty UTDs in operation and he could not understand why some were not operating in his plants. Orders were passed down through the organization and the beginning of one of the two disasters mentioned above was born. Because of the haste, every criteria listed in this paper was either misapplied or unknown. We were asked to assist, but too late and all we could do at the time was find another place in the company where the UTDs could be used successfully.

MIDDLE MANAGEMENT

The middle management level requires much the same information and education as presented to the top divisional level, but with much more technical detail. Each of the remaining criteria must be ingrained into their thinking and planning. The consideration of training of the engineering staff is very important at this level. We have sometimes found this to be rather hard - more excuses than action. Getting the lower level engineering people to take an active part in implementation has still not completely taken place in our own division even though we have had UTDs in operation for the last seven years. This is the group of people that require week long training sessions at the manufacturers of the equipment and later actual hands-on practice. Good engineering and planning cannot take place without such training.

PLANT MANAGEMENT

In each plant there are three areas of management which must be educated to the principles of UTD applications before the "first" installation is even considered. Again each area is informed in a different manner and content.

We must start with plant, assistant plant manager, operations and engineering manager level. If this group is not fully informed and fully agree to take an "Active" part in the planning, installation and launch of the first application then this is the point to stop. To proceed at the lower levels of the plant management (and this is possible) some form of failure is almost assured. The important point here is that an "Active" interest must be displayed, just as it must be displayed on the divisional top management level. Signing an appropriation request for the project, prepared by the lower levels, is not necessarily an active interest. It is necessary at this upper plant management level for you to be explicit as to the abilities and limitations of the robot including the importance of the ancillary equipment. We find that at this level no matter what goes wrong (ie: tooling, welding equipment, interface panels, etc.) the statement is always "that damn robot is down again". An objective analytic report is seldom produced. When we are called to assist or investigate any of the problems, we generally find that the criteria as listed was not properly implemented. The consideration of training is the one most often neglected followed closely by poor ancillary tooling. These are the two considerations which cost money and are the easiest to short change. This upper plant level should also be watched for the same "overzealousness" mentioned before. Full and proper planning and engineering are shorted for speed. In either case when the installation is made and a failure does occur (to any degree) getting a second project started in that particular plant becomes practically impossible. No matter what improvements are made by the manufacturers of the robots the people involved with that "First" installation failure will inevitably say that they had tried it and it did not work and don't want to try again.

The second area of management which must be educated is that of production supervision. They must be included in all planning and engineering. They must not be surprised by finding one day the UTD installed in their production area without any idea as to why it's there. In an instance where this did happen, it was the production foreman who pointed out the errors in the engineering and application. Time and money were wasted in making the required changes. Nobody knows his area better than the foreman, therefore, use him.

The third area of management is the engineering staff. They must be completely trained at the UTD manufacturers facility before the installation is made. Several people from each engineering department should attend the training sessions. In a classic example a plant sent only one man to be trained and after several weeks of the installation he left the company. Needless to say the operation turned sour for lack of knowledgeable people. The basic problem of the lack of trained people still has not been resolved at this plant and until it is, 100% success is not possible.

We have conducted information sessions on the divisional level and found them to be successful insofar as familiarizing the plant engineering staffs with the latest technology, problems encountered, solutions and general exchange of ideas. These sessions we plan to continue.

THE "RIGHT FIRST" APPLICATION

The importance of the first application cannot be over emphasized. As stated earlier, if it fails, it may also be the last. Our purpose is to find an application which is simple and does not require elaborate engineering, tooling and/or facilities to accomplish. Our goal is to develop confidence in the UTD by all the people involved. Later installations will then be easier to launch even though they may be more complicated. If at all possible, try to find an operation where the installation of the UTD will benefit an hourly person. Our first successful UTD application in 1970 was on a spot welding operation which had a health and safety grievance written against it. The spot welding was of very heavy metal which caused a large amount of metal expulsion and burned the operator. The installation while not exactly simple did make the operator happy and the Company and Union settled the grievance. Later installations at this location were made easier and it is now the plant with the most UTDs in operation in the Ford Motor Company.

If at all possible, do not be too quick to install the UTD on its intended operation. Make a temporary installation in the maintenance shop and allow the skilled trades, the engineering people to use it for "hands on" experience. The time used will not be wasted.

When the actual installation is made and the UTD is started up and the system is being debugged you will find that people, all kinds of people, will become curious. These people will appear from anywhere and everywhere to see the new machine in motion. It is at this point in time that safety requirements must be enforced, the details of which are covered later.

It is our opinion that the introduction of a different kind or make of UTD should be treated as a "First" installation and the same methods as mentioned in this paper be applied. It is not unusual for the second type of UTD to find it somewhat harder to be acceptable in a plant which has had a competitive make in operation. In the case where the UTD is of a new design either mechanically or the controls it is almost imperative that an application be found where production cannot be adversely affected. It is also just as important that adequate time be allowed for the UTD manufacturer to make corrections to problems which are only found in actual production tryouts. Expect such a situation to cost you some time and money (as it will the UTD manufacturer) therefore, include this in your planning stage.

As the problems with the first or new UTD present themselves, be careful to separate the problems which are caused by the UTD itself and those caused by the supporting equipment. Be sure that one or more of the considerations listed in this paper was not the cause of most of the problems. We repeat an earlier statement that no matter what goes wrong with the new device or technology, it is usually what is accused first. The lack of proper tooling, facilities, trained people and proper maintenance procedures are not generally admitted nor do we find the problems properly and objectively documented. Complete documentation of every event during the installation, debugging, launch and first months of actual production of the first application is of great value in the planning and engineering of later applications. Don't let the "First" be the "Last".

WHICH ROBOT WILL BEST SUIT THE APPLICATION

A robot is a robot. This statement is not necessarily true. A Cincinnati 6CH is not an Auto-Place which is not a Unimate which is not a Versatran and so forth. Each machine has its place and its application. There are, of course, overlaps where more than one type can do the same job. It is here that you must become the final authority and to do so you must become familiar with the many different makes of robots. After analysis of the requirements of the application call upon the manufacturers for advice, however, the responsibility of choice is still yours. Remember the technical advice provided by the manufacturers (especially related to performance of their equipment) must be considered carefully since vendors have a natural tendency to oversell their products.

Keep in mind that a robot should be flexible in its applications. Consider what would happen if the operation being performed became obsolete, would the robot being reviewed be able to be reallocated to another application? Also, what is the cost of such flexibility and do you really need it? We in the Automotive Assembly Division feel we do need such flexibility in most cases and are willing to pay for it, but do you need it? An example of such a situation is where we were asked if a robot could remove built-up automotive rear seats from an assembly conveyor and load them on to a delivery conveyor. The answer was yes, however, since seat build-up operations do not change much year to year or model to model, the transfer could be accomplished with a simple two axis device costing one fifth of what we normally pay for a robot. The device was designed and built by our own people and is operating successfully.

A good method of becoming familiar with the different types of robots is to spend some time and money to visit the users of each kind of robot. Ask the manufacturer to make the arrangements or make your own contacts, if possible.

PRODUCTION "BACK-UP" CAPABILITY

Your decision to make the UTD or robot installation has been made, the supporting tooling and facilities determined, but what about the method of continuing production when the UTD goes down. They will and they do. We have found and documented that robots are reliable and with normal maintenance and experience less than 2% downtime. The same documentation also indicates that with spot welding equipment attached to the robot downtime increased to 6%. The more equipment involved in the application adds to the downtime potential, nevertheless the robot is out of operation. The back-up system can be as elaborate as a duplicate system or as simple as a rack of finished parts in storage next to the robot operation. In many cases this back-up capability can be attained by leaving the manual system which the robot may have replaced intact. For example, in a transfer operation the manually operated hoist equipment may be left in place. Of course, availability of manpower to operate the standby equipment must also be considered.

SPECIAL TOOLING REQUIREMENTS

When a potential UTD application has been found orientation and fixturing requirements must be determined. If not properly engineered this aspect of the intended application can cause most of the launch problems. Tooling changes necessary to locate the work price or assembly relative to the robot within plus or minus .06 of an inch consistently is a must. Until sensory feedback devices are developed this added cost of tooling will remain with us and is as important as the robot itself.

In spot welding applications we have found that the use of a self-equalizing spot weld gun will simplify installation of the gun to the UTD. If the gun itself is not self-equalized then

the interface mounting device should be designed to fulfill
this requirement. The design of the interface bracket or mount
must be such as not to add excessive weight to the end of the
robot arm. We have learned that such interface mounts should
not be castings or made of aluminum but designed to be steel
weldments. No matter how careful the design may be made, nine
out of ten such mounts have to be reworked during launch of the
system and the steel weldment is easier to rework on site. We
have also learned that "balancers" spring or air should not be
used to help the UTD carry the weight of the tooling or welding
gun. Balancers only add to the inertia load on the robot arm,
particularly side to side motions. A small balancer to hold
the spot weld gun cable and hoses up and out of the way of the
robot, fixture or work piece is advisable.

Most installations will require some sort of sequence or interface control panel between the tooling, facilities and the UTD.
Included in the interface are signals to the robots from the
tooling or facilities - that the part is in place, welding complete, operator clear of the area or conveyor location. Signals from the UTD may include those required to initiate a weld
sequence or clamping of hook details, etc., and the end of the
robot program. These interface controls can also be very complicated when several model selections plus different part locations are to be engineered. Design them with care.

We have been using proximity switches to indicate the exact location of parts and tool mechanical components and found them
to be very reliable.

We must state again that part or work piece orientation and location repeatability is very important. Robots are only programmed for a point in space and could care less where the
part may or may not be.

PLANT LAYOUT AND FACILITIES

This consideration really goes hand in hand with the tooling
and sequence controls. Early in the engineering stage exact
layouts of the proposed installation should be made and then
field checked for interferences and accuracy. Working height
of the UTD and access to the job may require elevation above
the floor or changes to the part location. We do not recommend that the UTD be located in a pit, however, we sometimes
have found it necessary in order to accomplish the installation.

Guard rails should be provided to protect the robot and system
from material handling equipment and keep people away from the
robot.

For in-line, stop station applications, such as respot welding
on automobile bodies, speed up rolls may be required before and
after the work station with the part locationing tool station
in between. But remember any downtime in such a straight line

system will develop a "gap" or "hole" in the progress of parts down the assembly line. This gap must be expected and planned into the total system. Too often production people neglect to accept this aspect of a synchronous system, therefore, avoid such situations where possible.

Utilities and service requirements for the various robots can be obtained from the manufacturers.

SAFETY

This is an area which is fraught with opinion. Depending on where the installation is made or which group of people you deal with the instructions regarding safety and robots will be different. The following suggestions are based on our experience and the fact that to date we have had only one injury directly related to robots which happened during servicing.

As stated in other parts of this paper, guard rails (at least 42 inches high) must be installed at the same time as the robot. People will become curious and will walk within the working range of the robot whether it is powered or not. A robot may be standing still, but that does not indicate it will not move just as someone gets within its range. It may be waiting for "go or start" signal. Robots do not get into the way of people - people get into the way of robots. Keep people out of the area.

We also insist that all robots and their components be electrically grounded particularly when welding equipment is part of the robot tooling, also the spot welding gun should not be insulated from the robot arm.

There are two different times when a safety problem will most likely come about. The first is during teaching of the robot when the "teacher" must get right into the work area of the robot and the other is during service of the machine.

The teaching problem can be elevated by having an "emergency stop button" on the teach cable and hydraulic restricting valves which limit the speed of the robot arm. Such restricting valves are used only during teaching.

Each robot should have an emergency stop button on the control console (plus the teach cable). This stop button should have contacts which are externally connected to the conveyor, fixture or transfer device thereby stopping the complete system when the button is used.

We have also learned that when such an emergency stop is activated on one robot, in a system employing a group of robots, it is not always wise to power the other robots down. Let such stop action put the other robots into a "Hold" or "Freeze" state. Powering down the other robots may cause more safety

problems as they lose their electronic memory of arm location
or go limp and sag down under the weight of the tooling attached to the arms. The only machine which should be powered down
is the robot in question and any transfer or tooling devices
in the system.

The safety problem during maintenance is a strict matter of
personnel training on the procedures of safe practices. A robot is like any other piece of moving machinery and should be
regarded with the same respect. All power should be locked out,
all hydraulic and air power removed (accumulators relieved of
pressure) and the arm blocked up on a holding device designed
for the purpose before any service is started.

Some people have suggested that the robot be restricted in
movement with the installation of steel posts in the floor.
After years of working with robots, I do not like the suggestion at all. I would rather have the moving arm push me over
than pin me to a post or any other restrictive device.

The only other thing we can say regarding safety is use good
common sense in all aspects of your application and check each
part of the engineering for safe practices just as you would
for any other piece of automated equipment.

TECHNICAL TRAINING, SPARE PARTS AND TEST EQUIPMENT

The key to minimizing robot downtime is a well trained maintenance force, provided with adequate tools and spare parts. Robots as we stated before have a high reliability factor, but
they do go down and when they do you won't be making production
with the machine while waiting for a service man or spare parts
from the manufacturer. We have found that the best system is
to train our people on all shifts and have spare parts on hand.
We know of a case at the present time where the robot installed
only operates on the day shift because the only people trained
in the operation and service of the robot are on that shift.
The afternoon shift cannot use the machine because they do not
know how to keep the equipment running.

It may become apparent after close examination of some types of
robots, that your skilled trades people may not have sufficient
background to handle the sophisticated control systems. If not,
then take the time and money to train them because if you do
not spend the time and money at the beginning of the project -
you will pay for it later in equipment downtime.

Since the mechanical and hydraulic systems are relatively
straight forward, the major training requirement is for the
electricians and electrical engineering people. Our experience
has shown that it is best to assign the primary maintenance and
programming responsibility to the electricians with other
skills assisting as required. We have also learned that it is
better to have the manufacturer conduct training sessions on

your plant site. The engineering people should be included. This may take several weeks and appear to be expensive but the cost is worth it if you intend to use robots in your operations. Initial complements of spare parts and test equipment must be on hand during any training. This allows your people "hands on" training with such equipment and the robot you "temporarily" installed in the maintenance area.

If the robot you have chosen has available (at extra cost) a recorder for permanently recording programs - buy it. This permanent record can be used to quickly restore programs in the robot control memory as required. Robots have been known to lose their minds.

CONCLUSION

The success of any first robot application is dependent on the effort made to apply the above considerations. Anything less than maximum dedication to all of the above will result in some degree to failure.

We would like at this point to pass on some statements of fact based on our past experience with robots.

Don't ever use a robot to pace a manual operation nor threaten people with the use of robots. The reasons are, we think, self evident.

Do not expect a robot to do what a person can do - it can only replace a man who is performing an operation like a robot.

We repeat, get your management to back you up - total commitment on all levels is required for success.

We are positive that other considerations will be developed as the technology of robots and their use advances. However, we feel that those presented above will always remain basic to robots in industry.

Presented at the Robots III Conference, November, 1978

Robot Interface: Switch Closure And Beyond

By Brian J. Resnick
Cincinnati Milacron

Applying a robot involves an interface between the robot and the application. Depending on limitations within the robot, this interface could be a mere START/STOP contact, or it could mean a data communications channel with an external device or higher level computer.

A computer-controlled robot allows for the full spectrum of interfacing, from switch closure to data communications and beyond.

INTRODUCTION

A computer-controlled Industrial Robot (such as the one shown in Figure 1) can greatly simplify the teaching process. Using coordinate transformation matrices and other mathematical wizardry, the computer translates the operator's commands of UP, DOWN, IN, OUT, etc. into the desired motion at the tool. Thus, the computer allows the operator to manipulate the tool without concern for the individual robot axes.

In addition to simplifying the teaching process, computer control also offers certain important advantages in interfacing the robot to an application. Its extreme flexibility can minimize interfacing hardware costs by performing switching logic internally, and it also provides a whole new bag of tricks for solving the more difficult problems. Keeping abreast with robot technology is a difficult but important task. This paper attempts to acquaint the reader with a variety of these interfacing capabilities provided by computer-controlled robots.

Input signals, determined by the presence or absence of a specific voltage at the input signal terminals, tell the robot when or when not to do something. A typical condition is "If Input Signal Three is present, put the part in chute one." The robot makes a programmed decision to perform the routine of placing the part in the chute based on the state of Input Signal Three.

More complex applications may require decisions to be based on more than one condition. For example, "If a part is ready, and if the oven is up to temperature, and if the oven is empty, then put the part in the oven." Three conditions are required to make this decision, which normally means wiring three switches in series to an input signal terminal, or first wiring them to a relay panel or programmable controller if additional decisions are to be made with those same switches. It is this type of application that can bring out the cost effectiveness of a computer-controlled robot, because the additional switching logic hardware is not required. The robot's computer can perform the tasks of the relay panel or programmable controller and allow the robot to make decisions based on <u>combinations</u> of input signals. It also allows decisions at different points in a routine to be made from the <u>same</u> input signal. Thus, a computer-controlled robot typically requires <u>less</u> additional hardware and uses fewer input signals to perform the same task as a non-computer-controlled robot which is generally limited to one routine per input signal terminal.

OUTPUT CONTACTS

Output contacts (switch contacts operated by the robot) provide the robot with some control over the application, such as: turning on or off motors, heaters, grippers, welding equipment, etc. Controlling output contacts becomes part of the robot's routine. The robot is taught to close (or open) a contact at a particular point in the routine. Whenever that routine is replayed, that contact is closed (opened) at the same point at which it was taught.

In addition to turning on or off individual output contacts, computer-controlled robots allow simultaneous operation of several contacts, pulsing of contacts, and "handshaking". The later term means that after an output contact is turned on or off, the robot waits for a specific input signal to acknowledge that action before performing the rest of the routine. This is especially useful when controlling a tool. It is very reassuring to <u>know</u> the tool has actually opened or closed, as commanded, before any other motion occurs.

An example using inputs and outputs is shown in Figure 2. The robot is to feed two different types of parts (Part A and Part B) from respective part feeders into any one of three ovens provided an oven is empty and up to temperature. Each oven has two timers, one for each type of part. The robot sets the corresponding timer after depositing the part and, later when the timer indicates the oven has completed its cycle, the robot removes the part and places it into the appropriate chute depending on which type of part it is.

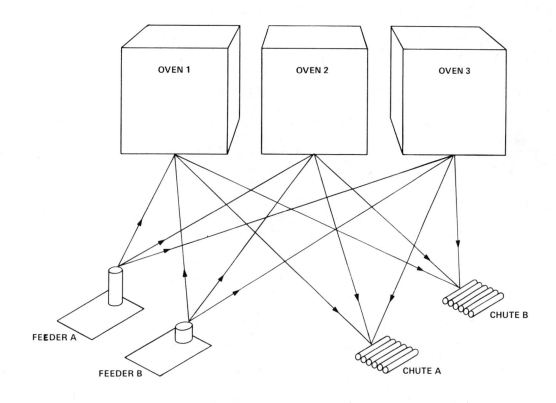

FIGURE 2: EXAMPLE USING INPUTS AND OUTPUTS

Figure 2 illustrates one solution that uses twelve robot routines to move the parts from feeder to oven and oven to chute. Now examine the signals and contacts involved. Each oven has a temperature switch (T1, T2, T3), an EMPTY switch (E1, E2, E3), two timers (S1A, S1B, S2A, S2B, S3A, S3B) and a OVEN CYCLE FINISHED switch (D1, D2, D3). Each feeder has a PART PRESENT switch (FA, FB). With the information from these contacts, everything is known except into which chute, A or B, to place the part. Some form of memory is required. One simple method is to close an output contact when it's from feeder "A", and open it when it's from feeder "B". Since there are three ovens, three output contacts (C1, C2, C3) will be used in this fashion.

Now determine what occurs in each routine, and what the decision process is for selecting each of the twelve routines.

If there is a part in feeder A and oven 1 is empty and oven 1 is up to temperature, then move the part from feeder A to oven 1, close contact C1 indicating the part is from feeder A, and pulse contact S1A which starts the timer for parts from feeder A.

Similarly, test all three ovens and both feeders.

If oven 1's cycle is finished and it contains a part from feeder A, then move that part from oven 1 to chute A.

Similarly, test all ovens and place the parts in the appropriate chutes.

This simple application involves twelve routines, nine output contacts and the derivation of twelve input signals from fourteen switches. Without computer control, the input signals would have to be created by wiring switches together in the proper sequence or external switching logic hardware. With computer control, each switch is wired to a separate input signal terminal, and the logic combinations are taught to the robot. (See Figure 3.) It is less expensive, faster to set up, and more flexible should a change of plan occur.

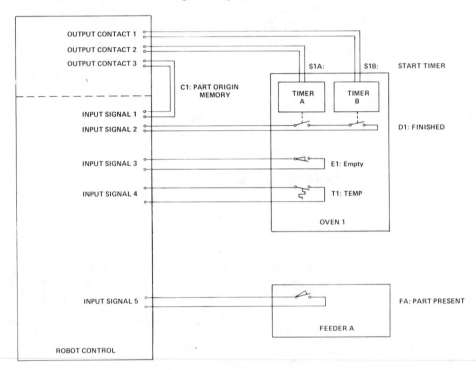

FIGURE 3: PARTIAL SCHEMATIC FOR FIGURE 2

ANALOG OUTPUTS

Not all processes can be controlled with contact closures. Some processes are of an analog nature and require a variable type control, such as: wire feed speed and arc voltage in the welding process. If the equipment is suitably adapted or manufactured to accept an external control voltage, then a computer-controlled robot with a programmable voltage source could be taught the appropriate settings, and they would become a part of the robot's routine. This greatly increases the versatility of the robot by eliminating the undesirable single or limited multiple set-point compromise.

A CHANGING ENVIRONMENT

A computer-controlled robot can solve problems in a changing environment that are impractical to solve with other types of robots. How does it handle a stack of parts? How does it work on a moving assembly line? Without special fixturing and alignments to particular axes, a non-computer-controlled robot can not readily solve these problems because it is incapable of self-modifying a taught maneuver with respect to a desired change in the robot's position relative to the workpiece.

A computer-controlled robot can be taught a routine at one position in space and told to perform it at another position in space. This is very powerful because it allows the robot to interface with a changing environment. Take the stack of parts for instance, as the height varies a different set of joint positions is required to maintain relative position and orientation between part and tool. All that is necessary, is to teach the robot the routine for only one part in the stack, and the computer handles the rest. By sensing the stack height with a gripper mounted switch, the robot "knows" when to stop moving towards the stack, and "knows" which coordinates to modify in order to work with the part at the sensed height.

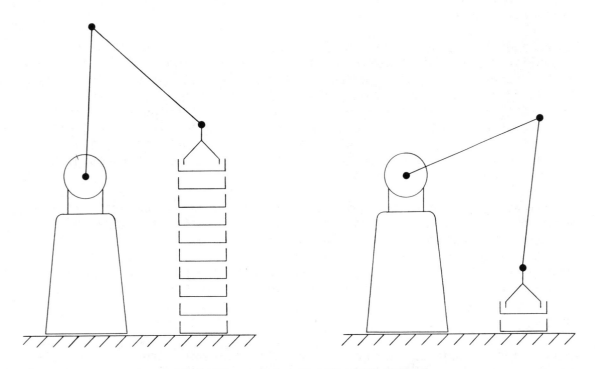

FIGURE 4: HANDLING STACKED PARTS

Another changing environment application, is performing operations on moving objects, such as spot welding on an automotive assembly line. Here, the computer-controlled robot constantly senses the car's position and responds by altering its routine so that the weld gun maintains the programmed positions relative to the car. Again, the computer-controlled robot interfaces to a changing environment with a minimum of special fixturing and additional hardware.

EXTERNAL DEVICES

As applications grow increasingly more sophisticated, sooner or later they will exceed the computer-controlled robot's capacity for making decisions, or its capacity for programmed points. A feature that helps solve these problems is the ability to communicate with external devices. This feature provides an external device (such as another computer) with the full capabilities for supplemental program management. Position, orientation, velocity and the function associated with each point of a routine may be modified by the external device, and then sent to the robot over a data communications link for execution.

Certain applications, such as intermixed small batch processing, would require the robot to remember hundreds of routines. That many routines would exceed the memory and decision making capacities of the robot. One solution is to store all the routines in another computer with sufficient memory. That computer would detect which type of part is presented to the robot and provide the robot with the appropriate routine. This external computer need not be dedicated to just communicating with one robot. It could be a computer that already exists within the application. In which case, its duties are expanded to include communicating with the robot.

There are other applications, such as part pick-up, which could be accomplished with only a few routines, if the randomness of the situation can be conquered. Consider a rectangular block. It can rest on any of its six faces, but requires only three pickup routines due to symmetry. The difficulty arises in the infinite variations in orientation and the resulting problem of gripper alignment. One solution is to use a vision system to detect position and orientation of the block, modify the block pick-up routine, and communicate it to the robot. Now the robot with vision is conquering a real world situation.

Communicating with external devices is a very powerful feature. It is important, in terms of convenience, to use an industrially accepted method of interfacing, such as the RS232C standard. It is also important, in terms of safety, that these communications are performed with an error prevention protocol. Without an established protocol, erroneous information could enter the robot and cause catastrophic results.

CONCLUSIONS

Interfacing a robot to an application is a problem of great concern when purchasing a robot. A computer-controlled robot can minimize additional interfacing hardware and provide solutions to problems that are impractical to solve with non-computer-controlled robots.

The robot capabilities discussed in this paper represent commercially available computer-controlled robot technology. Applications similar to those described in the text are presently operating in the field.

Presented at the Robots II Conference, October, 1977

Robot Decision Making

By H. Randolph Holt
Cincinnati Milacron

Decision making allows a robot system to deviate from its normal path program and perform other tasks based on changes in external events. Starting with rudimentary branching capability in a computer-controlled industrial robot system, more sophisticated decision-making capabilities can be added to facilitate application of the robot system to a wide variety of areas. Having a computer in the control allows these enhancements to be easily incorporated, and only minimal hardware changes are required. Conditional branches, offset branches, and search function are features which dramatically increase the decision-making capability possible in a commercially available robot system. Several other such features can be added to provide increased decision-making power in industrial robot systems.

Introduction

Few, if any, industrial robots perform tasks that require no connections to equipment surrounding them. Typically it is necessary for the robot to receive such signals so that it can interface to a working environment. Examples would include limit-switch closures to the robot when a part is in position, electrical signals to denote that a piece of auxiliary equipment has failed, or timing signals to synchronize the robot arm to a particular process. Relay contacts may also be provided by the robot to the external equipment to interlock the two systems.

To illustrate this interconnection of equipment, a diagram of a simple robot system is shown in Figure 1. This example is given to illustrate interconnections and does not represent an economical robot application; two or more machine tools in the machining cell would be economically justifiable. An industrial robot is servicing a milling machine, with raw (unfinished) parts arriving on one conveyor and finished parts exiting on another conveyor. A "ready for part" signal from the machine tool control to the robot control would inform the robot to pick up a raw part and load a fixture on the table. A "raw part available" signal to the robot system would permit the arm to move to that conveyor and transfer the part to the milling machine. The robot then signals the machine tool control to begin its operations on the part. While waiting for completion, the robot could perform housekeeping functions such as chip removal, inspection of finished parts, staging of tools in the tool changer, and inspection of tools for breakage or excessive wear; in all of these chores, other interconnect signals could alter the functions performed by the robot depending on the outcome of these tests or the presence of any unusual situations during this housekeeping. If during these functions the milling machine control detects a malfunction or a tool breakage during a machine operation, the robot must abandon these routine tasks and take some action to either remedy the problem or

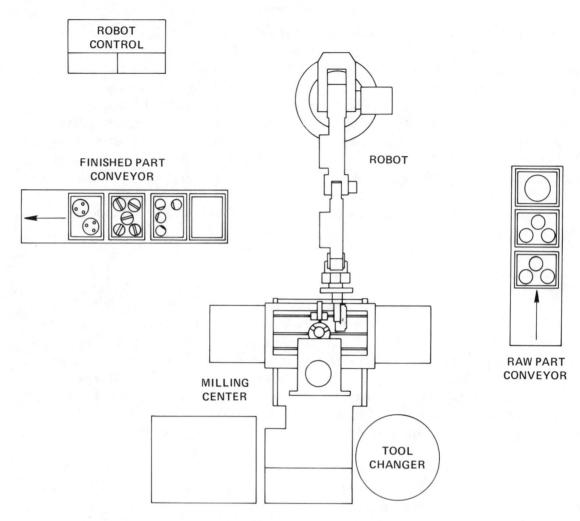

FIGURE 1. MILLING MACHINE/ROBOT SYSTEM

initiate an emergency procedure for the total system. A "part finished" signal from the machine tool to the robot would request that the finished part be unloaded and transferred to the outgoing conveyor. The cycle could then be repeated.

In the above example, the robot must respond to signals from the surrounding equipment and perhaps alter its movements and operation based on these inputs. Robot decision making can be defined as the ways an industrial robot system can respond to these changes in its working environment.

With the advent of computer-controlled industrial robots, significant improvements are possible in the way the robot is interlocked to the application. First, the computer's inherent decision-making power allows very complex decisions to be made in the robot control, thus minimizing external equipment that may have been necessary to implement this capability; this can have a significant effect on the total cost of the robot installation. Secondly, software programmability in the computer control allows for flexibility in the type of decision making included in a specific robot control; that is, the decision process can be tailored to the application to further reduce total system cost and/or to simplify the way an operator can teach the robot program.

This paper will discuss several decision-making components that are either already implemented or possible in computer-controlled robot systems. Specifics and terminology will be based on the Cincinnati Milacron robot, but the intent of this paper is to consider these components as being only examples of the kinds of decision-making elements that are possible with any computer-controlled robot system.

Standard Branching

Truly general purpose industrial robot systems must have some way of selecting or altering the programmed path and function based on changes in the environment around them. The name given to such a facility may vary with each manufacturer and the details may differ, but the purpose is the same: The robot reaches some point and interrogates an input signal to determine whether it is electrically active, or the robot is interrupted by activation of another input signal. In either case, the robot path "branches" to a section of the path/function program; if no signal is present at this decision point, or no interrupt occurs, the robot continues in a normal path sequence.

Figure 2 displays a top view of a simplified version of the diagram given in Figure 1. A "mainline" cycle (M001 through M005) has been programmed to execute a sequence of events that performs routine housekeeping, inspection, etc. The "branch" routine symbolizes the actions the robot must take to pick up a new part for machining. A "part present" signal is shown coming into the robot control that may originate as a limit switch closure when the part is present at the raw part conveyor. When the robot arm reaches point M003, (the decision point), the control examines the "part present" signal; if the signal is on, the path sequence passes through 0A01, 0A02, 0A03, and then returns to M004; if no part is present, the robot continues in the mainline routine directly to point M004, and the housekeeping cycle continues through M005 and then back to M001.

In an actual application, a number of these standard branches may be included in the program. For instance, if during inspection of a machined part some defect was found, the robot might branch to another part of the program that would dispose of the part, alert an operator of the defect,

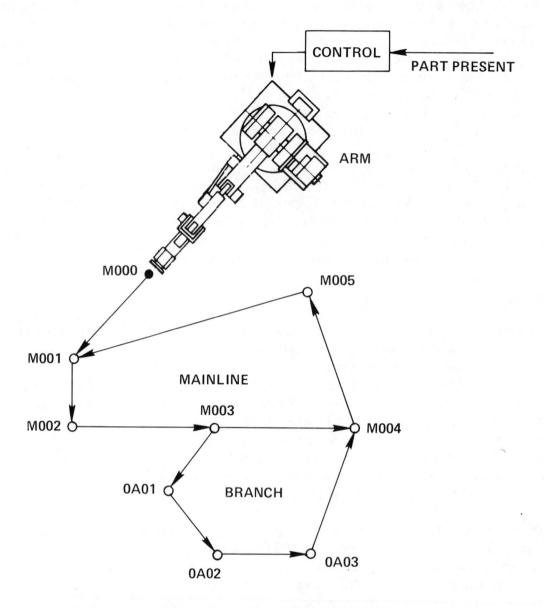

CONTROL

PART PRESENT

ARM

M000

M005

M001

MAINLINE

M003

M002

M004

0A01

BRANCH

0A02

0A03

FIGURE 2. SIMPLIFIED MACHINE LOADING SEQUENCE

and possibly even shut down the total system.

A number of these standard branches may be programmed at the same physical position, thus permitting the robot to "scan" them quickly. The first signal to be detected would direct the robot into that particular branch sequence.

Conditional Branching

Standard branches are the simplest way to alter the robot's path program, but they are limited in the sense that each branch is associated with one particular input signal to the robot control; if 32 such lines are provided in a control, any robot program is limited to 32 standard branches. Some applications can also have simpler installations and lower wiring costs if an external decision-making element is available to look at the number of signals simultaneously as a condition for branching; that is, if another computer system is supervising the total robot system installation, then 8 lines can provide 256 conditions for branching. This extension of the standard branch in the Milacron robot is called a conditional branch.

Using such a feature, a decision point can now be programmed to branch to a path sequence only if a particular combination of input signals occur at that point. For example, if the following branch function is programmed at point M007:

BRANCH 0A   1, -5, 10, -24

the branch sequence called 0A will be entered by the robot system only if input signal #1 is "on", AND input signal #5 is "off", AND input signal #10 is "on", AND input signal #24 is "off". If any of these signals differ from these conditions, the robot will proceed to the next mainline point. Programming a second branch function after such a point, say at M008:

BRANCH 0A   2, 5, -7

will branch to 0A if signals #2 AND #5 are "on", AND signal #7 is "off". In other words, branch 0A will be entered if the signal conditions at M007 are met OR if the signal conditions at M008 are satisfied. With conditional branching, therefore, complete Boolean expressions (AND-OR) are possible with the input signals to the robot control.

To illustrate one use of conditional branches, suppose that in a material handling application a supervisory computer is being used in the factory for other automation functions, and that capacity in the computer is still available for other tasks such as providing some level of input to the robot system. If 16 different branches are necessary to perform the robot tasks and if all necessary signals relating to these tasks are monitored by the supervisory computer, only 4 input signals representing a 4-bit word need to be communicated to the robot control from this computer to identify the various branches. Using standard branches, 16 input signals would have to be wired into the robot control to accomplish the same functions.

If other robot applications require external relays or electronic logic to implement Boolean expressions for robot decision making, conditional branches allow these functions to now be performed within the robot system, thus simplifying and reducing the cost of the peripheral equipment.

Offset Branching

In many applications it is desirable for the robot system to have a method for creating a branch that can be used at a number of points in the robot cycle, a branch that can effect its sequence relative to the physical location of the robot arm when the branch is requested. In other words, the entire branch is offset about the robot's position and/or the wrist orientation.

A _standard offset branch_ is programmed into the cycle in similar fashion to a standard branch: At each point where the offset branch is program-ed, the proper input signal is checked and the branch is entered if the input is active. Standard offset branch 0A in Figure 3 is programmed

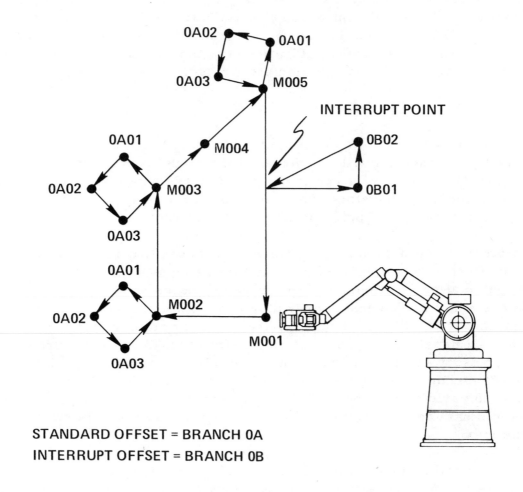

STANDARD OFFSET = BRANCH 0A
INTERRUPT OFFSET = BRANCH 0B

FIGURE 3. OFFSET BRANCHING

at M002, M003, and M005. Such a branch type could be used where the same sequence of operations is to be performed at a number of points in the application, two examples being drilling and material pickup from a palletized configuration.

Some applications require another kind of offset branch where the branch can be entered not only at specifically programmed points but anywhere in the cycle. An input signal from the surrounding equipment causes the robot to enter this interrupt offset branch whenever the input is activated; branch OB in Figure 3 is such a branch. One possible use of this branch type would be in spot welding to allow the robot system to recover from instances where the welding tips have welded themselves to the workpiece; the interrupt offset branch would be programmed as a set of twisting movements that would loosen the spot weld gun tips from the stuck position. The interrupt signal would be generated by external equipment supplied by the user when this condition occurs.

Search Function

In addition to altering its cycle by entering branches, a robot system can adjust the position of data points within an existing cycle based on changes in external equipment and workpieces. The search function, depicted in Figure 4, is one example of this type of robot decision making. Three points, 0A05, 0A06, and 0A07, have been taught to the industrial robot.

In a "normal" cycle, the center of the robot end effector (hand) would move in a straight-line path at a defined velocity from 0A05 to 0A06 with controlled acceleration and deceleration at the ends of the segment. After stopping and performing the programmed function at 0A06, the robot would similarly move to point 0A07.

In the example of Figure 4, however, a search function has been programmed at 0A05. The robot end effector would accelerate from 0A05 and proceed as in the normal cycle above, but in this case the robot control monitors an input signal that causes the robot to decelerate and stop when it is activated. This new stop point (called 0A06' in the figure) is treated as the "new" 0A06 point and the robot end effector moves from 0A06' to 0A07 during the next part of the program cycle.

One obvious use of the search function is in stacking operations, especially when the stacked items are fragile or have irregular thicknesses. The time delay inherent in decelerating from the input signal activation will permit some movement beyond the robot's receipt of the signal; so if the signal originates through a limit switch that is closed upon contact with the stack, some compliancy must be built into the robot gripper. A fragile workpiece would also require a slow velocity during the search segment.

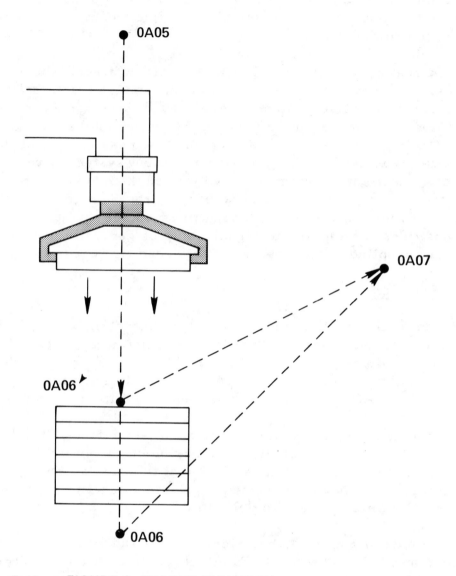

FIGURE 4. SEARCH FUNCTION

Adaptive Branching

Computer-controlled robot systems can be easily connected to other
computer-based systems since facilities to accomplish this communication
are provided in all computer system architectures. This more complex
type of external equipment in an industrial robot system can then easily
supply more than the "on/off" signals that are typically available from
other equipment in the system configuration. A higher level of robot
decision making is therefore possible with computers that are part of
a supervisory system or are an element in more complex sensor systems
employing vision, force, or tactile transducers.

The data comprising a complete branch sequence in the robot control can be rewritten with data from the external computer system. If the position or orientation of a workpiece varies, the robot movements and functions can be adjusted by the external device to properly pick up or perform the required operations on the work. If an emergency situation is detected, a branch can be replaced with other branch data to handle the specific condition. The speed with which this data can be transmitted to the robot control is fast enough to complete the communication in a fraction of a second, so this type of branching alteration could be called <u>adaptive branching</u>.

Conclusions

The flexibility inherent in computer-controlled industrial robot systems permits a wide range of decision-making capabilities, thus creating a very effective use of robots in a variety of application areas. The examples of robot decision making in this paper -- standard branching, conditional branching, offset branching, the search function, and adaptive branching -- have been given as illustrations of the kinds of decision-making functions that are possible in computerized robot systems.

References

1. Corwin, M., "The Benefits of a Computer Controlled Robot," Society of Manufacturing Engineers, MS75-273.

2. Hohn, R., "Application Flexibility of a Computer-Controlled Industrial Robot", Society of Manufacturing Engineers, MR76-603.

3. Dawson, B., "A Computerized Robot Joins the Ranks of Advanced Manufacturing Technology", Numerical Control Society, 1976, Proceedings, Thirteenth Annual Meeting and Technical Conference, Cincinnati, Ohio.

4. Dawson, B., "Moving Line Applications with a Computer Controlled Robot", Society of Manufacturing Engineers, Robots II Conference and Exposition, Detroit, Michigan, November 1-3, 1977.

Presented at the Robots IV Conference, October, 1979

Safety Equipment For Industrial Robots

By Dr. -Ing. H. Worn
KUKA

Comfortable safety equipment for the protection of men
and the industrial robot is being demanded increasingly
by users.

The high speed of procedure, the large number of axes and
the heavy weights increase the danger of accidents for the
operator, programmer or the service man. Collisions of
the industrial roboter with equipment or the robot gets
out of the standard control values could lead to destruction of the components of the robot.

In order to protect, during the automatic functioning,
the operating or monitoring personnel just as the personnel at adjacent working positions or inquisitive persons
as well, the operation sequence, basically, has .to be
organized in a manner to avoid any spatial and chronolo-
gical interference of personnel and industrial robot.

Since some of the protective devices used within the
automatic functioning have to be by-passed for programming and set-up purposes, an important function to fulfill is to design the overall plant such as to allow
intervention only by authorized personnel.

Ideal conditions, therefore, will include safety equipment being active within each mode of operation and
able to protect all human beings against injury just as
the robot against any destruction.

2. Technical safety standards and realized examples

In general, figure 1 represents the technical safety
requirements for industrial robots.

A sufficient mechanical strength has to allow either a
so-called safe-life or a fail-safe behaviour. Especially,
an adequate weight balance system just as mechanic rest
positions have to be provided.

The external configuration of robots has to be designed
in a way to avoid squeezing and shearing points just as
sharp edges and corners.

Control equipment for industrial robots, in accordance to the generally recognized rules of safety technique have to allow a fail-safe functioning as well as disconnection by means of emergency cut-off devices;

upon activation of emergency cut-off means for instance, displacement will be possible only at creep speed or according finger-tip control.

On account of the increased acceleration and mass powers, adequate braking means have to be provided according to the size of the control unit, device able to immobilize the industrial robot as fast as possible. The braking ratio has to be maintained even in case of lack of energy in order to avoid any collapsible effect of the unit.

It has to be ascertained that no failure within the energy transmission line results in unwanted movements; the only effect to release is to immobilize the unit.Especially, it has to be ascertained that the robot will not be started automatically in case of energy return.

Whenever, within a fault or danger condition, the industrial robot is stopped by means of actuating of the emergency shut-off device, it has to be possible, without drive energy and after having opened the unit via several valves or brakes, manually or after having suppressed the emergency shut-off position, to move the unit by finger-tip control or at creep speed.

Without regard to the type of drive energy (normally as a hydraulic, electric or pneumatic type) these conditions have to be satisfied.

To mention as well that the configuration of the grippers has to avoid any sharp edges or corners whenever possible, that means following a design to reach adequate tractive power in accordance to the weight of workpieces and possible dynamic forces. It is wanted to reach an optimum positive or conjugate gripping action. On the other hand, it has to be determined, according to the labor safety regulations, if, in case of danger, either the workpiece remains within the gripper means (positive holding action or maintained by means of springs) or releasing action of the workpiece (for instance hot-working piece) under safeguarded conditions.

Interfaces easy to be recognized, a decentralized structure of system just as aids for the operating and maintenance staff will allow to reduce the involved danger of operating errors.

By means of remote indication of any weak points the latter are able to be suppressed just as avoided during subsequent operations.

In order to increase the labour protection potential, elements just as maintenance or working schedules for overhaul, special tool sets just as adequate instruction of the operating and set-up personnel can be helpful.

The security measures or rules and equipments for industrial robots and work places will comprise mainly the so-called traditional safety techniques (see figure 2) comprising

- safety fences
- pull cord
- safety platforms
- safety equipments equipped with proximity detector

Still, as usual, the fence system remains the safest, most efficient and relatively low-cost protective measure for no admittance-type sections (mostly projected parts), measure realized for all plants. Within the operating mode 'automatic', any opening of the doors will be followed by immobilizing the plant by means of the emergency cut-off circuit. For the operation mode 'set-up and programming', this protective measure will be by-passed. Pull cords have been used in order to prevent, in case of interlaced IR's any access to the transport means equipped for instance with body fittings. Any excitation of the pull cord will release disconnection of the overall transport plant.

These measures are completed by the use of safety platforms the use of which is required whenever an operator is called to load the equipment (for instance a revolving table). During the loading action, a given number of plant functions have to be by-passed (for instance transfer of revolving table). By using light barriers, it can be possible to introduce a 'no admittance' condition at defined points of the plant.

All these protective equipments are characterized by simple and cost-effective realization while offering an effective protection for the personnel and during the 'automatic sequence' of functioning.

Nevertheless, no protection will be offered during the set-up, programming, overhaul and maintenance sequence because of the need to by-pass any safety equipments repeatedly. Because of unfavourable light distribution, for instance during spot welding within body parts, the programmer, during the programming and testing sequence, has to remain in direct proximity of the tool in order to appreciate positional precision. Very often, access of the personnel will be required for fault localization, remedy just as during maintenance work.

On the other hand, these measures will remain without any protection on behalf of avoiding any destructive action for the robot.

This further safeguard therefore will require additive protective measures.

A further possibility is offered by means of a safety equipment for the IR-oriented range, equipment fastened at the IR and following the movement of the latter (figure 3).This equipment comprises a pressure-sensitive material folded around the equipment parts and acting as an area contactor able to disconnect the plant in case of obstacle contact.

Protection against collisions of the tool element or gripper means is offered by a safety equipment represented by the figure 4.

The sixth axis of the industrial robot (IR) is flanged to the tool element or gripper, by means of a Belleville spring package. This configuration allows disconnection in case of overload conditions and at lateral approach.

Whenever the tool element (gripper or welding tongs) gets in touch with an obstacle (f.i. equipment), there will be unilateral lifting of the flange unit at a given transverse load F. As a result, the center of the flange is displaced in axial direction. This value of displacement, by means of a spring-loaded ram way is transmitted to the limit switch which, whenever operated, disconnects the industrial robot.

Actually, by means of a Research Institute, a space-monitoring sensor system will be designed, system destinated for integration within the robot control unit. In this case, an optical sensor performs a permanent position measuring of lattice-type installed 'measuring points' at the robot and this in order to monitor dynamically the executed robot movements. In addition, it will be possible to localize any obstruction within the operating room.
The disadvantage of this system solution is the expenditure of finance. On the other hand, availability of the system will not be reached earlier then after a 2 to 3 years design period.
This improvement concerns only safety equipments to install additionnally at the IR; except the expensive sensor system, they will offer only a partial protection in case of collision.
To protect the operating personnel inside the IR operating room just as in order to protect the IR against destruction, immediate recognition of any movements beyond control of the IR just as disconnection of the plant, whenever possible prior to any collision, will be required in addition.
Furthermore, the goal of the following description is to present a security system able to satisfy sufficiently these requirements. The named system has been designed by Ets. KUKA
to be used for the type KUKA-IR 601/60. The safety system has been integrated within the software section of the control unit with the advantage to present a particularly cost-effective solution (see figure 5).
Preventive measures are intented to protect the robot, the programmer just as the maintenance personnel against false robot control and falsified data input.

The operator has the ability to monitor the display of the different operational steps given per plain language. The operation sequence is shown by LED displays. Input data just as program number, speed, **accuracy**, output/input data are checked for reasonableness. All instructions and errors are indicated to the operator in a plain language; as a result, fault controls are mainly avoided.

For the operation mode 'set-up', the motor supply voltage is automatically limited such as to allow operation with only 10% of the set speed. This feature is obtained by means of switching the transformer feeding the power section.

In case of 'emergency cut-off', it will be possible to set all output lines of the IR to the relevant states. Therefore, the gripper state (open/closed) for instance can be determined or introduction can be allowed for a retractive motion of the gripper leaving an oven.

By means of steadily control-monitored hardware and software limit switches it will be possible to limit the working area in accordance to the task in order to avoid any overshooting of the working area limitations.

As mentioned before, the unlimited starting of the IR that means recognition of dynamic movements is of an essential importance. The function to fulfill is reached by monitoring of the different motion cycles, by means of the used travel and speed measuring devices provided for the control section. Figure 6 represents the different motion phases.

Generally, for the P-type position control a setting variation is given in proportion to the speed and to the mechanical circumstances.

In a number of applications, this setting variation is monitored in accordance to a maximum amount set rigidly at time of putting-into-operation.

Only upon exceeding the maximum allowable limit, an error, for instance as a result of a failure within the drive section, position sensor system or tachometer, is able to be detected.

Since the industrial robots (IR) allow very high speeds, an error indication of this kind is classified to be too slow. As a result, the limit value of the tow distance will not be given as an absolut limit value but determined permanently at the base of the actual speed and according to the machine parameters (cf. figure 5) within the control computer.

For each of the different motion phases, an upper and a lower limit value, relative to the given speed is able to be introduced as machine-type parameters and within each motion phase. As a result, a tolerance range is set around the position area. Whenever the actual value of this range is situated below or beyond the given level, an error indication is given per plain language to the

operator. In parallel, an error signal reduces to zero
the nominal value for the drive section and introduces a
short-circuit braking action, followed by the action of
the existing brake systems for all axles. This procedure
offers the possibility of most rapidly braking action and
shortest after-running path.

By situating the tolerance range in a narrow or large
condition referred to the nominal value, it will be
possible to set the monitoring control with a smooth or
sensitive ratio according to the function to perform. In
a similar manner, it will be possible to introduce the
flanks of the acceleration or deceleration graphs (cf.
figure 6a) in terms of machine parameters, with ability
to set the given machine, tool or load parameters.
Deviation of the analoguous members of the control loop
is able to present a deviation eventually beyond the
tolerance scheme.
After putting-into-operation, and after each program,the
control section will set automatically such a deviation
with subsequent compensation.
As a result, a high precision will be reached for the
long-term behaviour.
By adopting these kind of measures, unlimited movements
of the IR just as collisions of IR, failures of drive
sections, of range and speed measuring systems are able
to be recognized immediately because of overshooting of
the tolerance band by each of these causes.
Because of the motion-sequence monitorings, state monitoring controls are combined with the control section
offering the possibility of simultaneous on-line diagnostic.
This configuration allows a steady monitoring of error
conditions of different components such as the path
measuring control system, the tachometer, the power
supply or indication of a temperature error or any failure of the CPU, monitoring combined with display features.
Peripherical equipment such as magnetic tape cartridge
or punched tape are steadily monitored on behalf of their
performance characteristics. Therefore, the program to
transfer from the cartridge into the control section is
transmitted three times with simultaneous check for
parity and presence of CRC-type errors just as on behalf
of coincidence. Only upon three subsequent transmissions
without error, the program will be finally accepted.
Therefore, it is guaranteed that no falsified position
data will cause any irregular path configuration.
A RAM-CHECK device is provided for monitoring of the
machine-type data stock. Activation of monitoring is
performed in each case and in an cyclic manner after

putting-into-operation of the control section.
Any failure of isolated bit positions is able to be
localized by constitution of a check sum.
This check sum is composed by addition of all cell contents of the used machine data. This sum will than be
stored as an internal machine date.
This function is performed at the time of initialization
and first switching-in of the plant.
At each program sequence, the check sum is renewed and
steadily compared on behalf of the check sum with the
stored machine date. Any difference localized will disconnect the plant; on the other hand, a failure message is
indicated at the teaching controller.

A so-called routine for system program storage (EPROM)/-
check firstly recognizes EPROM-type connector board
errors just as the failure of different bit positions of
an EPROM. The routine will than be called after the
switching-in during the initialization routine and at
each program sequence.
Any connection error produced by replacing of EROM's
are recognized or localized by means of a given numerical
combination situated at the first cell of the slide-in
EPROM that means combination law interrogated and compared with a value determined after a determined formation
formula.

Failure of different bit positions can be localized by
constitution of a check sum. This check sum is composed
by addition of all cell contents of an EPROM; in this
case, the resulting sum is limited to two positions. The
sum of digits is determined externally and introduced
into the last cells of the EROM's. The control section is
constituting this sum of digits at each passage and
compares this sum with the check sum compiled externally.

All these measures are able to avoid any irregular behaviour and irregular path configurations of the industrial
robot (IR).

Most of accidents will occur especially during service
and maintenance works. As a result, adequate guidance of
operator and error localization aids are essential on
behalf of work position safety, measures able in addition
to shorten perturbation intervals.
In this time, all error indications are firstly transmitted per plain language, for example, during the approach
towards the final switch ;there will be display of the
specific axle final switch and of the path direction. On
the other hand, the operator responsible for maintenance
for the user has at his disposition a graduated diagnostic system allowing to localize per subassemblies and
card levels the indicated errors.
In case of failure, the operator uses a diagnostic card
for the control section (see figure 7), card indicating
the different diagnostic programs in EPROM's. By means
of a selector switch, the different test sequences are

able to be called-up.

The first check to perform consists of the CPU-type test since a fail-safe CPU will be the preparative condition for any further tests to perform. To do this, within the CPU have been activated for instance several computer operations controlled by the diagnostic program (operations just as arithmetics, transfer and comparison operations) the results of which are compared with the values stored within the diagnostic programs. The test result is displayed for the operator by means of a LED at the CPU. Whenever the CPU is defective, the latter has to be replaced.

The diagnostic program "RAM Test" controls a random generator within the CPU delivering bit samples by means of which the storage is loaded. Subsequently, these random-type bit samples are transmitted and checked on behalf of their conformity.

The test of the programming set comprises a test for the alpha-numerical displays, the LED's, selector switches and operation push-buttons. During the display test (see figure 8), the displays just as the lamps are controlled according to a given sequence. For the function push-buttons test and for the test of the selector switches, each pushing action or switch control is displayed by means of a plain language. The operator has in this case to verify these displays.

The state of to inputs/outputs towards the peripherical units is able to be visualized during the programming sequence.

Nevertheless, only the test of the momentary state of the inputs/outputs will be possible. The interface to the peripherical equipment is able to be tested in its integrality and this by switching the outputs to the inputs, by means of a coupling connector. In this case, the diagnostic program will than place in a cyclic manner the outputs together with a verification of all values.

The path control system will be checked in a similar manner. At an immobilized axle, the control will deliver a nominal value and elaborates the back information for the error voltage.

Conclusion

Safety equipment for industrial robots (IR) intented to protect as well the human being as the IR will condition increasingly the utilization of an industrial robot.

Beside any protection against unwanted penetration of the plant or against projected parts, by means of a protecting fence, special care or protection has to be given on behalf of the putting-into-operation and maintenance personnel obliged to by-pass frequently the external measures intented for protection. The goal to reach is firstly to protect this personnel against unlimited movements of the IR as a consequence of operating errors or malfunctions of components. On the other hand, the task of the maintenance personnel has to be supported by special diagnostic aids and this in order to improve safety of work.

This intention is realized here by adopting within the control unit an integral movement monitoring system just as by means of the ease of a diagnostic unit and, finally, by especially-orientated diagnostic programs.

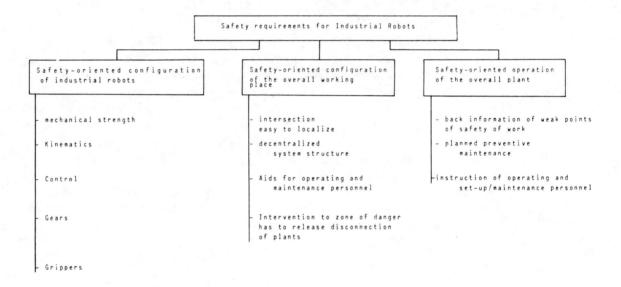

Figure 1:   Safety requirements for Industrial Robots

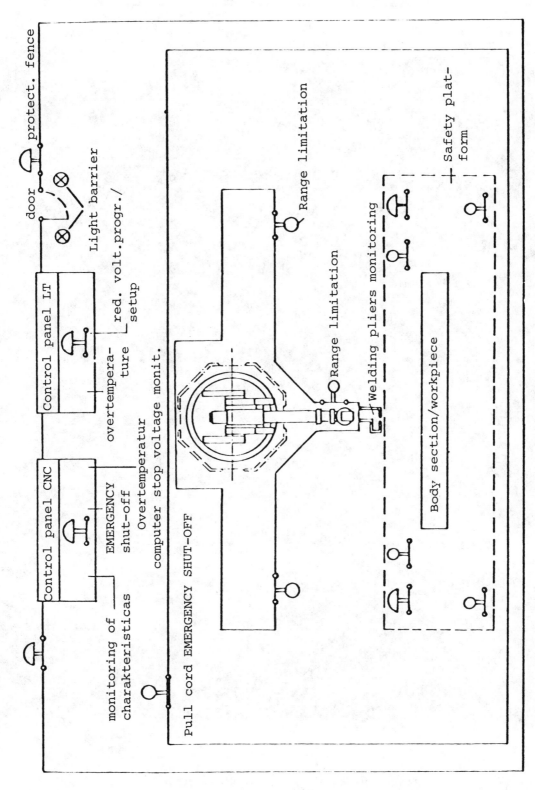

Figure 2: Realized safety functions for flexible transfer lines

Safety equipment (pressure-sensitive material)

Figure 3: Safety equipment (pressure-sensitive material)

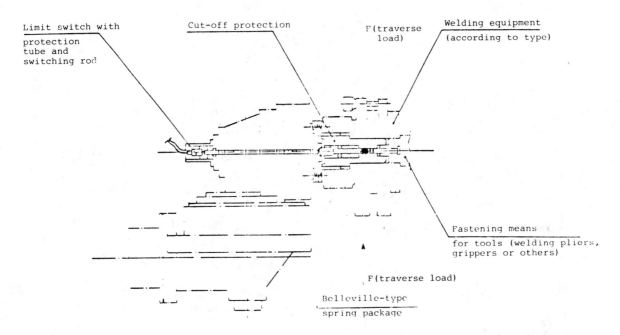

Figure 4: Cut-off protection means, 6th axle for industrial robot KUKA 601/6.

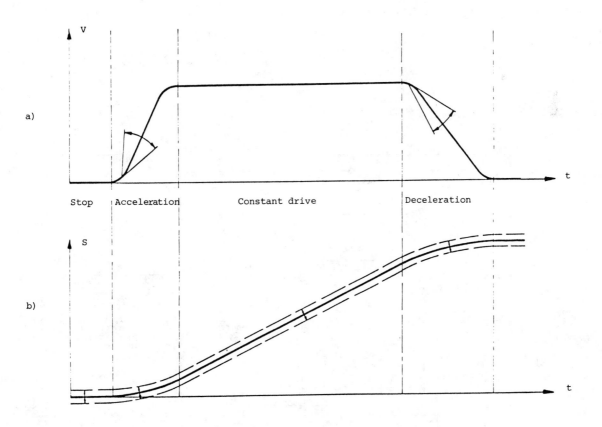

Figure 5: Monitoring of the industrial roboter motion phases

111

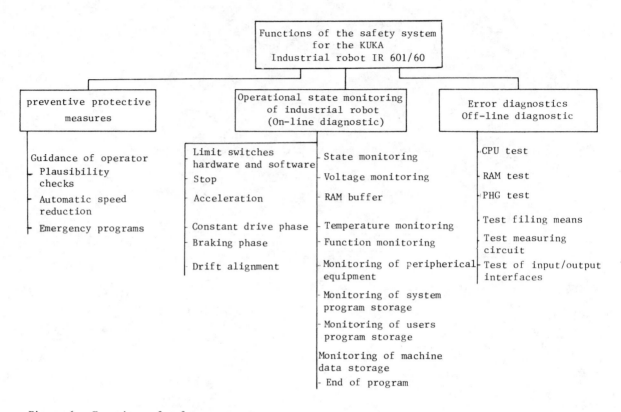

Figure 6: Functions of safety system

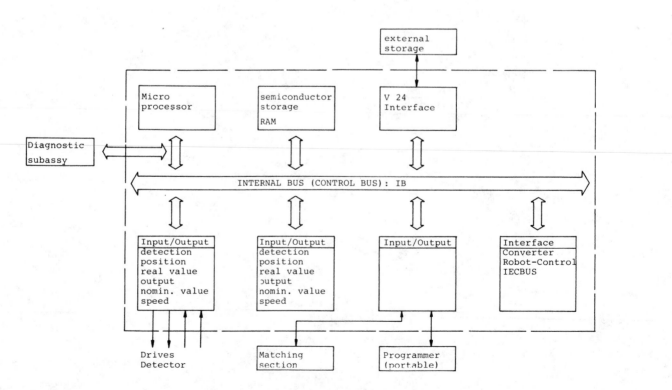

Figure 7: Diagnostic subassembly et control for Industrial Robot

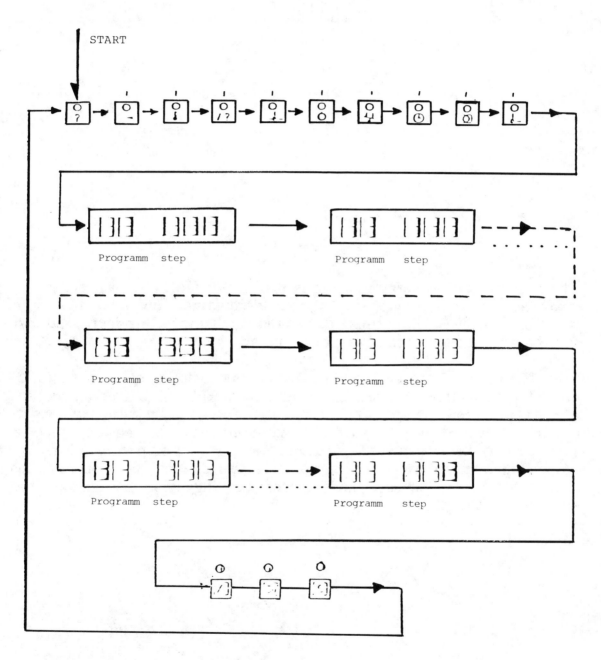

Figure 8: Test of displays

Presented at the Robots IV Conference, October, 1979

Safety, Training and Maintenance: Their Influence Of The Success Of Your Robot Application

By Robert R. Trouteaud
Prab Conveyors, Inc.

ABSTRACT

The utilization of industrial robots in production facilities has increased dramatically in recent years. The use of industrial robots has proved to be a sound business investment due to the requirement for increased production while coping with rising material and labor costs.

Very often those factors which influence the decision to select a specific robot for an application are not the same factors that insure the success of the application. Proper application of the robot is the primary consideration. However safety, training, and maintenance are equally important. The influence of proper safety considerations, training programs and maintenance procedures can make or break a well-designed robot application. This paper discusses these aspects and suggests guidelines for the success of robot applications.

SAFETY, TRAINING AND MAINTENANCE:
THEIR INFLUENCE ON THE SUCCESS OF YOUR ROBOT APPLICATION

Your company has decided to use industrial robots in its production facilities. After investigating available robot equipment you determined which robot best suited the application. You calculated that the use of the industrial robot will increase the productivity of this application and will provide acceptable payback, if it is a success.

Very often those factors which influence your decision to select a specific robot for your application are not the same factors that insure success of the application. Proper application of the robot is the primary consideration. However, safety, training, and maintenance are equally important. The influence of proper safety considerations, training programs and maintenance procedures can make or break a well-designed robot application.

TRAINING PROGRAM

As an essential aspect of a successful robot application, training should be considered an investment. As with any investment you expect a payback. The payback for training comes in reduced downtime and increased production. Training is one of the most neglected aspects of a robot project. Although training is always undertaken in one form or another, it is often inadequate.

An effective robot training program will consist of two phases: 1) programming and operator training and 2) preventive maintenance and troubleshooting. The program must incorporate both mechanical and electrical training. A full understanding of the system operation is essential to a smooth start-up and quick response to system failures.

The need for adequate vendor training is more important than ever due to increased complexity of many of today's robot applications. The use of computers, solid state controllers, and sophisticated system interfaces require a thorough understanding of the total process by operating and maintenance personnel. The electronics should be stressed since most of the mechanical components which make up an industrial robot are not new to plant personnel. If present plant personnel can maintain programmable controllers and numerically controlled machine tools then the addition of a robot control will not be a new experience. Robot manufacturers have designed their controls to be operated and maintained by plant level personnel.

Due to the wide range of industrial robots on the market today, the time required for training will vary. For example, the Prab Industrial Robot can be handled with a one day training program. This includes all programming, maintenance, diagnostic, and operator training necessary for the utilization of the Prab Industrial Robot line. In comparison, the Versatran training program at Prab Conveyors, Inc. lasts approximately four and one-half days.

A well-trained maintenance force, provided with adequate tools and spare parts, will prevent dependency on a manufacturer's service representative. They are rarely available on less than a twenty-four hour basis. Waiting for a service man or parts to arrive from the manufacturer is expensive due to lost production time. As an extreme example, it costs auto manufacturers about $4,000 for every minute of lost production on a final assembly line. Calculating your lost production dollars will only reinforce what I have said about the need to include training in your robot plans. Provisions for manual back-up of robots on high production lines are necessary to avoid costly downtime.

Many times assistance from the manufacturer can be handled via the telephone. We at Prab have full-time service personnel who are available to handle your problems over the telephone. This is a very important capability in reducing the downtime on your robot application. Built-in diagnostic capabilities aid in this telephone assistance. Your maintenance personnel will be able to read the diagnostic indicators and relay that information to the expert at the vendor's facility. Using this diagnostic information the service expert will be able to respond to the problem. Built-in diagnostic capabilities available on the various control packages make this quite practical. Your ability to react quickly to minimizing robot downtime will be increased with proper training, built-in diagnostic aids, and manufacturer assistance over the phone.

The Prab Robot line features many diagnostic aids. For instance, all Prab Robots include indicators to show filter cleaniness, hydraulic oil temperature, running hours, and directional indicators on all valve modules (See Figure 1). A home positioning light indicates when both the program controller and the mechanical arm are in the home position. The standard mechanical stepping drum control includes a step position locator and removable control drum for easy program diagnostics. Prab's solid state controller includes L.E.D. readouts of the programmed steps and functions. These readouts are easily viewed from the control panel on the robot. (See Figure 2).

The Versatran line of industrial robots uses an advanced microcomputer-based controller which features many diagnostic aids. The controller monitors oil filter cleaniness, oil temperature, cabinet temperature, operating hours, plus much more. The step number, program number, program sequence, position number, and function values are displayed in L.E.D. readouts on the control face (Figure 3). If a program problem occurs a program fault light will automatically activate. When the program fault light is on the functional value readouts will convey an error message code. These codes indicate the problem in minute detail and the coding chart provides specific cures for problems. This, of course, is a definite aid if assistance is needed from the manufacturer via the telephone.

In addition to these diagnostic features, the Versatran 600 Control permits teaching in a variety of modes to eliminate many problems before they occur. These include restricted speeds while in the teach mode and single step and test run modes for checking programming and positioning during start-up. The greater the diagnostic capabilities of the robot controller the more effective your maintenance team will be when responding to problems on line.

An important part of any training program is the manuals that are provided for your use. (See Figure 4). Included in this manual will be information on the operation, programming and maintenance of your robot. In addition to this, electrical and hydraulic diagrams and specifications, and layout diagrams of the robot are provided. The manuals include operating, diagnostic and service information plus spare parts replacement lists. Information and specifications on standard components are also provided.

Combining maintenance training and operational and programming training is of the utmost importance because it will teach your personnel to use the built-in diagnostic capability of the robot control. In addition, the training will teach common troubleshooting points and suggestions gained from the manufacturer's experiences.

By the end of your company's training class, plant personnel will be able to install the robot, program it to fit the application, and maintain the robot for maximum uptime in production.

WHO SHOULD BE TRAINED?

A variety of personnel have been included in robot training programs. Our experience reflects that the personnel have come from many skilled areas such as plant foremen, production supervisors, plant engineers, and plant managers. Although the mix of personnel to handle the robot application can be varied, there are two important points to remember. First, train more than one person. Shifting of personnel today dictates this need. Secondly, train the people most directly involved in keeping the operation at its maximum productivity. Put the knowledge where the need is. The task of programming the robot should be handled by the primary operator or supervisor responsible for the robot installation. Few people know the production operation better then the foreman or his lead man. Including them in the training, programming and maintenance requirements will insure a successful application.

WHERE AND WHEN TO TRAIN

The best location for training of personnel is in the robot manufacturer's facility. There are three main advantages to this location. First, the trainees will have a concentrated, in-depth training program with factory personnel who are most knowledgeable about the operation and maintenance of robot equipment. Second, the trainees will not be interrupted during training as they might be in their own facility. Third, members of other companies purchasing robots will be in the class and provide insight into a variety of production situations using robots.

Robot suppliers who provide training programs have full-time instructors. They are experienced in field applications.

The robot equipment required for training will be available in the manufacturers facility. Training should not be taken too far ahead of the scheduled delivery. The subject matter should be fresh in the minds of plant personnel when the robot is ready to be installed.

To familiarize other operating personnel in your facility, consider setting up the robot in a temporary area. Plant personnel can then have some hands-on experience with their robot without affecting production. This temporary set-up can contribute to the acceptance of the robot in the shop. The engineering staff can also gain experience in applying robot technology to other applications in manufacturing during this temporary set-up.

After the robot has been installed additional training can be handled in a number of ways. Robot manufacturers encourage retraining at their facility. In addition, field classes can be taught in your plant.

Once again, training is an investment with real payback. Emphasize it in your robot program.

PREVENTIVE MAINTENANCE PROGRAM

Preventive maintenance training taught by the robot manufacturers includes a full program of scheduled maintenance. Regular maintenance check procedures and recommendations for worn parts replacement are detailed. Daily, monthly and yearly check procedures are outlined in the training manuals and stressed during the class. These recommendations are based on years of experience with hundreds of robots. They include scheduled maintenance on the mechanical and hydraulic systems such as filters, oil pressure, bearings and other mechanical components. Electrical system adjustments are also covered in detail.

The advantages to following regularly scheduled maintenance programs on industrial robots are many. It is estimated that last year American industry spent more than two hundred billion dollars to maintain plant equipment. Almost one-third of that was wasted or could have been saved by proper maintenance management. In addition to the cost of disrupted production, it has been estimated that it takes three times as much manpower, time, and money to do an emergency repair job than a scheduled repair. It is good economics to have a sound maintenance program resulting in the increased use of production equipment.

Preventive maintenance programs make sense. Unfortunately, there is a great deal of resistance to implementing them. It is not an easy decision to stop the production of a machine or production line for scheduled preventive maintenance. There is always pressure to keep production schedules, and it is difficult for a foreman to shut down a machine in order to replace a part if the machine is still producing. Usually it is only when a machine stops that it demands immediate attention. The result is three times the maintenance cost for unscheduled repairs.

If the production manager is responsible for meeting the production quota, and the maintenance manager for fixing equipment, then who is responsible for the cost of not maintaining the equipment?

SPARE PARTS

The following scenario is a frequent occurrence. Management has had the foresight to have the right people attend the right training at the right time. Maintenance people have been well versed in the operation and programming capabilities of the industrial robot. The robot has been installed and is working successfully at your application. One day it stops. Your well-trained maintenance staff is quick to determine the problem, in this case, a contaminated valve. Having used their training the staff discovers that the necessary part is not carried in the plant. The routine is familiar from this point on: the emergency call to the vendor, determining when production can be restarted, advising others of the delay in delivery, etc. The lesson to be learned is that spare parts are a must!

In order to properly support an industrial robot you must carry an inventory of spare parts as suggested by the manufacturer. The amount and the dollar value of the spare parts will vary depending on the robot and its complexity.

A medium technology robot with its majority of standard components will have very few spare parts suggested by the manufacturer and carry a relatively low price tag. Most higher technology machines with their sophisticated servo controls will require a more costly inventory of spare parts. Prab's Versatran Robots are the exception due once again to standard components. As a general rule, spare parts will cost about ten percent of the robot system cost.

In case of temporary breakdown, having spare parts at the plant compared to waiting for their shipment from the vendor will result in savings. Keeping spare parts available when needed will pay for the cost of purchasing and holding spare parts in inventory. In setting up the recommended spare parts list the manufacturer can provide a system with regular return of any defective parts that can be handled on an exchange or repair basis.

A sound preventive maintainence program is vitally important to today's manufacturer. Increasing equipment uptime presents one of the most important solutions to increasing production profits.

SAFETY

Adequate safety procedures can impact the success of your robot application in three areas: 1) the safety of your work force, 2) the safety of the robot, and 3) the safety of other machines. Robots are great attention-getters. It is necessary, therefore, to provide adequate hazard guarding around the robot application so that no person can accidentally walk into the robot's work envelope. The hazard enclosure is important because a robot may not be moving while waiting for a signal from the machine it is tending so that a worker could think that it is not operating and walk into the area. Also, the robot's overall reach may be much greater than the area it's been programmed to move in. A malfunction could send it to its limits of travel and injure an innocent onlooker.

In order to limit the work envelope encompassing the robot arm, mechanical stops can be added to many of the servo controlled robots. This limits the maximum reach of the robots arm in all axes. Prab's medium technology robot line already utilizes programmable fixed mechanical stops for its positioning, so additional mechanical stops are not required.

The hazard guarding can be handled in a variety of ways. Methods that have been used include standard guard post and railing, (Figure 5), wire mesh fence, (Figure 6), or safety chain, (Figure 7). Sometimes the layout of the application is guarding protection itself. (Figure 8). The hazard guarding should be interlocked electrically so that when entrance is made through the gate the motion of the robot is halted.

Prab Conveyors, Inc./Robot Division has standardized hazard guarding on all of its industrial robots (Figure 9). Provided as part of the robot is a chain link fence with interlocking controls for use in surrounding the robot working envelope. A big advantage in using a chain link fence is that not only does it keep the passerby out of the work envelope, it also is capable of stopping a part which may be accidentally released by the robot's gripper during a sweep motion. A part weighing only a few pounds, moving at speeds capable with the robot rotary motion, would act as a projectile that is hard to stop.

A worker is more likely to be in danger during programming or teaching of the robot and during service. There are various methods of teaching the robot depending on the robot involved. Teaching may be accomplished from either the control console or a portable teach pendant. If the control unit is enclosed as part of the robot body then the hazard guarding must allow easy access to the control panel (Figure 9).

When a control unit is remote mounted, an ancillary teaching control must be used to get close to the robot for teaching. Caution must be taken as the teaching unit allows you to get inside the work envelope of the robot. Prab and Versatran Robots only move in a restricted slow speed mode during teaching. There is an emergency stop pushbutton in easy reach at all times.

While the robot is being serviced, extra measures should be taken to provide for the safety of the serviceman. One method is to install blocking posts set into the floor to halt the motion of the robot arm should it swing toward the serviceman. Service personnel must be properly trained and fully aware of the functions of the machine if safe working conditions are to be insured.

Most robot controls are designed to avoid the improper use of the programmable robot. Interlocking controls should be designed so that robots cannot be started in automatic modes unless they are first homed to the original start position. This keeps the robot from damaging surrounding equipment by improperly starting in mid-cycle.

In order the protect the industrial robot and the machinery it services, you will want to interlock all equipment through one common source. The interlock control sends signals to and from the robot and the appropriate machinery for transmitting such information as: when a cycle is finished, when a part has been cleared from the dye, mold or fixture, and when a cycle should begin. The interlock description or hardware is generally supplied by the robot manufacturer as part of the application.

The outstanding safety records of industrial robots have not come easily. Do not endanger that record by allowing a worker to walk into a robot set-up because of inadequate safety measures. Manufacturers and users must do everything they can to insure that the use of industrial robots is carried out in as safe a manner as possible.

SUMMARY

Without the attention to safety, training and maintenance, all the time and effort devoted to your robot project may be in vain. Unless you have taken the steps necessary to insure a successful application, the end result will be less than satisfactory no matter what robot was chosen for the job.

Be sure that a valid training program is available with your robot equipment. Training which covers programming, operation, and maintenance of your robot equipment in detail is essential to minimizing downtime. Pick the proper people to be trained and train them at the manufacturer's facility to optimize your investment in training.

Pay attention to preventive maintenance and scheduled maintenance procedures outlined by the manufacturers, and be sure that spare parts are carried as part of the robot project. The primary responsibility of maintaining your production is with your maintenance personnel.

Safety measures are essential for the protection of your equipment, your personnel, and your robot. Do not skimp on safety measures. You must do everything possible to insure a safe application.

Your industrial robot was installed to aid in increased production and profitability of your operation. The success of the application depends on your ability to maintain proper uptime. Your attention to proper safety considerations, training programs, and maintenance procedures will make your well-designed robot application a successful robot application.

BIBLIOGRAPHY

"Analysis of First UTD Installation Failures, Gennero C. Mari
Society of Manufacturing Engineers, MS77-733

"A Users Guide to Robot Applications, William R. Tanner
Society of Manufacturing Engineers, MR76-601

"Building the NC Team is a Management Job",   Ken Gettelman
                                                                and Karl Shultz
Society of Manufacturing Engineers, MS78-145

CNC Maintenance Training
Tooling and Production, July 1979

"Machine Loading with Robots", Walter E. Fritz
Society of Manufacturing Engineers, MS77-739

Preventive Maintenance: An Essential Tool for Profit
Production, July 1979

The High Cost of Bad Maintenance
Dun's Review, August 1979

Figure 1

Figure 2

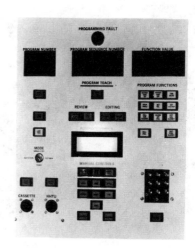

Figure 3

Figure 4

Figure 5

Figure 7

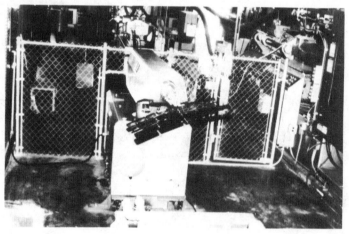

Figure 6

Figure 8

Figure 9

CHAPTER 3

JUSTIFICATION

Commentary

Justification of a robot installation may involve a number of factors, including increased productivity, improved quality, reduction of scrap or rework, performance of hazardous or undesirable tasks, avoidance of obsolescence, reduction or elimination of changeover time and, of course, reduction of costs.

The most important justification factor cited by robot users is economic—either through reduction of operation costs (primarily labor), an increase of profits, or both. This chapter presents a number of economic analyses of specific robot applications. It provides some experience-based guidelines for estimating the value of "non-economic" factors such as productivity, quality, flexibility and safety, and touches upon the social issues, as well. Typical installation, operating and maintenance costs for representative robot installations are also provided.

Presented at the Robots III Conference, November, 1978

"Selling" The Robot—Justification For Robot Installations

By William R. Tanner
Robotics Technology, Inc.

There are numerous factors cited for justification of the use of industrial robots, including: unavailability of labor for undesirable jobs, productivity and quality improvements, advancement of technology, competitive position, management direction, safety, flexibility and economics. The major factor is economics. Economic considerations generally fall into one of two classes, cost avoidance or cost savings. Economics can be measured in several ways, including return on investment, depreciated cash flow and payback period. Simple "rules of thumb" and a streamlined cost analysis method can be applied to determine the potential economic return of a contemplated robot installation.

INTRODUCTION

In the justification of an industrial robot in a manufacturing environment, the major factor is economics. The robot is considered capital equipment and it is therefore expected to provide a return of capital and, hopefully, a profit within its useful lifetime. Analysis of the economics of a robot installation generally falls into one of two categories--cost avoidance or cost savings. These will be described in detail later. Factors other than economics are also sometimes cited for robot justification.

"NON-ECONOMIC" FACTORS

The most common non-economic factors are increased productivity, improved quality, reduction of scrap, performance of undesirable jobs, advancement of technology, competitive position, management direction and safety. However, as will be seen in the following discussion of these factors, even the "non-economic" considerations are, in the final analysis, basically economic in nature.

Increased Productivity

Increased productivity through the use of a robot is, in many cases, more the result of the constant pace of the robot than of faster operation than a man. For many tasks, a human is capable of working as rapidly as or more rapidly than a robot, particularly where complex motions or adaptive movements are required. However, the human's pace tends to vary from cycle to cycle and, particularly on high-speed, repetitive tasks, fatigue will eventually reduce the person's work rate. The robot, on the other hand, will operate at a constant pace at all times. Thus, over the duration of a normal work shift, the robot's average cycle time may be less than a human's. The robot's higher productivity, measured in terms of increased parts per day, represents an economic gain.

Quality

Improved quality and reduction of scrap can result from the consistent operation of a robot. These gains are often cited in the die casting industry where the constant cycle time of the robot allows die temperatures to stabilize and operation without shut-down for breaks and shift changes avoids the production of cold shuts. Also, application of die lubricant by the robot is more regular and more uniform than by a human, which results in fewer bad parts as well as improved die life. Again, the greater number of good parts produced per shift by the robot, avoidance of material waste due to scrap and the reduction of die maintenance costs are measurable economic gains.

Undesirable Tasks

Operation of a robot on an undesirable task also has economic advantages. Worker's complaints about poor working conditions such as noise, dust, fumes, heat, dirt, heavy loads, fast pace or monotony, if unresolved, often lead to work stoppages or slow-downs, uncompleted operations, poor workmanship, high labor turnover, absenteeism, grievences or sabotage. All of these occurrences will be reflected in higher than normal operating costs. Overtime may be required to make up production losses, rework and repair necessitate additional labor and administrative costs are involved in processing grievances, hiring replacement personnel and training new workers.

Advancement of Technology

"Advancement of technology" may be cited as a reason for introduction of a robot. However, such a use is commonly limited to only one or a limited number of robots in more or less developmental applications. The motivation here is to gain the knowledge required to implement similar robot applications in a true production setting where the usual economic criteria and measurements apply. In fact, the cost of developmental trials is often factored into the cost of the follow-on production applications.

Competitive Position

Another consideration is competitive position, which has both direct and indirect economic implications. The direct economics are obvious--if a manufacturer can reduce the cost of producing his goods by using robots, he has a profit or pricing advantage over his competitors who do not use them. The indirect economic advantage lies in the inherent flexibility of robots. This flexibility gives a user the capability to meet shifts in market demands by increasing or decreasing production rates on various products without increasing or decreasing the size of the work force or to introduce new products into his manufacturing system quickly and easily, often with little change to production facilities.

Management Direction

A robot is sometimes justified on the basis of "management direction" without regard for economic considerations. In such cases, the success rate tends to be low. In an effort to comply with the management directive, neither the robot nor the application may be carefully chosen. The application may turn out to be more complex than first anticipated or the robot may not have the necessary capabilities to perform the chosen task. Aside from the obvious waste of capital, a bad experience may discourage management from further consideration of robots, even where other potentially successful (and therefore economically advantageous) applications may exist.

Hazardous Operations

A robot is frequently used on an operation which is potentially hazardous to the human worker, particularly as a means of compliance with safety regulations. Such tasks include press loading or unloading or operation in toxic atmospheres or extremes of ambient environment, for example. Even under these strong pressures for applying robots, the astute prospective user will consider the economics of guarding or protection for the human as an alternative to using a robot and, other factors being equal, will usually choose the least costly approach.

ECONOMIC FACTORS

Generally, a potential robot application is evaluated on the economics involved, with the factors described previously as well as projected labor cost reductions included in the financial analysis of the proposed installation. Of the two categories of economic justification--cost savings and cost avoidance--the latter usually involves less stringent economic objectives.

Cost Avoidance

A cost avoidance economic analysis is made to determine the least costly of several alternatives. In the case of new production equipment (new model tooling) for an automobile plant, for example, a comparison may be made between a labor-intensive approach, special-purpose automation and robots. The choice to be made is not "do we or don't we" retool the job, but which approach to retooling will result in the lowest lifetime cost. The labor-intensive approach would require less capital outlay and have a relatively low obsolescence cost but would result in high operating expenses. The special-purpose automation approach would require relatively high capital expenditures with moderate to low operating expenses and high obsolescence and changeover costs. The robot approach would also involve a relatively high initial capital outlay and low to moderate operating expenses but would have lower obsolescence and changeover costs than special purpose automation. An analysis of lifetime costs and expenses for each of these alternatives would be prepared and the least-cost method chosen

Cost avoidance economics would also be applicable in determining the best method of meeting safety requirements or reducing worker hazards. In a press loading operation, for instance, the cost of a robot installation would be compared to the cost of installing safety devices on the press. Here, the production rate of the robot approach versus the manual-with-guarding approach and the cost of part orienters for the robot would also have to be considered in the analysis, as well as the potential labor cost reduction with the robot.

Cost Savings

Cost savings economics, in contrast to cost avoidance where a capital expenditure is mandatory, apply in situations where there is the option to do nothing. In this case, the potential economic gain, as measured by return on investment or payback period, usually has to meet or exceed some established level to qualify for consideration. Since the demand for such discretionary capital usually exceeds the supply, a "target" is established to assist in screening requests for these funds.

A simplified cost analysis method is useful for initial screening of potential robot applications to determine if they will meet the established return on investment or payback period target. The Cost Analysis Form shown in Figure 1 may be used for this purpose. An explanation of each line entry on the form is shown in Appendix A. When completed, an estimated long-term (10 year) average after-tax return on investment will be obtained. An example of the use of this form is shown in Figure 2.

If return on investment is to be calculated on the basis of depreciated cash flow, the form may still be used as a convenient means of tabulating data. An example of the use of the Cost Analysis Form and the calculation of return on investment by the depreciated cash flow method is shown in Figures 3 and 4. Likewise, the form can be used to tabulate data for payback period calculation, as shown in Figure 5.

It is often useful to employ a "rule of thumb" for preliminary screening of potential cost savings robot applications prior to preparation of a detailed cost analysis. Such a "rule of thumb" can be established by working backwards through a cost analysis, starting with a desired return on investment and established direct labor rates and assuming some average operating costs for indirect labor, operating supplies, maintenance and repair, launching costs, etc. The objective is to project expenditure levels which can be supported by various anticipated direct labor savings at the desired return on investment. Once established, a "rule of thumb" will be applicable until there are significant changes in direct labor rates or return on investment targets.

For example, this process may establish a "rule of thumb" such as, "For a return on investment of 40%, 2 men per day savings will support an initial expenditure of $ 80,000." This "rule of thumb" can then be easily applied during cost savings robot application surveys by quickly estimating the potential manpower savings, robot and end-of-arm device costs and installation and rearrangement costs. By this means, low-return robot applications are eliminated from further consideration without the need for a detailed cost analysis.

In addition to determining the savings resulting from direct labor reduction with a robot application, the "non-economic" factors previously discussed should be considered and quantified in the cost savings analysis. If an increase in productivity can be projected for the robot, the number of parts per day can be estimated and, based upon their value, an annual savings can be determined. Similarly, if quality improvement and/or scrap reduction is anticipated, the potential annual reduction in scrap or rework costs can be calculated. When a robot is applied on an undesirable operation prone to high labor turn over, a savings equivalent to the annual cost of hiring and training new personnel can be credited to the robot. Avoidance of overtime to make up for production losses would also be attributable to the robot.

Other areas of potential savings include normal costs of protective clothing, safety equipment, lighting and ventilation levels, parking, dining, washroom and locker room facilities, supervisory work loads, etc. which are inherent with human labor. The determination of these potential savings and their quantification may be difficult and each factor may not contribute much in itself to the overall savings. However, the total of all of them may very well be sufficient to make a marginal project financially attractive or move a potential application from the unacceptable to the acceptable side of the required return on investment level.

While many of the foregoing considerations apply only to cost savings economics, some should also be taken into account when developing cost avoidance alternatives. The people-related factors, such as productivity, quality and reaction to undesirable tasks, as well as costs other than wages and fringes associated with human labor such as protective clothing, safety equipment, area lighting, ventilation, etc. all represent additional expenses to be assessed against the labor-intensive approach.

CONCLUSION

Although there are a number of different factors which may be employed in the justification of robot applications, the most significant and widely used is economics. Even when the justification is proposed under the guise of altruistic motivation or when factors other than pure economics are cited, the underlying reason for using a robot is usually profit oriented. Successful justification, therefore, requires consideration of all of the potential costs and cost benefits of a robot installation and the quantification of the economics of factors other than direct labor replacement in addition to the cost reductions related to the labor replacement itself.

Many of the cost factors can only be estimated during the justification preparation. However, following the installation of the robot, its actual cost performance usually can be easily and accurately measured. It is imperative, therefore, that the original estimates be as accurate as possible. It is far better to be complimented when the robot's performance exceeds the original projections than to be criticized when it fails to meet its economic objective.

Expenditures

1.	Robot and Accessories Cost	$_____
2.	Installation Costs	$_____
3.	Related Rearrangements Costs	$_____
4.	Sub Total	$_____
5.	Special Tooling Costs	$_____
6.	Total Expenditures	$_____

Operating Savings (Costs)

Long-Term Average Dollars - 10-Year Life

7.	Direct Labor - Man Hours	_____	
8.	Direct Labor - Dollars		$_____
9.	Indirect Labor - Man Hours	_____	
10.	Indirect Labor - Dollars		$ (_____)
11.	Operating Supplies		$ (_____)
12.	Maintenance and Repair		$ (_____)
13.	Launching Costs		$ (_____)
14.	Taxes and Insurance		$ (_____)
15.	Special Tooling Amortization		$ (_____)
16.	Depreciation		$ (_____)
17a.	Other Savings		$_____
17b.	Other Costs		$ (_____)
18.	Profit Before Taxes		$_____
19.	Profit After Taxes		$_____
20.	Investment Base	$_____	
21.	Return on Investment	_____ %	

Figure 1. Cost Analysis Form

Expenditures

1.	Robot and Accessories Cost	$ 45,000
2.	Installation Costs	$ 5,000
3.	Related Rearrangements Costs	$ 10,000
4.	Sub Total	$ 60,000
5.	Special Tooling Costs	$ 17,500
6.	Total Expenditures	$ 77,500

Operating Savings (Costs)

Long-Term Average Dollars - 10-Year Life

7.	Direct Labor - Man Hours	18.24	
8.	Direct Labor - Dollars		$ 50,560
9.	Indirect Labor - Man Hours	2.0	
10.	Indirect Labor - Dollars		$ (6,000)
11.	Operating Supplies		$ (2,560)
12.	Maintenance and Repair		$ (2,000)
13.	Launching Costs		$ (500)
14.	Taxes and Insurance		$ (1,250)
15.	Special Tooling Amortization		$ (-0-)
16.	Depreciation		$ (6,000)
17a.	Other Savings		$ -0-
17b.	Other Costs		$ (-0-)
18.	Profit Before Taxes		$ 32,250
19.	Profit After Taxes		$ 16,770
20.	Investment Base	$ 38,750	
21.	Return on Investment	43.3 %	

Figure 2. Example of Use of Cost Analysis Form for Return on Investment Calculation

Expenditures

1.	Robot and Accessories Cost	$ 45,000
2.	Installation Costs	$ 2,000
3.	Related Rearrangements Costs	$ 5,000
4.	Sub Total	$ 52,000
5.	Special Tooling Costs	$ 3,000
6.	Total Expenditures	$ 55,000

Operating Savings (Costs)

Long-Term Average Dollars - 10-Year Life

7.	Direct Labor - Man Hours	16.5
8.	Direct Labor - Dollars	$ 51,800
9.	Indirect Labor - Man Hours	
10.	Indirect Labor - Dollars	$ ( )
11.	Operating Supplies	$ (-0-)
12.	Maintenance and Repair	$ (2,800)
13.	Launching Costs	$ (500)
14.	Taxes and Insurance	$ (1,175)
15.	Special Tooling Amortization	$ (-0-)
16.	Depreciation	$ (5,500)
17a.	Other Savings	$ 1,675
17b.	Other Costs	$ (-0-)
18.	Profit Before Taxes	$
19.	Profit After Taxes	$
20.	Investment Base	$
21.	Return on Investment	%

Figure 3. Example of Use of Cost Analysis Form for Data Tabulation for Depreciated Cash Flow Analysis

Using the data shown in Figure 3:

	Annual Labor Savings (Line 8)	$ 51,800
	Other Savings (Line 17a)	1,675
1.	Total Annual Savings	$ 53,475
	Annual Operating Costs -	
	Maintenance (Line 12)	$ 2,800
	Launching (Line 13)	500
	Taxes and Insurance (Line 14)	1,175
2.	Total Annual Costs	$ 4,475
3.	Net Annual Savings (1 minus 2)	$ 49,000
4.	Net Annual Savings After Taxes (52% of 3)	$ 25,500
5.	Depreciation Tax Credit (48% of Line 16)	$ 2,640
6.	Total Annual Net Future Savings (4 plus 5)	$ 28,140

Assuming these savings are constant over the life of the robot (n years), for a discount rate of r:

$$\text{Current Value of Savings} = \sum_{1}^{n} \frac{\text{Total Annual Net Future Savings}}{(1 + r)^n}$$

$$\text{Investment Level} = \text{Total Annual Net Future Savings} \sum_{n=1}^{10} \frac{1}{(1 + r)^n}$$

Solving for r for the current value equal to the investment of $ 55,000:

$$\frac{55,000}{28,140} = \sum_{n=1}^{10} \frac{1}{(1 + r)^n}$$

r = .51 ; Return on Investment = 51%

Figure 4. Example of Return on Investment Calculation for Depreciated Cash Flow Method

Using the data shown in Figure 3:

	Annual Labor Savings (Line 8)	$ 51,800
	Other Savings (Line 17a)	1,675
1.	Total Annual Savings	$ 53,475
	Annual Operating Costs -	
	Maintenance (Line 12)	$ 2,800
	Launching (Line 13)	500
	Taxes and Insurance (Line 14)	1,175
2.	Total Annual Costs	$ 4,475
3.	Net Annual Savings (1 minus 2)	$ 49,000
4.	Net Annual Savings After Taxes (52% of 3)	$ 25,500
5.	Depreciation Tax Credit (48% of Line 16)	$ 2,640
6.	Total Annual Net Future Savings (4 plus 5)	$ 28,140

$$\text{Payback} = \frac{\text{Total Expenditures}}{\text{Total Annual Net Future Savings}}$$

$$= \frac{55,000}{28,140}$$

$$= 1.95 \text{ years}$$

Figure 5. Example of Payback Period Calculation

Appendix A. Cost Analysis Form Line Items

1. **Robot and Accessories Cost** — Includes cost of robot, optional equipment, special tools, maintenance and test equipment, accessories, etc. (excludes special end-of-arm devices).

2. **Installation Costs** — Labor and materials for site preparation, floor or foundation work, utility drops and connections (air, water, power), interface devices between robot and fixture or conveyor.

3. **Related Rearrangements Costs** — Labor and materials, procurement and installation of all facilities (excluding special end-of-arm devices) related to the robot installation, such as conveyors, part feeders and transfers; stock rearrangements; machinery and equipment relocations; guard rails; etc.

4. **Sub Total** — Sum of Lines 1, 2 and 3 (used for depreciation computation).

5. **Special Tooling Costs** — Includes labor and materials for special end-of-arm devices; related changes to other fixtures and special tooling, such as power clamping devices, interface devices (limit switches, sensors, etc.), control interlocks, etc.

6. **Total Expenditures** — Sum of Lines 4 and 5.

7. **Direct Labor - Man Hours** — Total _daily_ man hours of anticipated manpower savings, including relief allowances, shift premiums, etc.

8. **Direct Labor - Dollars** — Line 7 times normal work days per year times hourly direct labor rate including fringe costs (projected ten-year average).

9. **Indirect Labor - Man Hours** — Estimated _daily_ man hours required for maintenance and repair of robot and related new equipment.

10. **Indirect Labor - Dollars** — Line 9 times normal work days per year times hourly indirect labor rate including fringe costs (projected ten-year average)

134

Note: If robot maintenance is to be handled by outside contract, enter service contract charges or estimated parts-and-labor costs on Line 12 instead of completing Lines 9 and 10.

11. Operating Supplies — Annual cost of utilities and services for robot and related new equipment.

12. Maintenance and Repair — Estimated annual cost for maintenance supplies, replacement parts, hydraulic fluid, lubricants, etc. or service contract charges (see note above).

13. Launching Costs — Costs attributable to new equipment launching (such as production losses, overtime premiums, robot service personnel for start up, etc.) and to training of maintenance and programming personnel. Divide total estimated launching costs by 10 years for annualized average.

14. Taxes and Insurance — Calculated as 2.5% of the sum of Lines 1 and 2.

15. Special Tooling Amortization — Usually zero, unless new special end-of-arm devices and related equipment changes due to change in application are anticipated within the 10-year life. Amortization rate for non-capital equipment (expense items) would apply.

16. Depreciation — Calculated as 10% of Line 4 (straight line) or by whatever other method is normally used.

17a. Other Savings — Includes any non-labor savings, such as reduced scrap, increased productivity, investment tax credit, etc. See text of paper for further potential savings.

17b. Other Costs — Include any unique costs not covered above, such as prorated share of general test equipment, applications development costs, spare parts inventory, etc.

18. Profit Before Taxes — Calculated as Line 8, minus Lines 10 through 16, plus Line 17a, minus Line 17b.

19. Profit After Taxes   Calculated as 52% of Line 18.

20. Investment Base   Calculated as 50% of Line 6

21. Return on
    Investment   Calculated as Line 19 divided by
    Line 20, expressed as a percentage.

Reprinted by permission from Atlanta Economic Review (now BUSINESS Magazine). "Robots Make Economic and Social Sense," by Joseph F. Engelberger, July-August, 1977

Joseph F. Engelberger

Robots Make Economic and Social Sense

Sidestepping science fiction lineage, a practical industrial robot is taking over factory jobs scorned by humans everywhere in the industrialized world. From Detroit to Moscow, acceptance is mounting on economic and social grounds.

EVERYBODY KNOWS that a robot is a mechanical man whose interior is festooned with a maze of piping, wires, transistors, whirring gears, and flashing lights, all ingeniously interconnected to impart superhuman powers. Hardly anybody realizes that such automatons actually exist and that they are joining the "blue collar" work force in a wide range of hot, hazardous, or tedious jobs. Shucking the image owing to science fiction and television, these robots are hulking, powerful, one-armed machines which boast of a computer memory, a hydraulic musculature, and a penchant

***

Mr. Engelberger *is President of Unimation Inc., Danbury, Connecticut.*

***

for harsh, routine work. Early interplay between human workers and their robot counterparts seems to presage the proliferation of this new slave class. The robots are doing the dirty work, and the upgraded human operators look upon their servile coworkers with tolerant superiority.

The word "robot," which means "worker" in Czech, entered the English language in 1922 when Karel Capek, a Czechoslovakian philosopher and playwright, presented his most successful stage effort, "R.U.R." or "Rossum's Universal Robots." Old man Rossum and his son discovered a chemical composition that simulated protoplasm. They elected to organize it in the form of man and, they fondly hoped, in the

service of man. Young Rossum said, "It's absurd to spend twenty years making a man. If you can't make him quicker than nature, you might as well shut up shop." The practical engineer Rossum overhauled the basic design to eliminate superfluous organs, dimensions, senses, and especially a soul. Rossum opined, "A man is something that feels happy, plays the piano, likes going for a walk, and, in fact, wants to do a whole lot of things that are really unnecessary. . . . But a working machine must not play the piano, must not feel happy, must not do a whole lot of other things. Everything that doesn't contribute directly to the progress of work should be eliminated."

Only 55 years after Capek's prognos-

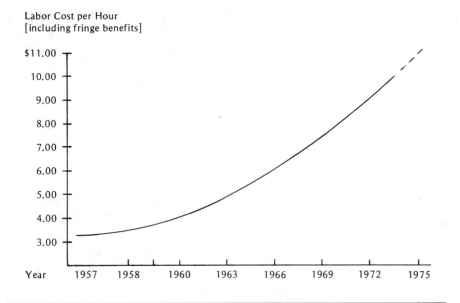

Exhibit 1: History of Labor Cost in the U.S. Automotive Industry

Labor Cost per Hour
[including fringe benefits]

tication, the aggregate advance of technology provides engineers with something akin to Rossum's artificial protoplasm. It is possible to build a machine that can do the work of a human; and, like the mountain that is climbed because it is there, robots have been built because it is possible. Indeed, close to 200 different industrial robots have been developed worldwide. Most of these efforts have aborted through lack of staying power, but a few manufacturers hang on tenaciously in the conviction that "robotics" is an "idea whose time has come."

Although it has been argued that a spectrum of available technology makes building a robot possible, the entire effort would be meaningless had not so much factory work been reduced to something grossly subhuman. There are few artisans in modern manufacturing plants. In the interest of production efficiency, work has been broken down into a series of simple repetitive tasks that can be quickly taught to unskilled labor. The principle applies not only to the assembly line laborers that Charlie Chaplin so poignantly championed in his classic, "Modern Times," but also to individual machine stations which have been automated except for loading and unloading.

Chaplin was telling us that unskilled and semiskilled workers are engaged in subhuman activities to serve the U.S. hard goods appetite. A noble goal would be the more human use of these human beings. But noble goals are not the only or the prime motivators of the captains of industry. The profit motive also must be served. In 1936, when "Modern Times" cried "foul," labor was cheap, plentiful, and intimidated. There was no economic pressure to foster the invention of a robot. A quarter of a century later a semiskilled factory laborer was costing $6,000 per year in pay and fringe benefits. And in the automotive industry, this cost was destined to rise to $20,000 per year by 1974. Exhibit 1 plots the history of labor cost in the U.S. automotive industry. The technique of designing work to eliminate skill had backfired. Workers demonstrated their aversion to dull, repetitive work by demanding higher pay than other workers received for skilled, more interesting jobs. Furthermore, turnover and absenteeism tended to follow job dissatisfaction and add to the total cost of so-called "cheap" labor.

The Industrial Robot

So now the industrial robot which became technically possible enjoyed economic pressure for its creation. It's time to consider just what is an industrial robot. Industrial robots tend to be stationary machines with a single powerful arm. One might say that a robot is a cross between a computer and a backhoe. It is taught a job by leading it around by the hand through its required motions. Its built-in memory records every action of the arm; it also remembers to turn off and on all surrounding machinery with which it works. The design is a failure, however, unless it is adaptable to a broad range of job assignments. Only then can the

Exhibit 2: A Sample of the Jobs Robots Can Do

Die Casting
Unload 1 or 2 DCM—quench—trim—die care—insert loading—palletizing

Forging
Drop forge—upsetter—roll forge—presses

Stamping Presses
Load/unload—press-to-press transfer

Welding
Spot welding—press welding—arc welding

Injection & Compression Molding
Unload 1 or 2 IMM—trim—insert loading—palletizing—packaging

Investment Casting
Wax tree processing—dipping—manipulating—transferring

Machine Tool Operations
Load—unload—palletizing—machining center—machine-to-machine transfer—operating with lathes, chuckers, drilling machines, broaches, grinders, multi-turrer machines, etc.

Spray Coat Application
Mold releases—undercoats—finish coats—highlighting—frit application—sealants

Material Transfer
Automotive assembly—automotive parts—glass—textile—ordnance—appliance manufacturing —molded products—heat treating—paper products—plating—conveyor and monorail loading and unloading—palletizing and depalletizing.

robot expect large series production, which is necessary to bring cost in line and to amortize the heavy R&D investment. Exhibit 2 tabulates jobs which have already succumbed to robotization, and Exhibit 3 is a montage of some of these jobs.

Return on Investment

Manufacturers use two common filters in evaluating capital investments. Exhibit 4 is a payback analysis. The payback period is nothing more than the cost of the machine divided by the annual labor savings less annual robot upkeep cost.

Evaluation of automation investments generally has leaned to the payback standard, and a typical cutoff is three years. Exhibit 4 shows that industrial robots on two-shift duty comfortably pass this test. But an industrial robot, because of its job flexibility, might better be classed as general purpose equipment which is not as obsolescence prone as is special purpose automation. Therefore, even more pressing justification for robotizing a job may be derived from looking at the return on investment (ROI). In Exhibit 5 this return is plotted versus hours worked per day for a range of likely hourly rates. Eight-year straight-line depreciation has been chosen, as

robots have already demonstrated the ability to survive two-shift operations for eight years. At any U.S. labor rate, the return on investment is quite handsome, and in some industries even single shift operation has become justifiable.

The Social Issue

The same ROI plot works pretty well throughout the industrialized world, as witnessed by the burgeoning use of robots in Western Europe and Japan. But there is a distinction. In Europe and Japan much more is made of the social issues. A robot pays for itself, yes; but the sheer issue of human use of human

Exhibit 3: A Montage of the Jobs Robots Can Do

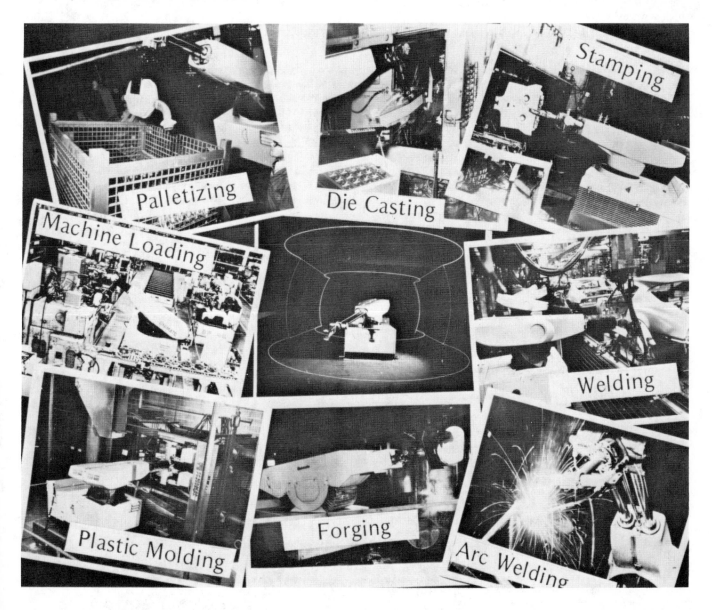

beings is a strong factor in the decision to employ robots.

The final social commentary comes from behind the Iron Curtain. The interest in robotics is intense, and large orders have been placed by both the Russians and the Poles. It has not been possible to discuss the applications in economic terms. Pay scales translated into dollars are ridiculously low. With labor at Socialist Republic rates, no American industrialist would consider the use of robots. Why then the USSR and Poland?

It seems that there is a genuine concern for the "worker's" lot. If a job is debilitating, or even demeaning, the socialist countries find adequate cause for displacing humans, irrespective of take-home pay. Ironically, we arrive at the same result. When the workers in a capitalistic country balk, they press for higher wages. When their price rises high enough, the robot becomes economically sound. The industrial world, socialistic or capitalistic, will find use for robotics. The justifications seem to be divergent, but the underlying pressures are identical.

The Robot's Future

The industrial robot working today lacks many attributes that would enable it to pervade man's productive capacity. However, robots are far from being at

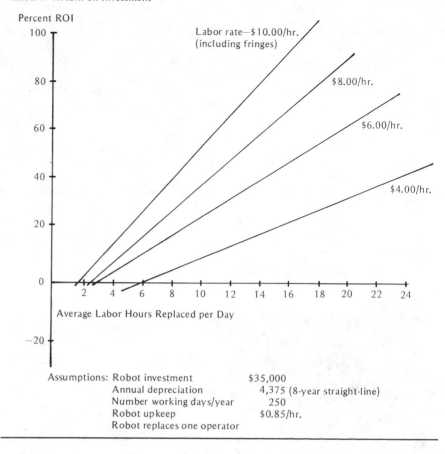

Exhibit 5: Return on Investment

Percent ROI

Labor rate—$10.00/hr. (including fringes)

$8.00/hr.

$6.00/hr.

$4.00/hr.

Average Labor Hours Replaced per Day

Assumptions: Robot investment $35,000
Annual depreciation   4,375 (8-year straight-line)
Number working days/year   250
Robot upkeep $0.85/hr.
Robot replaces one operator

Exhibit 4: Payback Analysis

$$P = \frac{I}{L - E}$$

where

P = payback period in years
I = total investment, robot and accessories
L = total annual labor saving
E = expense of robot upkeep

Typical Inputs

I = $35,000
L = $16,000 (automotive industry including fringes)
E = $2,000 one-shift use
    $3,000 two-shift use

One-Shift Use

$$\text{Payback} = \frac{35,000}{16,000 - 2,000} = 2.5 \text{ years}$$

Two-Shift Use

$$\text{Payback} = \frac{35,000}{32,000 - 3,000} = 1.2 \text{ years}$$

---

the end of their evolutionary tether. In the near future they will be made adaptable to a greater variety of manual tasks. The trick is to give robots at least one human sense, rudimentary eyesight, and engineers are hard at work on the development. Operating principles are similar to those used in such character-recognition jobs as check reading. A robot's eye, however, will be used to locate disoriented parts so that the arm can be directed to pick up parts which arrive random-scramble at the robot's work station. An otherwise puerile job is baffling to a robot if parts are not oriented at the pickup point.

So if a robot can be given an eye,

does not the gnawing question arise, "Will a free thinking robot be developed ultimately and will it prove to be super-human and therefore disinterested in servitude to mankind?" Presently there are scientists from several organizations who are working on the frontiers of artificial intelligence. They may very well think it possible to develop a truly sentient robot that might vie for the top rung in terrestrial evolution. As a roboticist of 20 years standing, I am convinced that a practical emulation of a man will remain an outright fantasy. It will always be far, far easier to make a robot *of* a man than to make a robot *like* a man.

Presented at the Finishing '77 Conference and Exposition, October, 1977

The Justification Of An Industrial Robot

By Timothy Bublick
DeVilbiss Company

Justification of an industrial robot requires investigation and analysis of the potential application, robot technical suitability, the economic aspect and humane considerations. This paper is concerned specifically with the rational application of robots to the field of spray coating and finishing which is a generally hostile human environment.

INTRODUCTION

Robots - as defined by Webster, are "heartless automatons, or an individual who works or acts mechanically." A practical definition for an industrial robot is "A machine that can duplicate human skills and flexibility with accuracy and precision." Independent of how robots are defined, the average worker subconsciously fears robots as management's ultimate scheme to eliminate man from his workplace. Contrary to this thinking, industrial robots are a long awaited advancement desperately needed by industry, a machine which has the ability to duplicate human movement with extreme accuracy and untiring consistency. Robots will ultimately free people from jobs which present serious health hazards, jobs that are mundane and repetitive but require human agility and mobility, and jobs that require human skill but which can not be performed effectively over long time periods because of fatigue.

Robots with a Purpose

Despite the obvious advantages, U.S. industry and especially the finishing industry has been slow to adopt robotics. This reluctance appears to have stemmed from a number of circumstances, foremost being the initial equipment investment as compared to the general availability of manual labor. Other areas of concern have been availability of suitable machines and reliability. Another factor influencing adoption of robots has risen from the promulgation of codes and regulations relating to employee health and safety. Now industry decisions are strongly shaped by EPA, OSHA, energy shortages, low profits, decreasing productivity, vigorous competition, both domestic and foreign, technical and sociological changes, and inflation. Now, more than ever before, industry must adopt automation to remain competitive and profitable. The fact is that painting robots today, based on ten years production experience in Europe and five years in the United States, are an economical, reliable and versatile form of off-the-shelf programmable automation. Robots are fast learners, they have an exemplary memory and more perseverence than the most intelligent, strongest human sprayer. Robots have a place and a purpose, they increase product quality, they generate paint and energy savings, they are the solution to problems encountered in hazardous environments and they increase flexibility for product mix or style changes. These qualities and benefits exert a highly positive influence on profits.

Robots in Finishing

It is most important to realize that modern painting robots can duplicate and improve on the best work performed by a skilled production spray painter. Once the robot is programmed it will repeatedly duplicate the motions of the sprayer and provide consistent results whether the application be finish top coat, primer, sealer, mold release, or the application of other types of material. The robot's computer memory also provides capability to interface with other equipment supportive to the finishing operations; such as: color changers, turntables, conveyors, lift and transfer tables or even other computers. In fact, using the robot's memory capabilities can alleviate many of the design and operating concerns which are associated with present day production finishing systems.

In order to justify the purchase of a robot painting system, economic feasibility must be established. The basis for justification can be established on any one of a combination of the above advantages.

Today, in the finishing industry robots are applying automotive exterior top coat to truck beds, underbody primer to truck cabs, top coat to steel office furniture, stains on wood furniture, sound deadener on appliances, porcelain on bathtubs, enamel on lighting fixtures, mold releases, primer on hoods and fenders and even the ablative coating on the booster rockets which will carry the space shuttle into orbit from the earth. These proven applications speak well for the versatility of the painting robot. In fact, an understanding of robot capability leads to the conclusion that the great majority of production painting operations can be economically converted to robotics.

ECONOMIC JUSTIFICATION OF A ROBOT SYSTEM

Justifying a Robot System

As an example we will take a typical robot application, applying baking enamel to household appliance components. Using this, we will proceed in establishing criteria for justification.

The preliminary investigation must establish how many hand sprayers are now applying the material, what is the material usage and cost of the material, the reject rate with hand sprayers and any considerations relating to health and safety regulations. The investigation determines that presently two men (one per shift) are spraying with a paint flow rate of one quart per minute and meeting production requirements. The reject rate due to finishing is averaging 15%. Thus, one robot will be required for the application and we will further assume that the spray booth and paint circulating system are suitable for the robot installation.

Let us examine economic justification from the viewpoint of both payback period and return on investment.

Payback Period (Refer to formula in Exhibit I)

The total first cost (C) shown as $80,000 includes the robot and controls, accessories required for spraying and interfacing. This initial cost will vary depending upon the simplicity or complexity of the overall system.

The wages and benefits (WS) have been estimated for one operator at $22,000 per year, for a two shift operation (WS) equals $44,000.00.

The yearly robot savings (RS) are estimated very conservatively at $11,200/shift X 2 shifts. This figure covers paint savings only, since no attempt has been made to project energy savings or reduction in reject rate costs. Assuming that an operator sprays at a deposition efficiency of 40%, we can project based on experience that with the robot programmed properly an increase of 15% to 20% could be attained in overall efficiency. We have elected to use a 15% figure for efficiency gain and a 50% gun use factor to arrive at our estimated material savings for the example.

Material Savings Calculation

Substrate area = 1.6 Sq. ft. (Average per hanger)

Conveyor speed = 21 F.P.M.

Part centers = 3'-0"

Flow rate = 32 oz./min.

Manual spray deposition efficiency = 40%

Robot spray deposition efficiency = 40% + 15% = 55%

Spray gun "on time" = 50%

Specified film build = 1 mil.

Material cost = $5.00 per gallon

MANUAL SPRAY

Material deposited on substrate:
  32 oz./min. X .5 "On time" X .4 deposition efficiency = 6.4 oz./min.

Overspray:
  (32 oz./min. X .5 "On time") - 6.4 oz./min. = 9.6 oz./min.

Yearly overspray (2 shift):
  9.6 oz./min. ÷ 128 oz./gal. X 60 min./hr. X 16 hr./day X 249 days/yr.
                                                = 17,928 gal./yr.

Yearly cost of overspray
  17,928 gal./yr. X $5.00/gal. = $89,640.00 per year

ROBOT SPRAY

Material deposited on substrate:

32 oz./min. X .5 "On time" X (.4 + .15) deposition efficiency = 8.8
Oz/Min

Overspray:

(32 oz./min. X .5 "On time") - 8.8 oz./min. = 7.2 oz./min.

Yearly overspray (2 shift):

7.2 oz./min. ÷ 128 oz./gal. X 60 min./hr. X 16 hr./day X 249 day/yr.
= 13,446 gal./yr.

Yearly cost of overspray:

13,446 gal./yr. X $5.00/gal. = $67,230.00 per year

MANUAL SPRAY VS. ROBOT SPRAY

$89,640.00 manual overspray cost - $67,320.00 robot overspray cost =
$22,410.00 paint saving/year with robot

For presentation example "round off" to $22,400.00.

Another factor to be considered is that specifications call for one mil. coverage; however, the majority of sprayers will apply a heavier coating to assure they are achieving coverage. With a robot properly programmed and with correct fluid flow and atomization air adjustments, it is possible to achieve and control the specified film deposition, thus saving additional paint. Recall that the $11,200.00 does not include savings for a reduction in the reject rate although robot users report as much as a 13% reduction. Another consideration is energy savings. It may be possible to reduce the required exhaust volume of a spray booth when an operator is not present. A reduction in exhaust volume can result in a substantial cost savings in energy requirements to heat the replacement air. Taking into account these savings, it is apparent that the $11,200 figure/shift is conservative.

For yearly savings (AS) in operating and maintenance, we are assuming $2,000.00. This figure is indicative of reduced spray booth maintenance, as well as average annual displaced automatic machine maintenance.

The robot depreciation expense (DE) is over a seven year period.

For yearly robot maintenance (RM) we are assuming $3,000.00 which consists of $2500.00 parts and $500.00 labor. The figure represents both preventive maintenance and maintenance due to failure. Robot reliability varies; however, operating time figures have been established as high as 98% with a mean time between failures of approximately 850 hours.

The robot operating and programming costs are estimated at $3,000.00. This figure represents the energy required to operated the unit, which with most robots is basically a hydraulic power supply in the five to ten horse-power range, and the cost incurred during programming time. The programs are established during production start-up; however, it may become necessary to re-program for style changes, different hanging configuration, new products, or even to improve the program to increase efficiency or finish.

By inserting the foregoing to values in formula (Exhibit I) the payback period is calculated at 1.17 years.

Many plants will wish to calculate return on investment. Such a calcula-tion is shown in Exhibit II, using the same values as for the payback cal-culation.

This process requires the aid of a computer programmed to calculate R.O.I. The calculated R.O.I. for this example is 121%.

The results of the calculations in Exhibits I and II show economic justifi-cation for this example. It is recognized that each potential robot user must evaluate the benefits in the perspective of his operations.

There are some applications where in the not too distant future it will be necessary to remove the hand sprayer from the atmosphere. These applica-tions require man's flexibility and agility but present serious health hazards due to the toxicity level of the material being sprayed. Here the robot is a logical solution.

There are also numerous applications where film build is extremely critical, such as in the Aerospace industry where weight and uniformity are closely controlled. It may not be possible for a man to meet specifications eight hours per day, five days per week. Other considerations may be a lack of qualified labor in certain geographic areas. Here robotics offer a viable solution to these problems and even with relatively low production rates prove to be an economical and practical solution.

Conclusion

We have set forth in this presentation basis for industrial robot justifi-cation within the finishing industry. No doubt, there are other advantages which may be peculiar to a particular plant or operation and which may assist in the justification of robotics in painting, welding, part handling, tool carrying etc. Here it is up to the prospective buyer with the aid of qualified robot suppliers to investigate all of the advantages which are pertinent to a particular operation. The finishing industry should, as an aid to American industry's competitiveness, human consideration, and pro-fitability, investigate automation of this type whenever and wherever possi-ble, whether it be an existing application or in the design of a new system.

EXHIBIT I

Payback time in years = cost/yearly savings

$$Y = \frac{C}{WS + RS + AS + DE(R) - RM - RO}$$

Y = Years payback

C = Total first cost (Robot, accessories, Control) = 80,000

WS = Wages & benefits - 2 workers - one year   = 44,000

RS = Yearly robot savings (Material, quality,
                                    energy, OSHA)   = 22,400

AS = Yearly savings (operating and maintenance)   = 2,000

DE = Robot depreciation expense 80,000/7   = 11,428

R = Tax rate   =   50%

RM = Yearly robot maintenance   = 3,000

RO = Operation and programming costs   = 3,000

$$Y = \frac{80,000}{44,000 + 22,400 + 2,000 + (11428)\,(.50) - 3,000 - 3,000}$$

Y = 1.17 years payback time

NOTE: ALL FIGURES CALCULATED ON A TWO SHIFT BASIS.

EXHIBIT II

CASH FLOW FOR ROBOT INVESTMENT

	YEAR 1	YEAR 2	YEAR 3	YEAR 4	YEAR 5	YEAR 6	YEAR 7
Initial investment	-80,000						
Investment tax credit	8,000						
Start-up costs	- 5,000						
Start-up tax credits	2,500						
Depreciation rate (.50)	11,429	8163	5831	3645	3645	3645	3645
Wages & benefits of replaced workers*	44,000	46,640	49,438	52,405	55,549	58,882	62,415
Tax cost*	-22,000	-23,320	-24,719	-26,202	-27,774	-29,441	-31,207
Maintenance cost of robot*	- 3,000	- 3,180	- 3,371	- 3,573	- 3,787	- 4,015	- 4,256
Tax credit for maintenance*	1500	1590	1686	1787	1894	2007	2128
Material savings*	22,400	23,744	25,169	26,678	28,279	29,976	31,775
Tax cost for savings*	- 4,500	- 4,770	- 5,056	- 5,359	- 5,681	- 6,022	- 6,383
Operating and programming expenses*	- 3,000	- 3,180	- 3,371	- 3,573	- 3,787	- 4,015	- 4,256
Tax credit for operating & programming*	1500	1590	1685	1786	1893	2007	2128
TOTALS	-26,171	47,277	47,292	47,594	50,231	53,024	55,989

*6% inflation added per year.

Return on investment = 121%

Presented at the Robots IV Conference, October, 1979

Economic Analysis Of Robot Applications

By John A. Behuniak
General Electric Company

Economics is the principle justification for robot application in manufacturing operations. This paper reviews one approach to economics in robots. Screening criteria is outlined for use in critical plant surveys to identify potential robot applications. Design of installations using economic criteria and more detailed analysis for both appropriation requests and later review of completed projects is presented.

INTRODUCTION

Economics is a principal driving force in manufacturing. Our competitive position as a business and a nation depend upon our productivity and our cost position. With limited resources in terms of both skilled personnel and money, our goal in manufacturing is to turn out quality products at the lowest possible cost. Robotics is a here and now technology that when properly applied can produce excellent economic results.

SURVEY

The first step in applying robots is to determine the number and types of potential robot applications in a manufacturing operation.

SCREEN

Analyze each potential application for both practicality and economics. The criteria in Table 1 is a useful guide to determine suitability of a proposed application.

THE TRAP

Robots have been designed to emulate the human, which is its strong point in terms of application and its weakest point in terms of economics.

Robots, unlike other forms of automation, usually only replace humans on a one for one basis. Any escalation in the cost of a proposed application easily renders the project unsound economically.

IDENTIFY AND PRIORITIZE

To set up a plan to implement robots, consider the leverage available from the number of similar robot applications with a suitable payback; four similar applications and a three-year payback is a good rough guide. Consider also the difficulty of the application and ease of implementation and prioritize the implementations.

SYSTEM DEVELOPMENT

To minimize risks in the implementation, suitable trials should be arranged to determine appropriate tooling, work place layout and cycle time for the proposed application. This is a critical step. Any problems noted can be addressed and corrected before any further commitment is made. The

TABLE 1

ROBOT SCREENING CRITERIA

I. Economics

     . Number of shifts per day

     . Number of setups per week

     . Major move of equipment required

     . Replace one or more persons per day

II. Practicality

     . Simple repetitive operations

     . Cycle time more than 5 seconds

     . Visual inspection not required

     . Part location and orientation
        suitable for robot acquisition

     . Part weight

testing also allows improvements to be made to obtain changes that allow optimization of the application in terms of cost and savings.

ECONOMIC ANALYSIS

Three robot applications are briefly described and analyzed to determine payback, return on investment (ROI), discounted rate of return (DCRR) and cash flow.

ROBOT SWAGING

Table 2 summarizes the data used in the economic analysis of a robot application for feeding and swaging a small part 1/4" dia. X 6" long. The unit operates on one shift replacing one man in an extremely noisy environment. In operation the robot picks a part from a hopper and feeds it into the swaging machine, then retracts and drops the part into a basket. An over-grip sensor is used to detect a no part condition and terminate the cycle.

ROBOT DIE CASTING

Approximately 450 robots are currently used in die casting applications. Table 3 outlines the economic analysis of one die casting application. Capital investment included the robot, a programmable controller and sensors. Labor savings are one-half man per shift, three shifts. A productivity improvement of 20% was included in the analysis. In operation the robot removes a part from the die casting machine, passes it through sensors to detect part presence, moves to a break-off fixture to separate the parts and places the sprue and gate in the furnace.

TABLE 2

ROBOT SWAGING

PAYBACK = 0.9 YEARS   ROI = 284.3%   DCRR = 284.28%

IMPUTED INTEREST = 11.50%   INSURANCE & TAX = 1.4%

INVESTMENT CREDIT = 10.0%   TAX = 46.0%

($ Thousands)

YEAR	1979	1980	1981	1982	1983
INVESTMENT	5.0				
EXPENSE	5.0				
NET SAVINGS	15.0	20.0	21.4	22.9	24.5
DEPRECIATION	0.0	.5	.5	.4	.4
INSURANCE & TAX	0.0	.1	.1	.1	.1
INVESTMENT CREDIT	0.5				
IMPUTED INTREST	0.0	0.7	2.1	3.5	5.1
INCOME BEFORE TAX	10.0	20.2	22.9	25.9	29.1
NET INCOME	5.9	10.9	12.4	14.0	15.7
ANNUAL CASH FLOW	0.9	11.4	12.9	14.4	16.2

TABLE 3

ROBOT DIE CASTING

PAYBACK = 3.6 YEARS   ROI = 29.2%   DCRR = 29.23%

YEAR	1979	1980	1981	1982	1983
INVESTMENT	60.0				
EXPENSE	20.0				
NET SAVINGS	18.0	36.0	38.5	41.2	44.1
ANNUAL CASH FLOW	-56.7	19.0	21.6	24.2	27.3

ROBOT PAINTING

Painting using a robot is one area where material as well as labor savings can be obtained. Table 4 presents an economic analysis for robot painting. One man is replaced per shift for three shifts, a 15% increase in overall efficiency and resultant paint savings of 4500 gal/yr was also included. (See ref. 1).

TABLE 4

ROBOT PAINTING

PAYBACK = 2.9 YEARS   ROI = 43.9%   DCRR = 42.55%

YEAR	1979	1980	1981	1982	1983
INVESTMENT	75.0				
EXPENSE	25.0				
NET SAVINGS	30.0	60.	64.2	68.7	73.5
ANNUAL CASH FLOW	-66.7	32.3	36.7	41.2	46.4

SUMMARY

Robots can be an attractive solution to certain classes of automation problems. Available robots are best suited for use in environments that are hostile to humans, hot, noisy, or where noxious fumes are present. They can be quickly implemented and provide satisfactory economic returns. However considerable care must be used to avoid the robot's economic trap, one robot usually replaces only one person.

References:

1. Economic Justification for Spray Coating Robots
    FINISHING HIGHLIGHTS May/June 1978.

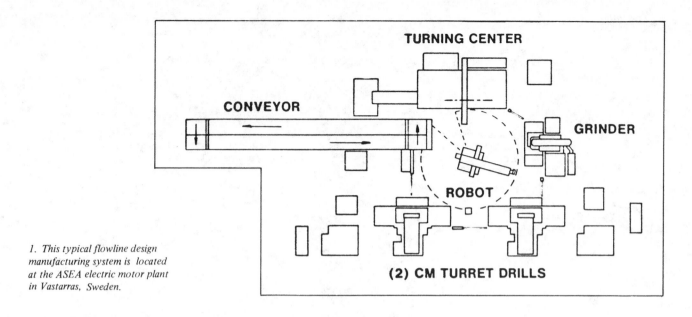

1. *This typical flowline design manufacturing system is located at the ASEA electric motor plant in Vastarras, Sweden.*

Reprinted from Robotics Today, Summer, 1979

Justifying a Robot Machining System in Batch Manufacturing

The use of functional and flowline design manufacturing systems in batch manufacturing helps increase efficiency and productivity

JOHN G. HOLMES
Project Engineer
Cincinnati Milacron, Inc.

Certain manufacturing fields, such as mass production and the continuous process industries, are highly automated and have high levels of productivity. Batch manufacturing, however, which will account for as much as 75% of all future production, is plagued with inefficiencies which contribute significantly to its relative low productivity. Some of the more evident problems occur in scheduling, setup, tooling and fixturing, and in-process inventories. Batch manufacturing is also very labor intensive, using higher skill level operators whose numbers are greatly diminishing. Machining operations are slow because they require a great deal of operator attention in the absence of automatic machine tools.

Production in a batch manufacturing facility consists of a relatively high part mix and low individual-part production volumes. The machines in the shops are arranged according to the functions they perform. This type of setup results in high in-process inventory and long lead times. It has been shown that 95% of the throughput time for the average workpiece is spent in storage, in transit, or in waiting. Even when the part is on an NC machine tool, the part is being worked on only 30% of the time. The result is that metal removal accounts for only 1.5% of the production process.

Productivity in the batch manufacturing facility can be increased significantly. Some of the solutions already being used include computer numerical control (CNC) machine tools, group technology, tool management, and robot machining systems.

A robot machining system is an assemblage of machine tools, robots, conveyors, and gaging equipment. The robot interacts with the CNC machine tools and peripheral equipment by controlling the entire sequence of operations. Extensive use of group technology and tool management is also incorporated for maximum efficiency. The machine tools become dependent on the robot for all required communication and signals. The robot also services the machining system as the prime source of material handling.

Batch-production shops employ basically two classes of manufacturing systems: a flowline design manufacturing system and a functional manufacturing system.

Flowline Design Manufacturing system. A flowline design has the machine tools grouped according to one well defined combination of processes and capacity required. Material handling within the system is reduced while manufacturing control is increased. The ultimate goal of such a system is the manufacture of a complete family of parts.

But there can be some disadvantages. In some cases, continued operation of any machine is dependent on the operation of its neighbors. Such a system is vulnerable if one of the machine tools breaks down, and a system of this type does have decreased production flexibility when compared to a functional manufacturing system. Another disadvantage is the difficulty of balancing the machining times of the various machine tools. Capital intensive machines may sit idle for extended periods of time or may produce at maximum efficiency.

There are many flowline systems beginning to appear in the manufacturing industry, especially in Europe. A typical European flowline system, Figure 1, is located at the ASEA Electric Motor Plant in Vastarras, Sweden. The system consists of a turning center, a rotary table surface grinder, two turret drills, a flip-over station, a parts conveyor, and a robot. This cell manufactures three different parts, each made in six sizes. Parts of one size are machined at each setup, minimizing production and delivery problems as well as the cost of flexible tooling. Some retooling is required between batch runs. This cell is manned for one shift and runs unmanned for the remaining two shifts. During the manned shift the raw stock conveyor is loaded, the finished parts conveyor is unloaded, and any required tooling changes are made. In order to justify a system of this type, batch sizes must equal the production of three or more shifts. This requires sizes of 200 or more parts for this particular family.

Research is being conducted at the University of Trondheim in Norway on a flowline system. Machines comprising the system are a vertical spindle NC milling machine, an NC turret drilling machine, a horizontal spindle machining center, a CNC lathe, and a Cincinnati Milacron T^3 (TM) industrial robot. This system is being studied to determine the maximum level of automation attainable along with maximum flexibility. The robot in the system will handle palletized parts. A DNC system will control both material handling by the robot and machining operations.

Functional Manufacturing System. A functional layout uses machines in specific groups, where the criterion for grouping is essentially the type of machining being done. This is the traditional organization of a batch-type manufacturing plant. When the part-family concept is applied, either partially or wholly, machines are occasionally regrouped into cells, each of which handles all operations required on parts of certain size and configuration. The part family thus created comprises a large number of more closely defined part families, sufficient to permit two or more like machines of each functional type required. Such a cell arrangement retains advantages of functional organization while facilitating application of group technology principles. Work travel distances are relatively short, and automation of the type offered by modern robots can be applied with still further improvements in work flow and productivity.

One of the most published functional robot automated manufacturing systems is a subset of a DNC system at Fujitsu Fanuc in Japan. Among the 18 NC machine tools are eight lathes with relatively short part machining times. Because of the frequency of loading and unloading, a robot was installed to handle the parts. The robot is mounted on rails above the machines and is allowed to move between the machines to perform its designated duties.

A stand-alone functional manufacturing system has recently been introduced by Cincinnati Milacron, Figure 2. Called a CINTURN/T^3 Industrial Robot Flexible Manufacturing Cell, this system consists of two 12″ (304.8 mm) universal type turning centers with Acramatic CNC controls; a T^3 industrial robot with an Acramatic computer robot control (CRC), which is similar to a standard CNC control; a gaging station; and a method for presenting raw stock parts and removing finished parts.

The system is specifically designed to increase the overall efficiency of a specific turning operation by forced pacing or loading of the turning centers. The chosen family of parts allows for maximum utilization of the cutting tools, the fixturing within the turning centers, and the robot's gripper. Within the cell, the robot controls the sequences of operation and is the prime source of communication. The system also maintains the correct part

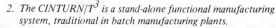

2. The CINTURN/T^3 is a stand-alone functional manufacturing system, traditional in batch manufacturing plants.

size by automatic tool compensation, using electronic feedback of the gage-measured value to the turning center Acramatic CNC control.

Typical Cycle. In a typical cycle, the robot moves to the conveyor carrying parts in random order and signals the conveyor to start, *Figure* 3. When a part is present in the pickup station, a signal is received from a device such as a limit switch. Using internal memory, the robot identifies the part in its gripper. The robot then moves to the available turning center, *Figure* 4, and with appropriate input signals checks to see that the machining cycle is completed and that the door is opened.

The robot then enters the work area and grasps the finished part, signals the tailstock quill to retract, checks to see which part has just been machined, and waits for the "quill retracted" signal. When everything is satisfactory, the robot removes the finished part and loads the raw stock into the holding device, *Figure* 5. For minimum load/unload time a double gripper is mandatory.

The robot then signals the quill to advance and signals the turning center CNC control to call up the appropriate part program. When the "quill advanced" signal is received from a limit switch and a check is made to assure that the part in the machine corresponds to the program selected, the robot arm then retracts from the turning center and the door automatically closes. Again, the robot waits for a combination of redundant signals assuring that everything is satisfactory. It then starts the cycle and moves to the finished part gaging station.

At the gaging station, *Figure* 6, the robot sets the part into the gage and initiates a "gage read" instruction. When the gaging is complete, the data is then directed by the robot to the turning center on which the part has just been turned. The data is compared

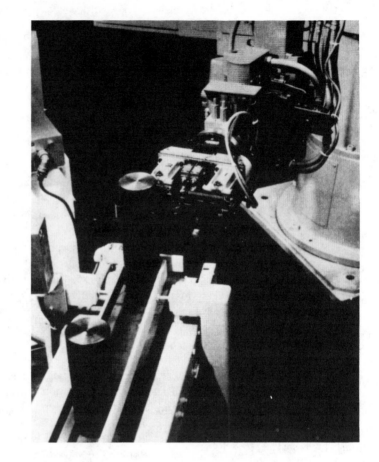

3. The robot picks up a raw part from the conveyor just after identifying that particular part.

4. The robot prepares to load the machine after checking to see that the door is open.

5. *Interchanging finished part and raw stock in the holding device.*

to a reference value for that specific part stored in the turning center control and the deviation from this value is calculated and used for the tool compensation value.

The robot also receives a go/no-go signal from the gage. If the part is within the tolerances prescribed, a number of take-off situations are available. The part can be placed in a bin, palletized, or placed on another conveyor. If the part is outside the specified tolerances, the robot places it in a reject bin. The robot then returns to the conveyor to begin the loading cycle on the next machine.

Advantages and Disadvantages. In a functional layout, as compared to a flowline arrangement, machine tools of similar functions are close to one another, making the system very flexible both in handling variations in part production requirements and in re-routing work when necessary. This allows for high machine tool utilization. If one of the turning centers in the system should fail, production can continue on the other unit. Another major advantage of a functional system is the marked productivity increase as the batch sizes increase.

Some of the major disadvantages in functional systems are the random material routes, high material costs, slow work movements, and high investments in work-in-progress. These problems can be minimized through use of the computer, both through simulation of the systems prior to construction and—more importantly— through implementation of computer aided process planning.

Productivity Benefits. The effectiveness of any system is measured by its output per hour or its productivity. Factors which influence the effective utilization of a system include the variety of parts being machined, setup time of the machines, process time of the parts, and the number of parts being machined per year.

6. *Gaging the finished part diameter. Data sent to the machine control is used for tool compensation.*

TABLE 1: PRODUCTIVITY COMPARISON OF MANNED AND ROBOT SYSTEM		
	Manned	**Robot**
Available Cut Time	120 Min/Hour	120 Min/Hour
System Attention	9 Min/Hour	12 Min/Hour
Efficiency	80%	90%
Total Utilization	(120 – 9) 0.8 = 88.8 Min/Hour	(120 – 12) 0.9 = 97.2 Min/Hour
Part Cycle Time		
Load/Unload	1.29 Min	0.37 Min
Cut Time	2.50 Min	2.50 Min
	3.79 Min/Piece	2.87 Min/Piece
Fatigue Factor (40 lb Part)	1.04	1.00
	3.79 x 1.04	2.87 x 1.00
	3.94 Min/Piece	2.87 Min/Piece
Throughput	$\frac{88.8}{3.94}$ = 22.5 Pieces/Hour	$\frac{97.2}{2.87}$ = 33.9 Pieces/Hour
Increased Productivity		$\frac{33.9 - 22.5}{22.5}$ = 50.7%

A comparison was made between the CINTURN/T³ manufacturing cell and two identical manned turning centers, based on projected total utilization of the systems for one year on a two-shift basis. As shown in *Table* 1, the productivity of the unmanned cell was more than 50% greater than that of the manned turning centers, showing that the cell's productivity was equal to that of three identical manned turning centers. When the operational costs per year, including power, taxes and insurance, floor space, maintenance equipment, labor, and depreciation tax credit, are compared, *Table* 2, there is an after-tax savings of $16,455 with the CINTURN/T³

Economic Justification. The return on investment for this system can be computed from any number of accounting methods. When operating savings per year and equipment costs of the CINTURN/T³ manufacturing cell with one full-time operator per shift are compared to those of three manned turning centers, the cash payback period is 1.5 years and the discounted cash flow rate of return on investment 65%, *Table* 3. These values are for the particular situation based upon the selected parts being machined. The direct labor rate chosen was $8.25 per hour. Other operating costs included taxes and insurance, maintenance, power cost, and tooling.

Substantial economic gains can be achieved by operating the system for more than one shift. The greatest difference occurs in the cost of labor per year, which doubles for two-shift operations. The other costs remain relatively constant or are insignificant in comparison. This results in a cash payback period of less than one year.

Several factors, such as scrap rework, reduced inventory costs, less material required because of less scrap generation, and reduction in operating costs because of better scheduling procedures were disregarded in this simplified economic analysis.

Some other intangible benefits which might be considered when justifying a robot system are the possible reduction of OSHA compliance costs, improved utilization of floor space, increased machine cycle rates, quicker run changeovers, and more system flexibility.

Application Problem Areas. The major problems to be considered with a typical robot automated system are in the areas of parts, manufacturing process, and economics. The parts to be used in this type of machining system should be geometrically similar. The parts in the turning application are parts of rotation. Unless the production volume is relatively large, parts to be machined must be grouped into families very similar in dimension and requiring approximately the same tooling to minimize setup costs. The dimensions of the part dictate the gripper design, fixture requirements on the machine tool, tooling, and the machine tool and robot specifications.

The "typical batch size" in batch manufacturing is uncertain. Although not much study has been done in the area, it seems obvious that small batches of 50 pieces or less increase the manufacturing problems to such an extent that this system would only be justifiable in a company which is extremely advanced in manufacturing automation technology. However, this system is geared to manufacturing batches in the 50 to 10,000 piece range. In this range the system can run for an extended period of time without additional setup. Part families and batch size should be determined so that the minimum time between setup changes is about 24 hours. Process control problems are also minimized with larger batch sizes.

Some of the most serious problems can arise when the parts are on the machine tool and the machining process begins. Probably the most serious problem is that of broken tools. If a tool insert should break, the feed on the machine tool must be halted rapidly to prevent damage to the machine. Much research is being done in universities and private industry throughout the world on measuring and predicting tool wear and breakage. Devices are available which monitor horsepower during metalcutting and stop the process when a value falls outside the allowable band. Adaptive control systems that sense changes in machining conditions and adjust machine feedrates accordingly, resulting in longer and more predictable tool life and/or faster metal removal rates, are appearing. More work is required in these areas, however. Sensors must be developed that give an instantaneous and accurate indication that the tool has broken. Software which will allow the machines to react to these signals is also needed.

Chip snarling is a less severe prob-

TABLE 2: FLEXIBLE AUTOMATED PRODUCTION CELL DETAILED COST REFERENCE SHEET		
	Three TC's Operator Attended	Two TC's Robot Attended
Installed Cost		
Equipment	$573,000	$600,000
Installation Cost	600	600
(80 Hours @ $7.50/Hour)		
	$573,600	$600,600
Investment Tax Credit (ITC)		
10% of Capital Amount	$ 57,300	$ 60,000
Net Installed Cost		
Equipment—ITC	$515,700	$540,000
600 x $0.52 After Tax (A.T.)	312	312
	$516,012	$540,312
Maintenance		
3.75% of Cost	$ 21,500	$ 22,500
Power Cost	$ 10,560	$ 8,800
Taxes and Insurance		
2.35% on Capital	$ 13,465	$ 14,100

cost of computers has been dropping and their performance increasing to a point now that they are accepted throughout the manufacturing industry. CNC is now common in machine tools and total computer controlled manufacturing systems are being developed. Constant software improvements simplify programming. Research is being conducted using robot vision for part identification, removing parts from bins, and tracking parts on moving conveyors. Higher level programming languages are being developed which will allow the logic portions of the robot program to be developed off-line in a more suitable environment, with only the arm positioning being performed in the manufacturing area.

These are only a few of the developments which will be available in the future. Productivity improvements and robot automated machining systems are here today, however, and should be seriously investigated when possible applications exist. ■

lem. In turning, some grades of steel have a tendency to have long snarling chips which wrap around parts. This becomes a major problem when parts are gripped and when measurements are being performed on them automatically. It is necessary to select the proper chip breaking inserts and the process parameters, feeds, speeds, and depth of cut that will minimize this problem. Adaptive control, including torque controlled drilling, also addresses this problem.

Another area of concern for many applications is economic. The initial investment in a system such as the CINTURN/T³ is quite substantial and it must be carefully justified. Other costs not directly involved with the purchase of the machine must be considered. Indirect labor costs will increase, especially in the maintenance area, since machines are worked harder and higher skills may be necessary to keep the system functioning. Even though the cell is capable of producing more parts per hour, productivity will depend greatly on both ends of the cell. Sufficient raw material inventory and scheduled production must be available to supply the cell, and the manufacturing system must be capable of absorbing the parts being manufactured. These problems necessitate more complete off-line support,

especially in the area of computer aided process planning.

The final area where more work and research are required is in machine diagnostics and adaptive control. Machine diagnostics is needed to lessen the downtime of the cell or one of the machines. Adaptive control optimizes the cutting operation and minimizes human intervention when process fluctuations are encountered.

The advantages of robot machining systems outweigh the difficulties. The

−Adapted from "An Automated Robot Machining System". Presented at the 9th International Symposium and Exposition on Industrial Robots, March 13-15, 1979, Washington, D.C., sponsored by the Society of Manufacturing Engineers and the Robot Institute of America.

TABLE 3: FLEXIBLE AUTOMATED PRODUCTION CELL ECONOMICS JUSTIFICATION			
Annual Operating Cost	Three TC Operator Attended	Two TC Robot Attended	Difference In Operating Cost
Operators Required/Shift	3	1	
Direct Labor Cost/Year ($8.25/Hour)	$ 46,530	$ 15,510	$31,020
Taxes and Insurance	13,465	14,100	(635)
Maintenance	21,500	22,500	(1,000)
Power Cost	10,560	8,800	1,760
Annual Fixed Tooling Cost	1,500	1,000	500
Total	$ 93,555	$ 61,910	$31,645
Operating Gain After Tax			$16,455
Equipment Cost	$516,012	$540,312	24,300
Cash Payback Period			1.5 Years
Discounted Cash Flow (DCF) Rate of Return on Investment (ROI)			65%

Presented at the Robots IV Conference, October, 1979

Robot Flexibility Allows Transfer To New Job

By Stan Gilbert
Goodyear Tire and Rubber

In late 1977, production requirements at the Luckey Plant of the Goodyear Tire and Rubber Company caused us to investigate the installation of a robot device. The requirements were as follows; a semi-finished styled wheel had to be picked off a mold that is mounted on an intermittantly advancing pour conveyor and deposited on a takeaway/trim conveyor approximately every 15 seconds. This was accomplished previously by a personnel unit who physically pryed the wheel loose, manually picked it up, took one or two steps while carrying the wheel, and placed it on the takeaway conveyor located approximately 5 feet away. As each wheel weighs approximately 25 pounds, and considering the rate of production, the individual lifted 100 pounds/minute for 440 minutes per shift for a total of 44,000 pounds per day - a dull, heavy, and boring job for anyone. A job perfectly suited for a robot.

Constraining parameters of the installation of a robot were determined to be as follows:

1. The robot unit had to be compatible with the speed of the production unit--that is, it had to take away one wheel each 15 seconds.

2. The robot system had to interlock with the production unit. In the event of a failure of the robot system, the pour conveyor had to stop producing to avoid a jam which could result in considerable production downtime.

3. Inasmuch as we have only general maintenance employees, the robot system had to be simple to repair and hopefully, practically maintenance free. Also, if necessary, repair parts had to be readily available.

4. Cost factors had to meet the company's established payout periods, accomplished through the anticipated elimination of a personnel unit per shift. The payout was good and well within our corporate guideline.

5. Machine size and connecting mechanisms had to be compatible with three slightly different production units. This parameter was very important as the customer requirements from time to time could necessitate the relocation of the robot to any of the three units.

6. The robot had to be safe for personnel in the overall production area. This presented no problem other than restricting personnel from the immediate working area. From the overall safety aspect, the robot eliminated two potential safety hazards--(1) The possibility of the personnel unit dropping the wheel on a foot or leg; and--(2) The claim for a strained or sprained back due to the amount of lifting and turning required manually.

A research of the suppliers to locate equipment which could meet all of the above requirements was made and the robot finally selected was the Prab model 4200. It was installed late in 1977 and was continuously operational on one production unit until several months ago, when a change in production schedules warranted the relocation of the robot to a more productive unit.

Originally, the cycle of the robot was: reach, drop, grasp, raise, retract, pivot, extend, release, retract and reposition to the starting point. The cycle time of the robot was just able to meet the cycle time requirements of the production unit. However, because of the design of the wheel being produced on the unit to which the robot was to be relocated, a new parameter was introduced. This required the inversion of the wheel before it was placed on the takeaway conveyor.

Because of the ease in reprogramming and the capability, with optional modifications, of turning the gripper 180 degrees, we did not anticipate any problem in the relocation. However, we did find the cycle time of the robot was now the determinate factor in unit cycle time as it exceeded the cycle time of the pour conveyor by a second or more. Likewise, this problem was able to be corrected due to the versatility of the robot's programmable drum controller. We changed the robot cycle to the following: reach, drop, grasp, pivot and rotate, release and reposition to starting point, resulting in the elimination of two steps. One of the robot's features which enabled us to perform these changes was the sufficient strength of the equipment which allowed for the pivot operation with a 25 pound object while the arm is in the extended position.

The ease of relocation was simplified because the robot is virtually a self-contained unit, ie: it has its own base, hydraulic package, and control components built into a single unit. Also, we were able to utilize a different gripper assembly enabling us to rotate the wheel without making changes to the robot. Boom itself, as the robot is designed by the factory to accept numerous hydraulically operated gripper or holding devices.

In summary, during the period the robot has been in operation on either unit, the performance has been totally satisfactory.

1. The timing is synchronous with the production unit without reducing production.

2. Downtime has been minimal and maintenance within the capability of our general maintenance mechanics.

3. It replaced a production personnel unit with no safety problems encountered.

We feel the addition of the robot has been very rewarding to our company in both productivity and personnel satisfaction.

Robots Reduce Exposure To Some Industrial Hazards

Reprinted by permission of the authors from National Safety News, July, 1974

Problems of machine guarding, heat, noise, fumes, and lifting of heavy loads related to metal presses, forging, and paint spraying are lessened by versatile devices, but *OSHAct* regulations still apply.

Iɴᴅᴜsᴛʀɪᴀʟ ʀᴏʙᴏᴛs ᴀʀᴇ ʜᴇʟᴘɪɴɢ to eliminate the hazards involved when workers put their fingers in metal-working presses, are subjected to long exposures of toxic materials, must load and unload hot parts from processing furnaces, or lift heavy loads.

The importance of industrial robots for risk control has been especially great since the *Occupational Safety and Health Act* went into effect in April, 1971. Before, industrial robots were viewed primarily as machines to improve productivity in manufacturing operations.

Industrial robots are not merely the curiosities they once were. Since 1962, more than 600 Unimate industrial robots alone have been installed by their manufacturer, Unimation, Inc., Danbury, CT and its licensee, Kawasaki Heavy Industries, Japan.

Productivity is still of major importance to manufacturing engineers contemplating new production processes, but safety engineers have a good reason for learning about the capabilities of industrial robots as an aid in evaluating potential applications where risk control is a necessity.

Nature of Industrial Robots

The industrial robot, as considered here, is a general-purpose, programmable, parts-handling machine that will also control and synchronize the equipment or production machinery with which it works. As with a human, it can be "taught" a job, can "remember" instructions it has been given, can be "retaught" when the job content changes, and can be transferred to a different job when the first job ends.

General-purpose industrial robots are most likely to be practical when the part or tool to be handled weighs up to 350 pounds, the job is repetitive even though complicated, and working conditions involve potential hazard to a worker.

The accompanying information explains what an industrial robot is —and isn't.

Capabilities of Robots

Capabilities of the industrial robot determine whether it is an appropriate method of controlling risk in a particular industrial environment.

The prevalent types of industrial robot have an articulated mechanical arm to which can be attached a variety of hand-like grippers or a tool, such as a welding gun or impact wrench. An electronic memory and control system directs the actions of the hydraulically powered arm, point-to-point in accordance with a preset program. In addition, the memory sends electrical signals to control equipment the robot is tending, and receives signals from the equipment.

From two through six programmable arm and wrist movements are available. For example, the three arm movements of a typical robot are up-down, extend-retract, and rotation (move left or right). The three wrist movements are twist (roll), bend (pitch), and yaw. For most applications, five movements are sufficient, with yaw needed infrequently.

Memory size is expandable in various modules from approximately 100 sequential steps to more than 1,000. Fewer than 100 steps will encompass the vast majority of jobs.

By **Norman M. Heroux,** Field Service Manager, Unimation, Inc., Danbury, CT.

And by **George Munson, Jr.,** Marketing Manager, Unimation, Inc., Danbury, CT.

Figure 1 shows an industrial robot positioning casting in a trim press, after quenching the hot casting in a tank of water. When the gripper is withdrawn from the press, the robot will signal press to cycle, removing superfluous metal from parts.

Alternative actions and repetition of sub-routine motions, as well as a straight sequential run-through and repeat of a program, are among the capabilities of the magnetic or solid-state memories of industrial robots.

A common method of teaching a job to an industrial robot is to switch its control system to a "recording" mode and move the hand of the robot through the operation to be performed. Once recorded, the movements and actions will be repeated endlessly and precisely, time after time, in synchronization with the external equipment with which the robot works.

The sightless robot's window to the world consists of electrical connections from its memory system to limit switches that signal the robot to perform—or not to perform—certain steps. For instance, a simple limit switch installed on a press ram will signal the robot when the ram is up. Only then will the robot continue, with programmed steps commanding it to reach into the pinch area to load a part. When the hand of the robot once again is in the clear, a signal recorded in the robot's memory will permit the press to continue and make another hit.

For teaching a second robot or for future use, programs can be stored on magnetic-tape cassettes.

Reach of industrial robots can span a few feet or up to 20 feet. Maximum load-handling capacities range from 150 to 350 pounds, with positioning accuracy of 0.05-inch.

Risk-Control Applications

Industrial robots are being employed to load parts into and then unload parts from production equipment, stack and palletize parts, assemble parts in a sequence, transfer parts from one machine to another, and spot-weld assemblies (See Figures 1 and 2).

The accompanying information lists selected Occupational Safety and Health Administration standards from Part 1910 that guide compliance safety and health officers inspecting workplaces in industry. The standards are ones for tasks or environments in which industrial robots are being found valuable for risk control.

Some specific tasks and environments, in which general-purpose robots or paint-spraying robots are

What Is a Robot

It was 1922 when the word, "robot" entered the English language, through the translation of a play written by the Czechoslovakian author, Karl Capek. The title of the play was "R.U.R." an abbreviation for "Rossom's Universal Robots." In Czechoslovakian, the word *robota* means *work* or *worker*.

The U.S. inventor George Devol began patenting his designs for an industrial robot in the early 1950's. He had noted the growing rate of machinery obsolescence and the large number of industrial workers who were assigned to repetitive machine tending tasks. His robot was to be a universal machine capable of doing a variety of repetitive machine-tending jobs, yet be immune from obsolescence when the nature of a job changed.

In 1958, Devol licensed a firm that was to become Unimation Inc., to produce the robots. Three prototypes were ready for operation in 1962. Seventy hand-built universal "Unimates" were working in factories by 1966. Then Unimation tooled for production of what today is the general-purpose Unimate industrial robot.

Unimation Inc. maintains that to be dubbed a "robot," an industrial device should be as general-purpose as the worker who would be hired as a machine tender.

Special-purpose automation machinery would not qualify, because it is built for a single purpose, customarily high-volume production, and is rarely adaptable for another job when the first is completed.

Nor are remote manipulators, which are extensions of a human operator's hands, classifiable as robots. They are often very sophisticated, but basically tools used by human operators. An example would be devices to assist an operator in manipulating radioactive materials and components.

Figure 2 shows how machine tools clustered around a single robot are loaded, unloaded, and cycled in sequence. Robot transfers part from vertical lathe to broach to drill, then deposits completed part on conveyor.

employed, are described in the applications that follow.

Press Loading

A number of years before *OSHAct,* a "big three" auto manufacturer had industrial robots doing press-loading work. Two of these presses, each tended by a robot, are illustrated (See Figure 3). Electrical interlocks synchronize operation of each press and robot. A failure results in stoppage of the operation because either the robot or the press will not receive the next signal it requires before proceeding.

Note the railings and chains around the work area to prevent unauthorized personnel from approaching too close.

Another use of stamping presses has pointed out that even new presses designed to help meet *OSHAct* requirements may occasionally make a double hit. Guarding provides extra insurance that there will be no hands in the die area when this happens—but it is an additional fail-safe procedure if the feeder is an industrial robot, instead of a man.

A simple pneumatic cylinder with mechanical stops is adequate for pushing, pulling, or lifting small, light-weight parts so that a man will never have his hands in the pinch area of a press. An industrial robot is frequently essential, however, where more complex actions and heavier parts are handled.

Spraying Paint

Spray painting of auto parts or other products exposes workers to solvent vapors and aerosols. Protective masks and clothing are encumbrances for workers, and can be uncomfortable. Robots were developed to eliminate the need for a man to spend prolonged periods in this kind of environment (See Figure 4). Programming this type of robot consists of having the operator guide the spray gun through the required (continuous path) pattern while the robot's memory is in the "recording" mode. When switched to the "play" mode, the robot will repeat the pattern for each new part presented.

More than 125 of these spray-painting robots currently are in operation.

Furnace Tending

Manual loading and unloading of a furnace with tongs is a task in which heat-induced fatigue can lead to carelessness and increased risk, regardless of protective clothing worn.

At one plant (See Figure 5), an industrial robot loads and unloads a rotary-hearth furnace for firing ceramic coatings on interiors of valve bodies. Formerly, a man was stationed in front of the 1,800 F. furnace.

Valves and fittings of various sizes and shapes must be processed in this application; therefore, a library of programs has been established. A tape containing the appropriate program is withdrawn from the library and electronically transferred to the robot's memory before each different lot of parts is processed. Program transfer takes a maximum of three minutes.

When the parts are completely new, the robot is taught the new program in the conventional manner. An operator plugs a hand-held teach control into the industrial robot and uses it to lead the arm and hand of the robot through the required steps. Later, the new program can be put on tape and filed for future reference.

Die Casting

More than 100 industrial robots are in use at die casting plants where they extract die castings from

Figure 3 illustrates two press-tending robots loading and unloading parts, thus eliminating the possibility of workers placing their hands in hazardous areas. Each robot controls the press with which it works.

between the open platens of the casting machines (See Figure 6). The heat, molten-metal splatters, and hazards of reaching into the mold area to pull out a casting make this an unsuitable task for a man.

Although the robot illustrated simply extracts the casting and places it on a conveyor, at other plants such units perform additional operations while the next shot is injected into the mold, and solidifies. These operations include periodically picking up a spray gun and using it to lubricate the mold; quenching each casting in a tank of water; placing the casting in a trim press, which removes the sprue and excess metal flash that occurs along the mold parting line; alternately unloading two die-casting machines, and loading inserts into the mold (See Figure 1).

Plastics Molding

At a number of plastics products manufacturing plants, general-purpose industrial robots are tending compression-molding and injection-molding machines.

In a Southwestern plant, each of two industrial robots alternately unloads parts from a pair of eight-cavity compression-molding machines. Employee turnover was high because of the heat and vapors. Now, two robots are working in the hazardous clamp area of the machines, rather than the nine employees who were there formerly. A third industrial robot is soon to be installed. Besides unloading the machines, the 120-step programs in the robots' memories also control a magazine holding pre-heated parts fed one at a time to the molding machines.

Elsewhere, elastomeric parts five-feet-long are being unloaded alternately from two injection molding machines. The robot performing the job reaches into the press and strips the part from 395 F dies, relieving two operators per shift from exposure to vapors as well as the heat. Previously, operators relieved each other at intervals to keep their exposure to the vapors within allowable time limits.

Each elastomeric part consists of two usable pieces joined by a run-

Figure 4 shows how spray painting robot limits worker exposure to solvent vapors and aerosols. Such robots are used in enameling of bath tubs, washing machines, other equipment, as well as spray painting of automobile parts and lighting fixtures. A worker must first "walk" a robot through a work pattern to "teach" it its operation; subsequently, parts are sprayed automatically.

ner and sprue. After stripping the part from the die, the unit draws the part over two parallel blades on a cutting table, severing the pieces from the central sprue. Finally, the robot drops the usable pieces on a conveyor and turns to the second injection molding machine as it opens. On the way, the robot drops the sprue in a box.

Forging Work

A manufacturer of hammer forging equipment in 1971 began selling automated forging systems, consisting of a die forger controlled by an industrial robot.

The robot (See Figure 7) removes a hot piece of stock from an electrically heated magazine, transfers the stock to the first of two die stations, triggers the forging hammer, which delivers a series of blows, while the robot holds the workpiece steady in the first impression, moves the workpiece to the second impression, triggers the hammer again, may transfer the workpiece to other impressions, and then deposits the forged "platter" of parts on a conveyor. Noise levels can be well in excess of 90 dBA, with blows equivalent to a ton of force being delivered at rates up to 100 per minute. One man monitors the operation from a remote, quieter area where he is not exposed to

Figure 5 illustrates an industrial robot loading and unloading a rotary hearth furnace. Formerly, a worker with tongs was stationed in front of the 1,800 F. furnace.

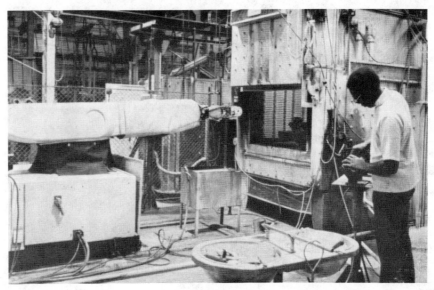

the heat and fumes customary in forging.

For the production of crankshafts at another plant, a large version of the Unimate takes 2,200 degree F. forgings from a nest, loads each into a twisting machine that does the preliminary forming of the crankshaft, then unloads the forgings onto pallets. Weight of each forging is in excess of 200 pounds —some types weigh more than 300 pounds. This would be a dangerous task for a man.

Munitions Work

Industrial robots are transferring projectiles from conveyors to one of 15 positions in storage carts at an ammunition plant. Weight of the unfilled projectiles is up to 200 pounds each. After the transfer, the robot unscrews a lifiting lug from the nose of each projectile. This particular palletizing task is not excessively hazardous, but explosives are present elsewhere in the plant. Also, as in the case of handling 200-pound crankshafts, the projectiles are heavy and, in the normal operation, must be transferred with the help of a hoist. This exposes the operator to a risk situation due to human error or equipment failure.

Safety of Industrial Robots

When a robot is in operation, safety chains or guard rails prevent personnel from inadvertently walking into the operating area. Interlocks on production machinery will cause a shutdown or interrupt the programed task. The industrial robot is designed to be fail-safe. As an extra precaution, emergency stops on the robot can be adjusted to mechanically restrict arm traverse.

The robot is taught with the aid of a teach control, which enables the operator to move the robot's arm and wrist through the necessary manipulations. The teach control has an enabling trigger that the operator must squeeze to move the arm. All motion ceases when he releases the trigger. In addition, the hydraulic system is restricted (fail-safe) in the "teach" mode, so that the arm will move only at slow speed.

In some jobs, the industrial robot works with an operator who performs a complex operation on a hot workpiece while the robot holds it for him. Here, the robot would be programed to extend its arm to the maximum before presenting the workpiece to the operator. The operator would stand behind a chain or rail, facing the robot, and restricted to a location where the arm could never move closer.

Unimate industrial robots will work in ambient temperatures to 120 degrees F. and humidities as high as 90 per cent saturation. When ambient temperatures or humidity are higher, a duct can be used to supply fresh air to the cooling-air intake. When facing a furnace, the robot's wrist is protected with wire-on thermal insulation, and a heat shield is often installed on the side of the industrial robot's body facing the heat source.

Molten metal splatters are stopped by the standard metal-mesh filter on the air intake.

Spray-painting robots can be furnished to meet Factory Mutual System's intrinsically safe standards for Class I, Division 1, Groups C and D, pertaining to hazards from electrical arcing in explosive atmospheres.

Aside from spray-painting applications, concentrations of explosive vapors or dusts are virtually never present in the usual industrial environment for which such robots were designed. Similarly, corrosive materials that could damage or cause deterioration of the robot are rarely present in concentrations high enough to be a risk. Heat and hot metal particles are among the most severe conditions that an industrial robot is likely to face. Simple shields on arm, wrist, or hand, and special materials for the gripping hand and fingers of industrial robots are usually sufficient for such difficult environments.

Figure 6 shows how the die casting machine at the right is unloaded by robot, which reaches into the die area and removes the hot casting, then places it on conveyor. Note warning sign on robot console.

Maintenance of Robots

Unimation Inc. usually recommends that there be an eventual potential for at least three industrial robots at any single plant location. A trained maintenance man will not be effectively utilized until there are a minimum of three robots for which he is responsible. Exceptions, however, include cases in which risk-control concerns were dominant or in which the productivity gains achievable with one or two robots would more than off-set the expense of a specially trained employee.

Manufacturers of general-purpose industrial robots operate customer training programs or can arrange service contracts.

Reliability of machine tools is considered to be satisfactory if "up-time" is 90 per cent, and very good if 95 per cent. The comparable reliability figure for Unimate industrial robots is 97 per cent, and mean time between failures is more than 450 hours, at present. Thus, the need for repairs is infrequent.

Auto manufacturers with multiple industrial robots usually have spare robots available for production-line operations, such as spot welding. A fork-lift truck side-lines any robot needing repairs, and replaces it with a spare. The program is quickly extracted from the memory of the defective robot and written into the memory of the spare robot, or a duplicate program on file can be transferred into the spare robot's member, typically within three and one-half minutes.

Economic Considerations

A key difference between the usual safety equipment and the industrial-robot is that the industrial robot is designed actually to increase productivity, because it can work more rapidly than a worker could in many difficult applications. The consistency and uniformity of its work cycle eliminates the effects of worker-related time variations that affect the quality of products, such as die castings.

Consequently, industrial robots are capital equipment that can be evaluated in financial terms, justifying acquisition of the equipment. Although an industrial robot may be acquired primarily for risk-control purposes, the equipment is almost invariably an income-producer.

Investment in an industrial robot, plus accessories, ordinarily can be returned through savings within two years on a two-shift operation. Payback on a three-shift operation would be shorter, commonly 1.4 years.

In the compression-molding press application, with each industrial robot loading and unloading two presses, the robot-controlled presses

have an output double that of manually operated presses. Payback information is unavailable.

In automation of an operation in which industrial robots are interfaced with standard non-automated machine tools, one Southwestern manufacturer found that payback took less than a year on the three-shift operation. Each robot tends four machine tools.

One die caster reports a nine-month payback with industrial robots operating on two shifts. Another cites a six-month payback for a three-shift operation, while a third reports an 18-month payback on a single-shift operation in which industrial robots unload die-casting machines.

Meeting OSHAct Regulations

To help insure that particular manufacturing operations are free from recognized hazards to workers, industrial robots are being used and considered for a number of jobs covered by *OSHAct* standards. Chief among these are the jobs, machines, and conditions listed under Subpart G, *Occupational Health and Environmental Controls;* Subpart I, *Personal Protective Equipment;* Subpart O, *Machinery and Machine Guarding;* and Subpart P, *Hand and Portable Tools and Other Hand-Held Equipment.*

The accompanying information *Selected OSHAct Standards,* presents a broadly interpreted listing of *OSHAct* subparts for job-related and machine-related safety applications for which industrial robots could be considered or are being used.

A Look at the Future

According to the president of Unimation Inc., Joseph F. Engelberger, greater sophistication is the expected direction for future developments in industrial robots. The greater sophistication will make the robots still more flexible and will be the evolution following the robotics revolution of the last decade.

Greater sophistication will likely include more add-on accessories, such as program-storage and fast-teaching devices, increased memory

Figure 7 illustrates how a robot (reaching in from lower left) holds a hot forging in the second of three successive dies, as the forging hammer (blur above) descends to deliver a series of blows.

capacity, flexibility, and speed, and greater modularization to permit purchasers to begin with a relatively basic, less-expensive robot that can be upgraded as requirements change.

Development of visual sensors, to give industrial robots "sight" is not a high-priority development item for builders of general-purpose industrial robots. In many production operations, workpiece orientation is consistent and position is highly predictable at any given time during the operation, making optical sensors superfluous. Thus, demand is limited, and cost would be too high for many jobs.

Cost is crucial to purchasers who must be able to show that there is economic justification, in terms of payback or return on investment, when acquiring capital equipment.

Greater sophistication almost invariably means greater complexity and lower reliability. This has been offset so far by the increasing reliability of solid-state electronics. Because this situation cannot be expected to continue, it will become more difficult to keep industrial robot up-time high. Unimation is developing diagnostic systems that can be built into the robots to monitor operation and pin-point trouble spots the moment they occur. Faster reaction to impending problems will assist in controlling downtime.

It is evolutionary developments such as these that will make general-purpose industrial robots increasingly attractive for a growing number of manufacturing operations.—**End.**

Selected OSHAct Standards
Applicable to Use of Robots

Various subparts of part 1910, Occupational Safety and Health Standards, extend somewhat indirectly into the use of industrial robots.

As with all *OSHAct* standards, such applications are determined basically by the type of industry or nature of the operation involved. Hence, the industrial user of robots should be familiar with all of the many *OSHAct* references that pertain to his plant operation.

He should keep uppermost in mind the fact that even though robots may be used, *OSHAct* regulations still must be complied with to protect those employees entering robot station areas, as well as to safeguard employees working in areas adjacent to where the robots are located, and where, for instance, such hazards as noxious fumes, excessive noise, or extreme heat may be present.

Specific questions should be referred to *OSHAct* compliance officers or to the OSH-Administration area director.

A selected listing of pertinent *OSHAct* standards might include:

● Subpart G—Occupational Health and Environmental Control.

—Section 1910.93 Air contaminants.

—Section 1910.93a Asbestos.

—Section 1910.94 Ventilation.

—Section 1910.95 Occupational noise exposure.

● Subpart H — Hazardous Materials.

—Section 1910.106 Flammable and combustible liquids.

—Section 1910.107 Spray finishing using flammable and combustible materials.

—Section 1910.108 Dip tanks containing flammable or combustible materials.

—Section 1910.109 Explosive and blasting agents.

● Subpart I—Personal Protective Equipment.

—The various sections herein as they pertain to workers in the areas where robots are located, and to workers who would be subjected to environmental or job conditions requiring the use of such personal protective equipment.

● Subpart L—Fire Protection.

—Sections 1910.156 through 1910.165b, as they relate to specific types of manufacturing processes and structures, would also pertain to the use of robots therein.

● Subpart N — Materials Handling and Storage.

—Section 1910.176 Handling materials—general.

—Section 1910.177 Indoor general storage.

—Section 1910.178 Powered industrial trucks.

—Section 1910.179 Overhead and gantry cranes.

—Section 1910.180 Crawler, locomotive, and truck cranes.

—Section 1910.181 Derricks.

● Subpart O — Machinery and Machine Guarding.

—Section 1910.212 General requirements for all machines.

—Section 1910.213 Woodworking machinery requirements.

—Section 1910.214 Cooperage machinery.

—Section 1910.215 Abrasive wheel machinery.

—Section 1910.216 Mills and calenders in the rubber and plastics industries.

—Section 1910.217 Mechanical power presses.

—Section 1910.218 Forging machines.

—Section 1910.219 Mechanical power-transmission apparatus.

● Subpart P — Hand and Portable Powered Tools and Other Hand-Held Equipment.

—Section 1910.242 Hand and portable powered tools and equipment, general.

—Section 1910.243 Guarding of portable powered tools.

—Section 1910.244 Other portable tools and equipment.

● Subpart Q—Welding, Cutting, and Brazing.

—Section 1910.252 Welding, cutting, and brazing.

● Subpart S—Electrical.

—Section 1910.308 Applications.

—Section 1910.309 *National Electrical* Code.

The listing cited is not to be considered complete. In addition, in some of the named references the relationship between the job condition covered by *OSHAct* regulations and the use of robots may be quite remote, with clarification depending upon the specific situation under consideration. At present, robots are not involved in every one of the cited job areas, but their capabilities are such that they could serve in some capacity.—**End.**

CHAPTER 4

HUMAN FACTORS

Commentary

This chapter deals with the issue of resistance to industrial robots—resistance that can be found in management as well as on the factory floor.

The UAW's stand on industrial robots is presented, as are the characteristics of management attitudes toward this technology. Some approaches to overcoming worker and management resistance are suggested, as are some solutions to problems of worker displacement.

An industrial robot public relations checklist is included. This is a basic guide for industrial robot manufacturers and users to properly present robot information and to avoid the necessity of defending robots against attacks and criticism based on fears of the unknown.

Presented at the Robots IV Conference, October, 1979

Three Laws For Robotocists: An Approach To Overcoming Worker And Management Resistance To Industrial Robots

By Neale W. Clapp
Block Petrella Associates

Advances in industrial robot technology continue to exceed understanding of the dynamics of worker and management acceptance. Continued attention to developing equipment and applications with neglect or indifference to the "soft" science of management may result in sophistication without application, or installation "failure" wrongly attributed to technology, while the real problems of human interfaces remain unaddressed.

The crude basis for a beginning approach to confronting management and organizational issues may be the promulgation of "Three Laws of Industrial Robot Implementation." As a statement of management intention, they will be suspect. As a commitment to management action, they will prove themselves to both management and worker.

Furthermore, consistent with the laws, the robotocist would be well advised to explore innovative and social environmental benefits accruing from industrial robot application. Manufacturing management, currently "under siege" may, instead of resisting the inevitable, find reason for optimism in the demise of traditional patterns of work relationships.

INTRODUCTION

A review of the current literature in robotics reveals a dearth of data regarding what engineers are prone to call "human factors." When mentioned at all, the term usually relates to the human-robot interface, or the economics of manufacturing.

Yet, the history of technology is replete with examples of retarded application due to inattention to human and organizational dynamics. Frequently, this inattention is a function of the seduction of the proponent by the fascination with his own ideas. Sometimes, it is merely a desire to avoid the "soft" and intangible barriers to application, which being outside of the realm of the proponent's expertise, are a source of discomfort and even threat.

It appears that industrial robotics is at this crossroads. Opportunities for applications abound but implementation is, at best, cautious. Meanwhile, there is a vague hope that improvements in sensors and tooling will overcome the resistance. I suspect that lurking underneath all this, is the unexpressed fear that the roadblock is not technology nor economics, but something less tangible called worker and management reaction.

Investigation reveals that worker reaction is not nearly as resistant as anticipated, and in fact, management resistance may be more crucial. A recent study by Doris B. McLaughlin for the National Sciences Foundation concludes that:

> The most common response that this country's labor unions make to the introduction of new technology is willing acceptance. The next most common initial response is opposition. That is followed by adjustment, encouragement and, finally, competition.

> The report notes that while opposition ranks second on a short-term basis, 'that response is very often a temporary one and is usually followed by either a move to adjustment or willing acceptance. Thus,

in the long run, it is willing acceptance followed by adjustment that constitute the most common union reactions to technological innovation.'

When a union does respond negatively to the introduction of technological change, the report continues, the underlying reason is invariably that acceptance of it would have an adverse effect on a large and/or important segment of its membership. 'Once the employer can convince the union's leaders that their members either will not be adversely affected or that those who are will receive some offsetting benefit, union opposition evaporates.'[1]

Contrast this statement to the statements of Joe Engelberger, John Wallace and Gennaro Macri regarding management reaction:

Manufacturing is an extremely conservative activity, and its ramparts are already under siege of technological innovation capable of traumatically changing traditional patterns.[2]

Production people are generally conservative, and want to use tools and processes that are first of all, proven, secondly up-to-date. The reverse order does not have too much appeal to most people in production.[3]

It is at this upper level of management that basic policy regarding the implementation of any new technology must be made. Lower levels of management will rarely assume any of the risks involved. We must start with plant, Assistant Plant Manager, Operations and Engineering management level. If this group is not fully informed and fully agrees to take an 'active' part in the planning, installation and launch of the first application, then this is the point to stop.[4]

Resistance, then, is not the exclusive province of the workers, but if the evidence is to be believed, more likely to occur within the ranks of middle and lower management. Or, as the McLaughlin study concludes:

Both literature and interviews reveal that employer representatives, particularly at middle management level, more often constitute the real barrier to the introduction and effective use of technological innovations. Since no study now exists of the part played by management in slowing the rate and direction of technology, McLaughlin suggests that such a study be undertaken.[5]

Given "resistance," wherever its locus, it may be time for the industrial robotocist to take a moral and ethical stance which, if honored, will provide: 1) a roadmap for his behavior, 2) a system of governance and policy for the industry, and 3) a commitment to assure human and civilized treatment of those most directly impacted by the advent of the robot.

---

[1] Bureau of National Affairs, Inc., "Unions Not Prime Obstacle To Productivity, Study Says." (Washington, DC, 1979).

[2] Engelberger, Joseph F. "A Robotics Prognostication." (Danbury, Connecticut: Unimation, Inc.).

[3] Wallace, John J. "Creating a Robot Market," MS75-242. (Dearborn, Michigan: Society of Manufacturing Engineers, 1975).

[4] Macri, Gennaro C. "Analysis of First UTD Installation Failures," MS77-735. (Dearborn, Michigan: Society of Manufacturing Engineers, 1977).

[5] World of Work Report, "Labor's Answer to Technology is Often "Willing Acceptance." Volume 4, Number 6, June 1979.

THREE LAWS OF ROBOTICS

With lasting thanks to Isaac Asimov and no attempt at humility, I submit three basic "laws" which might well govern industrial robot installations.

I - ORGANIZATIONS MAY NOT INSTALL ROBOTS TO THE ECONOMIC, SOCIAL OR PHYSICAL DETRIMENT OF WORKERS OR MANAGEMENT.

Among other things, this law would mean that workers displaced by the robot would be guaranteed equal-rated jobs, that supervisors would not be penalized by "down-time" while the robot is being de-bugged, that the advanced manufacturing engineer would not be out on a limb by himself in promoting the robot. It would include setting rates that operators agreed were equitable, and recognizing that a supervisor might find it difficult to be less knowledgeable about the robot than the newly trained operator. It would acknowledge that old interface frictions between manufacturing units could easily be aggravated by the robot and win/lose battles ensue. In fact, it would require increased sensitivity to the workplace, the reward system and the management system.

II - ORGANIZATIONS MAY NOT INSTALL ROBOTS THROUGH DEVIOUS OR "CLOSED" STRATEGIES WHICH REFLECT DISTRUST OR DISREGARD FOR THE WORK FORCE, FOR SURELY THEY WILL FULFILL THEIR OWN PROPHECY.

Admittedly, this appears to be so obvious that it is trite, but it bespeaks the possibility of a management attitude toward the work force which, whether implicit or explicit, will be transparent. Acceptance of this law would foreclose the euphemisms of "universal transfer devices," or the "let's-get-it-on-the-floor-first-and-then-worry-about-the-reaction" school of management. Successful applications studied invariably revealed that management had consciously leveled with and listened to the workers. However, it is not only the work force which has a stake in the outcome. Middle and lower-level management must be included in early deliberations, particularly with regard to review of management measures and standards.

The issue for organizations is to assure a proper match between the _type_ of problems encountered and the _structure_ best suited to provide optimal solutions.

Ironically, the installation of the robot reveals examples of both appropriate and inappropriate matches. In _technology_, robot installation is an example of a well-structured problem. That is, the output is known, can be counted, and errors in program can be detected relatively quickly and precisely. Given the state of the art, technical information is relevant, available and complete. External standards such as costs, output targets, performance and rates can be estimated and reasonably well controlled. Feedback about results is fast, and can be attributed to the action of the robot or its operators. Thus, the manufacturing organization with clear levels of authority, division of labor and high use of rules and procedures is _appropriate_.

However, in _human system_ terms, robot installation is an example of an _inappropriate_ match between an ill-structured problem existing in an authority/production centered organization. Obviously, proponents of robot installations must be able to operate in a knowledge/problem centered organization, with high interchange of various skills, problem-solving capability, procedures and rules developed specifically for the purpose, and authority based on knowledge, not on hierarchical position. Furthermore, there are many required links to units within the total organization, and new communications channels must be developed.

An "open" strategy therefore, is essential for optimal implementation. "Open" in this context means clear articulation

of what's at stake, what legitimate conflicts can be expected
between sub-units and an attitude conducive to problem-solving
rather than fault-finding.

> *III - ORGANIZATIONS MAY ONLY INSTALL ROBOTS ON THOSE TASKS*
> *WHICH, WHILE CURRENTLY PERFORMED BY MEN, ARE TASKS*
> *WHERE THE MAN IS LIKE A ROBOT, NOT THE ROBOT LIKE A*
> *MAN.*

At this stage of evolution, the "blind, one-armed imbecile" of
Engelberger's description is performing noxious, unsafe, repe-
titious tasks. As robotics sophisticates, this may become
less true, although there are certainly a multitude of manufacturing tasks onerous enough to last for the foreseeable
future.

If the robot is to fulfill a role as man's servant, this law
demands adherence. Any robot takeover which further impover-
ishes a work culture already burdened with dull, unrewarding
work should be excluded.

Of course, the facts are that the robot does and can continue
to, provide enrichment for manufacturing tasks. Nonetheless,
claiming this, without demonstrating commitment to enrichment
will surely lead to disenchantment. One operator released
from an onerous, dead-end job by his training with the robot,
wistfully expressed a forlorn hope to train others when new
robots were installed. That's the motivation which management
should seek and encourage by its actions.

These three laws were originally written to be "provisional,"
but after reflection, they seem to have sufficient validity to
warrant permanence.

Consistent with the laws, the robotocist would be well-advised
to begin to explore innovative, and social environmental benefits accruing from industrial robot applications. Manufacturing is, as Engelberger states, "under siege" and traditional
patterns of organization and management, teeter on the brink
of an uncertain future. If the vision of the organizational
future can include optimism rather than merely a reluctance to
part with the tried-and-true, resistance will dwindle.

INNOVATIVE BENEFITS

One innovative benefit outweighs any other and demands most
attention.

Robotics requires interdisciplinary action. Installation cannot be successful without a high degree of coordination between
interfacing units. To consider robots is to raise issues where
the rigid vertical hierarchy of manufacturing must come under
scrutiny. The typical structured problems so well managed in
conventional manufacturing are suddenly uncertain, complex and
unyielding. Long established measurement standards are no
longer appropriate. The hierarchy is further jolted when the
operator is more knowledgeable through training than his foreman. The enthusiasm and energy of the robot proponent is tem-
pered by the lack of psychological "ownership" by other units
on whom he must rely.

Typically, three critical units must manage their interface to
assure productivity. Under the leadership of the Plant Manager, Shop Operations, Production Control and Manufacturing
Engineering coordinate their activities. The differentiation
in the sub-unit goals, time frame to measure performance and
rewards systems are sources of sustained conflict, negotiation
and hopefully, resolution.

One case in particular highlights the interface dilemmas.

The impetus for installation started with Manufacturing Engineering. With relatively long-term goals, the engineer was
prepared for the time required for robot re-tooling to equip it

to feed 800-ton presses, with one operator acting as set-up man. Previously, the operation had required three workers engaged in heavy, back-breaking manual operation. Even Engineering would admit that re-tooling and proper programming took longer than expected. For example, the robot "went crazy" (violated its program) for inexplicable reasons. After some time it was learned that a passing fork-lift truck was sending out signals which required additional suppressors on the robot. A special gripper had to be designed and built. Frequent new set-ups (programs) were required as volume of work diminished.

As the events proceeded, Engineering felt strong reluctance from the Operations group. Whenever down-time occurred, Operations would quickly return to manual production, except in those cases where the robot set-up precluded that possibility. As down-time mounted, a battle raged to convince Plant Management of the "rightness" of one group or the other.

According to the Engineering Group - Operations Management is the roadblock, they blame all down-time on the robot, while Engineering claims only three per cent is attributable to robot failure. Operations doesn't want to "make waves," but maintain the status quo. Therefore, they remain unimpressed despite the obvious qualities of the robot. They underestimate the workers and are more concerned with justifying the size of their "empire" (which affects Operations Management job classification) than improving cost-effectiveness. They are short-term oriented and unimaginative. Furthermore, OSHA legislation demanded a search for alternatives, and the robot was a way to minimize labor while conforming to federal standards.

According to Operations Management - Manufacturing Engineering "fell in love" with the robot and was blind to the problems, such as:

1) frequency and time required for set-ups,
2) costs of re-tooling,
3) requirement of electrification of other equipment to assure compatibility with the robot,
4) cost of training workers and the fact that only the Manufacturing Engineering group could solve problems, and
5) the foreman hated the robot, as they were ignorant regarding its operation and were being "killed" by down-time. Operations Management claims that appropriate application opportunity did exist in the plant; on high-volume repetitive work such as heli-arc welding, but would use humans first to see if a robot was warranted.

The potential for innovative benefit here is obvious. In order for "success," Plant Management must assure that Production Control, Manufacturing Engineering and Shop Operations address those central dilemmas which have always plagued them. Before installation, it is necessary to reassess management measurements, revamp customary standards of acceptable down-time, accurately forecast volume, set-up, time training, shift and maintenance requirements and review production rate to assure that new rates are realistic and not oppressive or demotivating to incentive workers. Attention must be paid to the foreman and his loss of status, as well as the selection process for blue-collar workers.

This reassessment holds the potential for enabling increased integration and coordination mechanisms. In brief, the industrial robots' utilization slices through the differentiated sub-units and calls to management attention the goal conflicts inherent in their interfaces.

A second innovative benefit is the identification of talent for robotics (and manufacturing in general) among the work force. As robots increase in number, multiple uses and sophistication, there is the opportunity for career development and movement towards skilled technician positions. In several locations, workers long respected but "stuck" in their present positions, found the robot as a way up the occupational ladder. The Manufacturing Engineer who seemed to have the best rapport with the work force, was himself a product of upward mobility. In the words of one engineer (and echoed by many others), "Don't sell the workers short."

SOCIAL ENVIRONMENT BENEFITS

Robot economics are well-described in existing literature. As a matter of fact, it can be said that they are frequently the sole management criterion for installation. Elaborate formulas are explained by Joseph F. Engelberger to provide classical cost-effective justifications.[6] Plant Managers invariably spoke of cutting labor costs and competition from other corners of the world. Some mentioned the impetus provided by OSHA legislation. Inevitably, the last named benefit was the demise of onerous work. Only one, the owner of a small company (and a former labor leader) spoke first to the unpleasant nature of many tasks. He clearly advocated an open stance with workers regarding profit and losses. However, he acknowledged that in larger companies, success would depend more on middle management support than on top management budgeting for capital expenditure on robots.

According to Engelberger,[7] Iron Curtain nations are intensely interested in robotics, and are hardly likely to use capitalistic economic justification. Not surprisingly, their spokesmen use the well-being of the worker as their motive. Propaganda aside, it does seem there is a greater recognition of the general malaise of the workplace, and the increasing resistance to sub-human work. Recent events in the United States, from labor unrest over job content, to the emerging "Quality of Work" movement, would indicate that unions may turn their attention to work content as the line hardens on wage demands.

Whatever the strength or vitality of this new force, it is evident that productivity decline is linked to job dissatisfaction. One need only visit manufacturing facilities to observe the dull glaze in the eyes of women on assembly lines, with peperback books propped up in front of them as their hands almost automatically perform rudimentary, totally repetitive tasks. Even the minor intrusion of the investigators was a source of more animation than might be expected. At least for a minute, the routine was interrupted. Work stations usually lacked even the most insignificant signs of an individual personality, other than an ashtray, or a favorite cartoon pinned up.

This dehumanizing picture was in sharp contrast to the women running a large bank of meter voltage testers. The costly automated equipment was her domain, and she was described by our guide as feeling like "the king" with her total responsibility.

Even more dramatic was the pride of the hourly workers associated with robots. They were pleased to be selected, and saw that as management recognition of their worth. Some personalized the robot and gave it names (one beige Unimate was known as "Nancy the Blonde" because "she's erratic and tempermental") while others disdained personalization and viewed it as just one more evolutionary step. Invariably, workers used the word "respect" in describing their appreciation of the robot. Workers' "respect" for the machine represented a dual meaning. On one level, it was a statement of regard for the power and capability of the industrial robot. At quite another level, it appeared to connote an alliance between man and machine.

---

[6]Engelberger, J.F. "Economic and Sociological Impact of Industrial Robots." (Danbury, Connecticut: Unimation, Inc., February 1970).

[7]Engelberger, J.F. op. cit.

Typical concerns about production rates still prevail, and there can be disgruntlement if rates are set which fail to account for the unexpected. Discussion over fair and justifiable rates will undoubtably continue, but it appears to lack the consuming anger and dissatisfaction of the perennial worker-management strife.

When questioned as to whether workers who viewed the robot on a plant tour were threatened, one Manufacturing Engineer responded: "No way, they were fascinated." This same engineer worked hand-in-hand with the blue-collar workers trying unsuccessfully to "de-bug" a robot. Eventually, the robot, derided as a "lemon," was replaced by another, and as the workers were part of the ultimate success, they were "tickled to death."

Finally, robot installation will continue at a very gradual pace. No massive displacement can occur given the limits of present day robots and the organization of manufacturing space. Thus the opportunity presents itself for careful study and control of application and concurrent management learning not only about the untapped qualities of the work force, but their own responsibility for the current state of affairs in dehumanizing jobs. Taking responsibility might commence with adhering to the Three Laws of Industrial Robotics.

Presented at the Robots IV Conference, October, 1979

A View Of The United Automobile, Aerospace And Agricultural Implement Workers Of America (UAW) Stand On Industrial Robots

By Thomas L. Weekley
UAW

To understand the UAW's* position concerning the affects of industrial robots, it will be necessary to first establish the fact that the UAW membership as a whole does not currently place a specific emphasis on robots, but rather views robots as just one of the technological advancements facing the union today.

It should be a pleasant surprise to know that the UAW membership, in general has favored the introduction of new technological advances, and has recognized them as an essential means of promoting economic progress through increased productivity. Further, labor in general has accepted and encouraged such changes.

Not only does increased productivity mean additional benefits to management but employees recognize that their wages, working hours, fringe benefits and safety will also improve as productivity increases. Productivity, therefore, is a recognized and accepted necessity for advancement of personal and corporate goals. The vast demand for quality low-cost goods makes increased productivity essential for economic progress. The UAW has included the following clause in major contracts - clearly accepting the principle of increased productivity.

> "The improvement factor provided herein recognizes
> the principle that a continuing improvement in the
> standard of living of employees depends upon
> technological progress, better tools, methods, pro-
> cesses and equipment and a cooperative attitude on the
> part of all parties in such progress. It further
> recognizes the principle that to produce more with
> the same amount of human effort is a sound economic
> and social objective."

In addition to benefits received from the increased productivity, monotonous and hazardous jobs are a cause of discontent and robots and/or automation offers an acceptable alternative.

These reasons provide the reasoning, willingness, and motivation behind employees acceptance of technological advances, however, it is necessary to outline some of the negative facts and casualties of automation that prevents blind acceptance.

There is an immediate negative impact on employees both in terms of security and mental awareness when automation/robots are introduced. Regardless of the overall view of automation there can be no denial of the fact that the benefits gained by employees through automation, will be applied to a smaller work force. Common sense and simple facts show that automation does replace workers, and though a certain amount of new work is shifted to a skilled maintenance group, the long term affect is a reduction in non-skilled labor.

Based on historical data, it is estimated that the hours of labor to produce a car will go from 146 in 1976, down to 103 by 1990. This leads to the conclusion that hours worked in the auto industry by 1990 will increase less than 5% to produce a 47% rise in production. These figures reflect all production workers who are considered to be in the auto industry, according to U.S. government statistics. Thus they include workers who

* The abbreviation UAW throughout this paper shall refer to The United Automobile, Aerospace and Agricultural Implement Workers of America.

make trucks, replacement parts and other automotive items different than cars. Nevertheless, they provide an approximate measure of the decreasing labor required to make a car.

Since the inception of automation this reduction has been offset by ever increasing product demands causing an increase in necessary manually operated work and/or an expanded work force. Such favorable conditions have helped create a climate of acceptance and have helped to mask the real impact of automation on the work force.

The increase in productivity at existing facilities further reduces the number of available jobs due to the fact that fewer facilities are needed to produce ever increasing amounts of products.

Robots not only replace manual production operations, but in combination with electronic optical inspection devices are gradually replacing the next higher labor grade of semi-skilled workers.

As you can see, the UAW's acceptance of automation/robots is not without a very serious basic consideration: <u>SECURITY</u>.

The following policy outlines the UAW's position concerning new techno- logy and was approved by the 1979 collective bargaining convention attend- ed by delegates from every local within the UAW.

TECHNOLOGICAL CHANGE

The pace of technological change, spurred by new scientific breakthroughs and the pressures of meeting society's energy and environmental needs, will be quickening markedly in the period ahead. The President of General Motors Corporation, for instance, has already projected that by the year 1988, ninety percent of GM's new equipment and machinery will be computer controlled. In a few years robots will increasingly displace human beings even on the auto assembly lines. The full development of the "World Auto" will permit massive interchangeability of parts, produced any- where throughout the world. Standardization will become so prevalent as to reduce the need for flex- ibility in the manufacturing process and increase the opportunities for management to produce subassemblies and parts across national boundaries for final assembly. Subcontracting - a complex, difficult problem endan- gering job security - will know no borders. The world itself will be a single geographic area for integrated manufacturing facilities.

It is estimated that about 5,000 industrial robots are in operation in the world today. Fewer than 500 are in use in the U.S. and Canada, performing a variety of operations including parts transfer and welding. Recent develop- ments have produced robots which have vision; robots can also be designed to respond to spoken commands. Thus, they will "see" and "hear." We can well ima- gine what this portends in the displacements of workers who sort out parts or perform visual inspection! We are on the threshold of a robot explosion.

Attention must be given to the introduction of computers that control lighting, starting and stopping machines or a production line, or which provide an alert to trouble spots that occur. We must see to it that the manual operations, now computerized, are not with- drawn from the bargaining unit. Of special signifi- cance to our TOP members are the disputes that occur over the control of cathode ray tube visual display units.

This must remain within the jurisdiction of our bargaining units and not "farmed out" to non-bargaining unit employes.

The end result of these vast and rapid changes in technology will bring about social consequences that must be reckoned with. Jobs will be eliminated by the thousands while the creation of new jobs may not keep pace. Adjustments will be needed to maintain the distribution of purchasing power required for a healthy national economy. Management executives must examine not only the advantages for profit making as technological breakthroughs reduce unit costs but they must address with equal concern the demands for social responsibility called for by their decisions and actions as they affect the workers, employment opportunities, consumer demand and the national economy.

The Union must devise collective bargaining programs to protect workers against the loss of skills and jobs. The UAW has a long established policy that reduced work time represents a key element in creating more job opportunities. We are pledged to alleviate the problem of technological job loss by continuing to reduce work hours.

The Union should simultaneously undertake educational programs among staff, local union leaders and the membership, not only to alert them to the technological explosion of the next decade, but to provide them with the knowledge needed to protect their job security and the integrity of their bargaining unit.

- Management must be required to give ample advance notice to the Union concerning the introduction of new technology (particularly computer controlled and/or robotized equipment) so that there is the opportunity for full discussion and negotiation of the effects upon the workers and the scope of the bargaining unit.

- The introduction of new technology should be managed to effect the least amount of displacement of workers by gearing it to normal attrition of the work force.

- Where increases in productivity outpace the rate of attrition, the protection of workers against displacement is a proper first charge against the gains in productivity.

- The integrity of the bargaining unit must be maintained; bargaining unit work must not be transferred to out-of-unit employes.

- In-unit employes must be given adequate training to perform the jobs introduced by new technology.

- Work time must be reduced to afford adequate job opportunities to all who want to work.

In this regard it is essential that the UAW keep abreast of research and development being undertaken in the basic industries of our jurisdiction so that, as a Union, we can plan to meet the future challenges to job security which technological breakthroughs will no dougt bring in their wake. The International Union is already in the process of seeking out the advice of outstanding experts in the field as consultants. Special effort must be directed, moreover, to educating local union leaders and the membership in the new developments in computerology so that they will be fully aware of the potential problems affecting their welfare.

As you can see, there is a real awareness of the consequences of technological advancement and its future impact. Security is the underlying theme of this statement and of the goals of the UAW in recent and upcoming contract negotiations. Natural attrition is a recognized and acceptable method of implementing changes that will result in a lowering of the work force, when coupled with advance discussion, upgrading possibilities, a sense of individual security and a reduction in working hours.

It is simple logic that dictates the fact that an ever increasing labor market, combined with a dwindling source of jobs, due to technological advances, will in the long term view alter our economy and create an economy in which employment will not keep pace with the available work force. If such a trend becomes reality the amount of paid work will have to be apportioned more widely to ensure sufficient jobs for the available work force. As stated by Howard Young, Administrative Assistant to UAW President Douglas Fraser:

> "- - - hours of employment are dependent on the paid work needed to produce desired output; the number of job slots are dependent on the need to allocate that output. To equate those two, we will have to reduce the number of work hours per job."

The UAW negotiated a major step in that direction as part of its 1976 contracts with the auto companies and has continued to make inroads in this concept. The Paid Personal Holiday plan provides workers with individually scheduled paid days off. The scheduling is designed to simultaneously assure additional leisure time for workers, and to allow the companies continued efficient use of their plants and equipment. The combined impact of those two effects is to increase the number of workers on the payroll; that is, to create new jobs.

Since that program -- or any other negotiated reduction in work time -- represents an alternative to bargaining for higher hourly wage rates, there must explicitly be pay for the hours that work time is reduced. Otherwise, groups that opt for reduced work time to create jobs would be adversely affected. It is not intended that the reduction in work time be a disguised form of layoff; instead, those workers who are fortunate enough to have a job and forego the opportunity to earn even more money -- by working their previously scheduled hours at higher hourly pay rates -- are entitled to do so without decrease in their total income.

If workers perceive robots, or technology generally, to be the cause of unemployment or loss of income, they will no longer have a cooperative and receptive stance toward technology introduced by their employers.

Thus far, in light of the information presented, it can be seen that technology is acceptable and in fact encouraged as long as the current work force is enlightened and assured of continuing job security. This must be the prime concern when automation/robots are placed in operation if an attitude of acceptance is to be continued by workers. The current in-plant skilled workforce should be used wherever possible to install such equipment to further assure acceptance.

Some specific advice to those involved in design or construction of automation/robots from a union's viewpoint would be:

1. To design automation so that current Skilled members can perform necessary maintenance functions with as little training as possible, and within current technological know-how or provide an on-the-job update program for training.

2. To use standard fittings etc., as much as possible, keeping the purchasing of specialized tools to a minimum.

3. Build in as many automatic safety measures as possible including a shut-off at the actual operation center where injury could be most severe. With robots, this would require a safety shut-off at the very end of the extension arm.

Such basic consideration would enhance the possibility of acceptance of such devices as a safe, familiar piece of equipment and would eliminate the aura of reluctance or awe that accompanies a totally new and radical concept of technological change.

Skilled Trades Training Program should be updated to include training in new or future fields, whether currently in use or in the future. Such training may only be available from the manufacturer or some other outside source. The main concern is to provide such training in anticipation of the eventual need to make new technology familiar and welcome.

The need for such mental preparation for change extends beyond the skilled work force and brings us to the subject of communications in a general sense. Advance discussion and preparation for any new piece of automated equipment not only helps to assure acceptance, but may provide helpful, practical suggestions for installation, and operation.

In summary: Automation has the potential to benefit all of us through increased productivity and by doing dangerous or undesirable jobs. Automation also has the potential to cause economic hardship for workers whose jobs are directly affected and eventually for others through its effect on overall employment. Therefore, a sense of security by workers and a sense of obligation by employers is a must for the successful implementation and acceptance of technological advancements.

These concerns and suggestions have been intended to convey to you the UAW's position on automation/robots, and the basic concerns that are expressed by its membership. These concerns are basic to all workers and will hopefully, be considered by manufacturers of automated equipment and users of such equipment in the years ahead, as automated devices become more and more an integrated part of our society.

Presented at the Robots V Conference, October, 1980

Management Resistance To Industrial Robots

By Neale W. Clapp
Block-Petrella-Weisbord

While much attention has been focused on the potential for worker resistance to the installation of industrial robots, the management constraints, other than economic, have been largely overlooked.

The catch-phrase, "management commitment" is frequently hollow and ill-defined. Specific definition and concrete action steps by the roboticist are required if the organization is to accept and support the installation of industrial robots.

This paper will attempt an operational definition of the nature of management commitment and suggest three areas of management resistance worthy of attention in assuring the optimal utilization of Robotics Technology.

INTRODUCTION

This paper derives from the belief that the accelerated growth curve of industrial robot applications in the United States is threatened by management resistance.

While engineers and manufacturers continue to provide improvements in <u>technological</u> systems, the changes to be wrought in the <u>social</u> system of the workplace receive scant attention. Furthermore, when addressed at all, the focus of concern is the worker and/or the Union. Infrequently, a passing reference is made to the need for management commitment and support.

- "Get your management to back you up. Total commitment by everyone is necessary for success."[1]

- "In my mind, the problem boils down to getting those thirty thousand or so uncommitted manufacturing executives to think the same way 3400 of them are thinking right now. For the life of me, I can't see why it is so hard to do. Are they too smart to buy robots?"[2]

[1] Tanner, William R., "A Users Guide to Robot Applications," (Dearborn, Michigan: Society of Manufacturing Engineers, 1976), page 8.

[2] Wallace, John J., "Creating a Robot Market," (Dearborn, Michigan: Society of Manufacturing Engineers, 1975), page 7.

- "It is at this upper level of management that basic policy regarding the implementation of any new technology must be made. Lower levels of management will rarely assume any of the risks involved."[3]

- "We must start with plant, assistant plant manager, operations and engineering level. If this group is not fully informed and fully agree to take an "active" part in the planning, installation and launch of the first application, then this is the point to stop."[4]

- "Productive people are generally conservative, and want to use tools and processes that are first all proven, secondly up-to-date. The reverse order does not have too much appeal to most people in production."[5]

Unfortunately, none of the above statements help much in defining what management commitment looks like, nor do they explain the sources of management resistance.

The following sections are an attempt to explore these complex and neglected concerns. The first section describes management commitment and support, while the second deals with some aspects of management resistance.

THE CHARACTERISTICS OF MANAGEMENT COMMITMENT AND SUPPORT

The following characteristics are necessary components of management commitment and support. They are intended to go beyond the hackneyed use of that phrase, and provide operational direction.

Valuing - Commitment and support are functions of "valuing." Valuing can be described as freely choosing from alternatives, prizing the choice by affirming it publicly and acting by doing something with the choice, repeatedly.

For example, it is not valuing to say one loves classical music when one opts for televised football games rather than concerts.

In the case of robotics, valuing would include selecting the industrial robot over other manufacturing options after thoughtful consideration of the consequences. Next, valuing would require affirmation of the choice - not the public attitude that "it's the pet project of our advanced manufacturing engineer," but, "we looked at the idea with our advanced manufacturing engineer and agreed that

3
Macri, Gennaro C., "Analysis of First UTD Installation Failures," (Dearborn, Michigan: Society of Manufacturing Engineers, 1977), page 3.

4Ibid., page 4.

5
Wallace, op. cit., page 8.

it was the best solution." This public statement of "ownership" is crucial to the notion of valuing.

Reinforcing - Reinforcing is the act of underlining one's commitment by active participation. Management demonstrating commitment would reward and encourage behavior which was consistent with the goal of implementation. This demands attention to possibilities where reinforcement is possible. For example, in one instance of successful implementation, the manufacturing engineer listened and utilized the input and ideas of the operator. He encouraged his suggestions and provided the opportunities for "hands-on" experience. In another case, an operator who had been challenged by the intricacy of the robot, and had gained mastery, was open in expressing his wish to train others. Demonstrating reinforcement would mean making that opportunity available.

Reinforcement which appears artificial or contrived will not be successful. If the manufacturing engineer mentioned above had no real confidence in the operator, but wanted to give him the "feeling of" participation, his strategy would backfire and be accurately perceived as patronizing.

Discouraging Undermining or Indifference - Of equal importance to the notion of reinforcing goal directed behavior, management must be aware of and active in discouraging behavior which obstructs goal accomplishment. In one failure, the operators grew increasingly lax in maintaining the hydraulic level for the robot. This slippage was noticed but not mentioned, probably because the supervisor did not

feel directly involved in the installation. Eventually, the robot failed. In the case of the operator, it was indifference. Passive resistance (or undermining) was suspected as the motive for the supervisor's behavior.

"Clues" to worker indifference are discernable to committed management. They may take the form of joking hostility, over-attention to details of other aspects of work, jealousy between units, or conflict over the priorities of time and resources. Ignoring the clues is equivalent to condoning the behavior.

Consistency and Repetition - Commitment is not a brief flurry of active support, but a sustained process in which all opportunities for reinforcement over an extended period of time are viewed as instrumental to goal achievement. One successful installation later failed when committed manufacturing manager was transferred to another location. His replacement did not "own" the criticality of the application, and within a short time, the original manual process had been restored.

Objective Search for Resistance - Once management has determined a course of action for robot installation, the strength of their belief may cause them to become overly judgmental and punishing about the lack of support and commitment down line. Thus, the tendency develops to blame, find fault or point fingers at the guilty parties.

This tendency interferes with rational problem-solving. For example, it may well be that the productivity measures used to evaluate a first-line supervisor are prejudicial when the robot is "down."

The possibilities for legitimate resistance are manifold, particularly in interfaces between manufacturing sub-units. If these instances are perceived by management as ignorance, stupidity or sabotage, management reactions can further impede robot acceptance.

A good rule of thumb: if the robot is resisted at any organizational level, search for the legitimacy of the resistance before prescribing corrective action. What things that people value highly do they believe are being threatened?

Monitoring - This does not mean acting as a watchdog, but scanning the organizational environment to find, in Weisbord's words, "blips on the radar screen."[6] This management task assures that a goal "fit" exists between the sub-units involved, and that goal clarity and agreement pervade the organization. It would also include the establishment of mechanisms or procedures to 1) smooth difficult interfaces, 2) provide channels for upward, downward, and lateral flow of vital information, and 3) function as a management gyroscope to redress imbalances.

[6] Weisbord, Marvin R., Organizational Diagnosis: A Workbook of Theory and Practice, (Reading, Massachusetts: Addison-Wesley Publishing Co., 1978).

WHY COMMITMENT AND SUPPORT ARE HARD TO FIND

OR

THREE ELEMENTS OF MANAGEMENT RESISTANCE

1. <u>The Legitimacy of Sub-Unit Conflict and the Requirement for Continual Conflict Management Without the Assurance of Permanent Resolution.</u>

Lawrence and Lorsch[7] in their research on organizations and environments, identified four critical features in which sub-units in organizations varied.

    a.   The degree of reliance on formalized rules and formal communication channels within the unit.

    b.   The time horizon of managers and professions in the group.

    c.   Their orientation toward goals, either diffuse or concentrated.

    d.   Their interpersonal style, either relationship or task oriented.

Without an overly long description of their findings and conclusions, it is fair to say that these four dimensions which create <u>differentiation</u> require mechanisms for <u>integration</u> if the units are dependent upon one another for work accomplishment.

One brief example will serve to illustrate their thesis. If manufacturing operations and manufacturing engineering are compared, some crucial differences become evident.

Manufacturing typically operates with high reliance on rules, set procedures, and many levels of hierarchy. Time orientation is short; if machines fail or stoppages occur, the impact is felt immediately. Performance feedback is clear and swift. Goals are generally clear and precise: so many units per hour or day. Relationships are task-oriented, and less interpersonal energy is expended on participation in decision making, or on whose feelings are likely to be ruffled. The environment has a high degree of certainty.

Engineering, while operating with rules and procedures, usually demonstrates more flexibility to adequately respond to uncertainty or emergent needs. There tends to be fewer levels of hierarchy; the pyramid is flatter. The time orientation is somewhat longer, and it may take months or even years to determine the success of their per-

7 Lawrence, Paul R. and Jay W. Lorsch, <u>Developing Organizations: Diagnosis and Action,</u> (Reading, Massachusetts: Addison-Wesley Publishing Co., 1969).

formance. That is, feedback is delayed. Goals are more diffuse, abstract and less easily quantified. While task orientation is high, more effort must be expended on assuming interpersonal relations are satisfactory. The environment has a moderate level of certainty.

Obviously, other sub-units with organizations such as R&D, Marketing, Personnel, etc. have even more widely disparate differences.

Given these different orientations, the sub-units are expected to integrate their activities towards some common overall mission. To further complicate the picture, manufacturing and engineering need one another; that is, they are highly interdependent. One cannot achieve its unit's goals without the cooperation of the other.

To assure that this integration is accomplished, organizations create "managers" or "coordinators" whose central job function is to smooth this potentially rough path. Other mechanisms are employed, such as joint meetings, control forms, and management information systems to facilitate this required interdependence.

While all of the above may seem obvious, it is infrequently articulated by the sub-units, despite the fact that both individual and sub-unit rewards are distributed according to how well sub-unit goals are met. Not surprisingly, acrimony and ill-feelings develop. The manufacturing people are perceived by engineers as "too conservative, reactionary, short-sighted" and a host of other perjorative adjectives, while manufacturing perceives engineers as "falling in love with the robot, dreamers, unsympathetic to the minute-to-minute problems of production." In one instance, the advanced manufacturing engineer attributed robot failure in these proportions: 5% to the robot, 0% to the worker and 95% to the indifference of the operations supervisors and management. The operations manager claimed the robot didn't fit the application, the rejection rate was 50-60%, according to him, even the "lab" people couldn't keep the robot running.

It doesn't take long for these disparate views of organizational reality to harden into differences and conflict. Unfortunately, the means of handling the conflict often becomes forcing, pushing tough decisions "Upstairs", or using political clout to achieve sub-unit ends.

2. Manufacturing Management as a Conservative Activity

Ivan Vernon anticipates the future manufacturing organization and the skills required to manage it. Among the points he stresses is the potential loss of the manufacturing manager's autonomy. "Indeed, he may become just another member of the team and perhaps not the most important one."[8]

No one yields power and authority willingly, and faced with this probability, it is predictable that management will try to install robots without disturbing the existing hierarchy.

However, authority will need to be vested with those with knowledge rather than on hierarchical position. As operators are trained and become skilled workers, their level of knowledge is likely to exceed that of their supervisors. Supervisors can be expected to resent this threat to their authority, and may be prone to watch the robot fail, welcoming a return to manual operation. The probability of this is increased by the fact that the supervisor's rewards are based on getting the product out, not on enhancing the sophistication of the operation.

Furthermore, robot technology is still primitive. As application opportunities grow, and new generations of robots appear, the likelihood of temporary disruption in operations will increase.

Recent research in the Social Sciences reveals that of all the cherished assumptions about change in organizations, only three can be correlated with actual results. While this is somewhat disheartening to social scientists, it is worth noting that two of the three hypotheses which receive at least low-to-moderate support are both applicable to robot installation. 1) Collaborative strategies are more successful than unilateral efforts; and 2) strategies in which the change agent has a participative orientation (as opposed to an "expert" orientation) are more successful.

The "change agent" may be the manufacturing engineer, and success- ful implementations often reveal a growing rapport between the engineer and the worker. This participation once encouraged and enjoyed, is likely to become an expected part of the job. To the extent that this occurs, the traditional boss-subordinate relationship will be endangered. The inroads into established chain-of-command relationships is not hardly likely to sit well with those in authority.

REASSESSMENT OF MANAGEMENT MEASURES AND STANDARDS

While industrial robots will provide one of the means to increase manufacturing productivity, management-by-exhortation will not make it so. There is a clear need to diagnose the impact of the robot on existing measures and standards applied to sub-unit management.

8
Vernon, Ivan R., editor, Organization for Manufacturing, (Dearborn, Michigan: Society of Manufacturing Engineers, 1970), pages 238- 239.

For example, an Operations manager who yields to the need of the engineer for time to de-bug, train, reprogram, etc., the robot and its operator, will surely be penalized in terms of labor costs, meeting schedules, avoiding rejects and down time. The Operations manager, in fact, is being asked to perform against his own best interests. On the other hand, the advanced manufacturing engineer who never suggests innovations, nor attempts new models to speed productivity, nor interrupts schedules, will just as surely be seen as useless and non-performing by the organization.

Asking managers to act in a fashion which is "good for the organization," but denies their own self-interest is sure to lead to self-justification and rationalizing behavior. On the other hand, that manager who is open enough to express his self-interest will be particularly vulnerable to his management.

Thus, top management must demonstrate leadership which enables sub-unit managers to take a fresh look at the various stakes each has in the installation. Top management must create the climate in which their managers may jointly examine costs and benefits to their self-interest. Innovative solutions to integrate disparate goals will be time consuming and may appear, at first glance to be less "productive" than the more tangible measures usually associated with manufacturing.

"Thus the design of every production system provides an opportunity to explore options concerning the relationship between people, technology and economic costs. Manufacturing engineers are as much the designers of social systems as technical systems, and this point cannot be overstated. We therefore recommend that a fresh look be taken at the way manufacturing operations are organized, knowing full well that unique solutions need to be evolved which are appropriate to the circumstances of each situation."[9]

CONCLUSION

The knowledge gap between technology and the organizational and management sciences is nowhere more apparent than in the implementation of industrial robots.

9
 Fadem, Joe A., "Socio-Technical Approaches to Assembly Design,"
    (Dearborn, Michigan: Society of Manufacturing Engineers,
    1976), page 16.

While robotics technology accelerates, our grasp of management strategies for implementation lag behind. Oftentimes "failure" attributed to the robot or its tooling may well have been avoided if the management issues had been addressed and resolved. The "blind, one-armed bandit" is an easy target for displacing responsibility.

Perhaps it is time that we scrutinize the social systems of manufacturing with the same blunt objectivity we apply to the robot.

Presented at the Fifth International Symposium on Industrial Robots, September, 1975

An Industrial Robot Public Relations Checklist

By Robert Skole
McGraw-Hill World News (Sweden)

If industrial robots are to be accepted on a large scale by industry, robot makers and users must prepare to present robot information properly to the "public" -- workers, unions, mass media, managers, public officials, laymen. Until now, industrial robot informational activities have mainly been directed to specialists and by specialists. Now, a totally new and different "audience" must be approached and informed.

The informational check-list includes a wide range of points that industrial robot makers and users should keep in mind when preparing press and informational material. This ranges from the basic facts that should be in press releases to those people who should be invited to participate in press showings, but who rarely are -- union officials, occupational safety officers, and ordinary workers.

Using this check-list as a basic guide, industrial robot makers and users will be able to get the most accurate, complete and positive coverage. The author urges industrial robot makers and users to take greater initiative in informing the public about robots in order to gain a general understanding, acceptance and realistic enthusiasm. This will help avoid being put into a position of having to defend industrial robots against attacks and criticism based on what might be sincere fears -- fears of the unknown.

Public relations: gaining understanding and appreciation

Industrial robots -- because of their unique abilities and their very name -- are emotion-laden machines. At least, to the public. As a consequence, public relations activities involving industrial robots must be approached and executed with more than normal thoroughness and sensitivity.

It requires very little imagination -- especially in view of the current depressed world economy -- to predict what fearful, negative reaction there would be if public information about industrial robots were not handled properly. We are all familiar with the "Luddites" violent attacks on the first automatic looms in England, and the modern fictionalized version in the classic Alec Guiness film "The Man in the White Suit". French workers used their wooden shoes, sabots, to destroy machines, and we have the remains of this action in the word "sabotage". With intelligent informational activities, we may avoid the introduction of a new word: robotage.

One cannot expect the public to understand and appreciate the many benefits that can be derived from industrial robots unless these benefits are explained clearly and thoroughly. By "public" I mean a very wide range of audiences and groups: workers in the plants and industries involved with specific robot installations; trade unions; local press, radio and TV; national mass media; trade and technical press; environment groups; governmental labor agencies, on national and local levels; schools; universities; technical and engineering societies; business groups; occupational health specialists; women's activist organizations; minority groups; the list can go on and on. Public relations activities -- which are basically informational activities -- must be carried out by both industrial robot makers and users in order to win the best possible appreciation, understanding and support of the "public".

The following check-list is not a complete "how-to-do-it" text. But by referring to it, the major points will be covered.

1) Bring in the workers

Workers must be placed first. It is essential that workers be brought into robot applications at the earliest possible time. This is so obsious that it is usually overlooked.

Worker involvement in production development or planning varies greatly from country to country. But when it comes to robot applications, there should be no variation: workers involved must be given the greatest opportunity to participate, even in those nations where the concept of industrial

democracy or co-determination is virtually inknown today.
When companies first start to consider buying and installing
robots, they should bring trade union officials and workers
into the earliest descussions.

The reason for this is simple, and I will modestly call
it the fourth law of robotics, to be added to Isaac Asimov's
classic Three Laws. This "Skole's Law" is: A robot may take
a human's job, but never make a human jobless.

The fear of being put out of work must be allayed at
the start -- and that's why workers must be brought in when
robot discussions begin. Indeed, this could occur at the
time when a robot manufacturer introduces a unit to a prospective customer: the maker should request that worker representatives be allowed to participate. This is highly unconventional, but robots are highly unconventional machines.

Once workers, their unions and works councils are
convinced of the great benefits that can be gained by the
employees themselves through the introduction of robots, a
major step has been taken in a positive public relations
program.

In those companies where there are works councils or
workermanagement consultative groups, it could be a good idea
to organize a special committee -- or sub-committee -- to
study robot applications. The more people seriously involved
with robots, the better.

2) Clarify the benefits

Benefits resulting from installation of robots might
be very obsious to a production engineer or plant physician,
but could be overshadowed completely in the view of a worker
who feels his job is threatened. Robots certainly can and
will take jobs from humans. But responsible managers will
see that no worker is put on the street because of a robot.
The thousands of jobs that can be taken over by robots are
those that are not fit for humans.

Workers whose hands are not threatened by presses because robots' grippers are tending presses will be able to
use their hands for safer tasks. Workers whose lungs are
saved because robots take over tasks in poor air will be
able to breathe fresh air in control rooms. Positive points
like these must be clearly presented to the public. The
benefits of robots are not automatically understood -- they
must be explained carefully, thoroughly, and honestly. Once
workers and their unions and works councils understand the
great benefits that can be gained by workers themselves
through introduction of robots, a major step has been taken
in a positive public relations program.

3) Plan for new jobs

A vital point in public relations involving robots is early planning for new employment opportunities. When a robot installation is first considered, responsible managers, working with employees and unions, should start making plans to reassign or retrain workers so that no-one is laid off. This is essential to gain early acceptance of robots by workers. A manager who moves toward a robot installation while leaving workers in the dark about their future is making a serious mistake that will be extremely hard to correct.

When information is eventually presented to the public that a robot installation is underway, or in operation, it will be extremely valuable to be able to say specifically what the new, safer, healthier, more responsible tasks that workers are doing. (Of course, in a brand new factory, where robots do not affect workers because no worker had ever been hired to handle the job, this whole problem is avoided.)

4) Inform the "outside public" early

Only after those workers directly involved have been brought into the robot picture, and are themselves fully aware and convinced of the advantages of the robot and its applications, should the public relations program be directed to the "outside". Those directly involved with the robot -- the workers, union officers, shop safety stewards, foremen -- will be the most important authorities to introduce to the public when the robot itself is introduced.

The first to be informed of the robot installation that's planned -- after those directly involved -- should be the other workers in the plant, or company. They should be told -- through regular in-house communications channels -- the reasons for the robot installation, how it will benefit some specific employees, and how the workers involved feel.

At this very early stage, the local press should be informed of the planned installation. This will enable the managers -- and workers -- to get the full facts out to the public before stories start to circulate about "robots taking jobs". Naturally, the benefits -- by the workers and the company as a whole -- should be stressed in these early announcements of the robot installation investment.

5) Press showings: make them worthwhile

When a new robot is to be presented publicly for the first time, or when a new robot installation is finally in full operation, "press showings" should be arranged. These should be held only after workers, trade unions, safety officers and other directly involved have been fully informed.

Whenever possible, a number of different press showings should be held, in order to satisfy the interests of various publications. The trade and technical press, union press, general mass media, radio and television, all have different interests and requirements. For example, spending time explaining full details of a robot control system will be worthwhile to technical reporters, but not especially valuable to the "popular" press or to labor reporters.

Press showings should be held to demonstrate the robot -- not to give the managing director or production chief a chance to make a speech. A little planning and effort in handling the press showings will produce excellent results. Press releases, photos, full background material should be readily available, and all specialists involved with the robot or installation should be introduced and prepared to answer questions or provide details.

Most essential, union representatives or workers involved with the robots should be invited to participate and should be introduced to the journalists. They should be given an opportunity to tell frankly how they experience the robot and the installation. Union officials or union and government occupational health and safety officers could also be invited to attend showings of new robot models. They could serve as valuable "outside experts" to provide an added dimension to the journalists' reporting. Bringing in union people and "outside" specialists at press showings is an essential point that unfortunately is generally overlooked. A worker who can stand up and tell the journalists that the robot is a great job improvement -- from safety, health, or job satisfaction viewpoints -- is far more impressive than any expensive and colorful sales brochure or articulate statements by the robot sales manager or plant production chief.

Press showings of robot installations should always include demonstrations or fotos or films of how a process or operation was carried out manually. (This, of course, is difficult for a new, multi-purpose robot that has not been installed. However, it can be done, even if a before-after demonstration must be arranged as a "sample.") Do not assume that journalists are familiar with all industrial production methods. Showing a robot in action is only half-telling the story. The full impact of the robots is clearly seen only when an operation by a robot can be compared against a manual operation

It might be pointed out here that robot manufacturers -- as well as users -- should make efforts to take movie films of operations carried out manually, before a robot is installed. These films can later be used to illustrate the benefits that were obtained. (There is a serious lack today of before-after film records of robot installations. Industrial historians in the future will be most unhappy.)

6) Press releases: make them informative

Press releases -- the written material presented to the press -- should tell the full story of the specific industrial robot or the robot installation. These releases should answer the basic journalistic questions: who, what, where, why and how?

This may seem like a most obsious point. But very rarely is press material really complete. There are generally gaping holes in the factual material, or unclear writing, or many basic questions left unanswered. A way to avoid much of this is to let an outside public relations specialist prepare press material. Engineers and technicians and managers get too close to their subject to write about it clearly. For example, I have never seen an industrial robot press release that includes a brief definition of exactly what is an industrial robot. People in the robot field know, and so do most technical writers, but few non-specialized journalists know, and these are most important.

Several different and complimentary press releases are usually better than only one. The first could be a general description, including all essential facts, and explaining technical points in plain language understood by non-specialized journalists. (Probably only one journalist in one thousand knows what "degree of freedom" means. This term -- and others - would have to be explained.) Another press release could be intended for the technical press. Here, more complete details of the robot or installation would be included. A third release could be intended for the trade union press, and here, labor questions should be answered.

A vital point: all press releases should include quotations or statements by workers or union leaders and occupational health and safety officers. This human aspect must not be ignored.

7) Illustration: make it illustrative

There is nothing as uninteresting as a picture of a dead robot -- at least to people other than robot specialists. For public relations purposes, always illustrate the robot -- or robot installation -- alive, in action. This is not really difficult. It requires only a little extra effort and imagination.

Movements of the robot arm and hand can be captured on a photograph through multi-exposure techniques that any competent photographer can handle. Best of all, a robot should be photographed in an actual installation. Before-after type photos should be offered to the press: before showing the actual manual operation, and after, the robot handling the job.

Do not be afraid to put humans into the picture. After all, humans are needed to supervise or maintain robot installations. If a robot installation reduces the work-force assigned to a job, this can be clearly state in the picture captions that workers have been assigned to better, safer, or more satisfying jobs in the plant.

Television, of course, can be an excellent media for informing the public about robots. To allow television to make most informative programs, however, requires that the TV reporters be advised of the robot installation well in advance. This will enable them to film workers handling a job manually, and later come back and film the robots handling the job. Naturally, workers involved should be given full opportunity to tell how they experience the robots.

8) <u>Select the right time</u>

Timing of introduction of new robots or showing robot installations is important.

Presenting a new unit for the first time at a major national or international trade show is not always best. Remember that your "news" will be competing against many other new machines or systems, and yours might be over-shadowed or over-looked. Journalists covering trade shows have limited time for reporting and limited space in their publications. Prospective customers at a trade show do not care if a unit is being shown for the very first time. If they haven't seen it before, it's new to them. If they read about the unit months before in the trade press, and are interested, they will not want to miss it at the show.

Greater publicity can be obtained by introducing robots at other times than trade shows. There does not have to be any special event linked to an introduction. However, there could be related events that fit nicely into the robot field. For example, if a trade union is holding a congress and a major discussion topic is health and safety, there could be no better time to introduce a robot or robot installation that offers decided occupational health advantages to workers.

9) Reaching others

As part of the public relations program, efforts should be made to tell the robot story to various groups. The listing of ideas is wide: school classes can be invited to see the robot in action; workers' families can visit it during plant "open house"; films or slide presentations could be available for loan by workers and managers for showing at civic clubs or church groups; special visits might be arranged for members of local technical societies; presentations of the installation or robot can be prepared for trade unions; occupational health officials' associations could be invited to study the robot installation.

The list of activities goes on and on.

The vital point is that all involved in robot development, manufacture and application must make every effort to spread information about robots.

Listed here are only some basic points. There are many details not taken up, and specialists in the industry have clearly demonstrated that they do not lack imagination. Many more ideas and projects for informational efforts should emerge.

The initiative and efforts devoted to informing the public about robots will help gain general understanding, acceptance and realistic enthusiasm. This will help the industry avoid being put into a position of having to defend industrial robots against attacks and criticism, based on what might be sincere fears -- fears of the unknown.

CHAPTER 5

CAPABILITIES

Commentary

This chapter expands upon the basics of robotics and presents information regarding the specific capabilities and limitations of certain robots. It begins with the status of today's technology and an overview of current robot products.

The capabilities of non-servo, medium-technology robots; microprocessor-and computer-controlled robots; servo-controlled, continuous-path robots and servo-controlled, electric-drive robots are presented. These represent a fair cross-section of the available technology.

A brief look at some potential future capabilities is provided, as is some insight into the robotics research and development efforts of a major research organization.

Presented at the 1978 International Engineering Conference, May, 1978

"Robot Automation"—Today's Status

By William E. Uhde
Unimation, Inc.

The growth in the robot industry has been the result of several factors. The 250% increase in U.S. labor cost over the past 10 years while average robot cost increases were closer to 50% is an important one. However, the robot industry's endeavors to develop a series of products more adaptable to the user's needs and to provide more sophisticated automation systems in order to integrate robots into the manufacturing process are seen as the major factors. Much of the progress has evolved from previous experience. Some of the more recent developments and their implementation is the scope of this paper.

INTRODUCTION

The past few years have shown substantial growth in the robot industry, not only in numbers of machines but in increasing sophistication of the applications selected. This has been due to confidence in the reliability of the robots and to the increasing awareness of the true capabilities of the machines.

Generally, the purchaser of a first robot selects an application that not only has economic return but exercises only the basic functions of the robot in order to first become acquainted with the machine.

This was particularly true in early die casting removal operations where the user envisioned future applications development but initially used the robot to only unload castings from one or two machines.

Die Casting - Figure 1

A substantial shift has occured in the die casting industry. The escalation in the price of zinc a few years ago forced a shift to the use of aluminum which accelerated a trend twoard automatic ladling of metal in cold chamber units and an increase in the tonnage capacity of the new machines delivered.

Several things affected robot utilization. The larger tonnage machines require robots of increased strength and linear stroke in order to manipulate large castings at high speed. Cycle times have increased also, allowing the use of slower, less sophisticated units for part removal without prohibitively affecting total production output. However, this also increased the market for the robots with greater capability, as the "lead it by the hand to teach it" methods allow quick setup for trimming or other secondary functions and die care lubrication programs have been developed to help cool specific problem areas in these larger dies. Long time users now find value in utilizing the robots for insert loading in multiple cavity dies and also find that the removal of 65# transmission case castings can be performed more quickly automatically than with manpower. Large castings must sometimes be manipulated carefully around obstacles in the die area or the tie bars, thereby requiring 4 or 5 axes of motion. Many die care programs also require 4 to 5 axes of motions to reach all die projections.

A major benefit was realized when a large die casting facility with approximately 50% of its die casting centers automated with 40 Unimate industrial robots was able to maintain just under 40% production during a strike.

Investment Casting - Figure 2

The modern investment casting facility is a relative newcomer to the industrial scene and the industry is growing rapidly. Wax patterns are processed through a series of slurry and sanding operations to build a rugged shell to serve as a mold during the metal casting operation. Previously, manpower had been considered indispensable for the delicate manipulation during dipping and sanding operations. As many as ten such operations are necessary to form a shell around the wax pattern. However, with the weight that can be processed by manpower restricted, other means had to be developed to increase the output of the shell building area. The introduction of robots capable of manipulating several hundred pounds at variable speeds and with carefully controlled acceleration and deceleration characteristics has increased the productivity of shell building by a factor of 5-8 times. Even where individual shell weights cannot be increased due to characteristics of the product to be made, multiple shells can be manipulated at one time to increase the throughput at a location.

Several benefits have resulted to the investment caster. He has been able to penetrate new markets which previously had been restricted to sand casting techniques. The efficiency of the shell building areas has increased dramatically as more useful shell weight per linear foot of floor space is processed. Delicate parts are now being run under carefully controlled shell building programs by utilizing the robot as the manipulative device with a resultant decrease in defective shells.

The majority of investment castings are built in small lots. The intelligent robot, possessing decision making capabilities, can identify and process each lot individually. Process controllers and computers can be employed to provide the backup information to the robot to enable it to construct a good shell around the mold. This enables each investment caster to decide what technique is to be employed to build a shell, and to exercise complete control over the method chosen. The robot gives the investment caster this capability with a standard maintainable product at a very reasonable price. It also provides a building block approach which allows the user to determine the extent of automation used at any phase of his expansion plans.

Future problems to be faced as the production volume increases will be the manipulation of the shell during pouring and finishing operations. One function of the investment casting process is to provide a part that requires little additional finishing. However, gates must still be removed and some parts will require machining. In future years the robot will expand into these areas as labor content goes up.

Machining Operations & Group Technology - Figures 3 & 4

Considerable labor goes into machining operations in the United States each year. High volume parts are machined on transfer lines. However, more than 75% of parts manufactured fall in a category that does not allow transfer line techniques.

Industrial robots are now employed in machining centers to load and unload parts in parallel or consecutive machining operations. The idea was to avoid the reliability problems of long transfer lines. It also provided flexibility to run a series of related parts through the same machines.

An example is a ring gear machining operation where 6 different parts are to be run in controlled lot sizes. Parts are run from machine to machine without utilizing intervening storage areas. The limitations of this approach are that the machines in the work area must be relatively well matched with regard to output and reliability and that tool changing requirements are kept to a minimum.

Unimation Inc. explored the feasibility of employing the robot in a flexible transfer line. Here as many as four robots could be assigned to nine or ten machine tools, thereby optimizing the group technology approach. Not only would direct labor be eliminated in the manufacturing process but in-process inventory could also be reduced.

As previously described, the robot's automatic program selection capability allows it to respond to processing conditions on the line. In a parallel machining operation where all machines in the work center perform the same operation, an interruption in one machine tool's cycle would simply eliminate that unit from the work process. The robot would automatically continue to service the machines still in operation. The throughput would be reduced temporarily. This work center is separated from upstream and downstream work centers by in-process storage facilities. These can consist of storage conveyors, towers, or pallets which are robot loaded. Part inspection can be handled by in-process gaging or automatic gaging which is robot loaded.

Auxiliary process controllers can be employed to monitor status of each work center and control the flow of parts from work center to work center. These units direct robot programs to change in accordance with the conditions on the line, thereby using the flexibility of the robot to position parts for the most orderly flow pattern. Where consecutive machining operations are organized into one work center, intermediate buffer storage is provided to allow for downtime at each machining station. The number of buffers and the number of parts stored at each is determined by the reliability and tooling of the machines, part lot size and total proportionate cost of operation of each machine in the work center.

Several of these systems have been installed and information gained from them will be used as models for future installations.

Forging - Figures 5 & 6

For several years various forging machines have been utilized with industrial robots. Upsetters, roll forges, programmable hammers, impacters and forge presses have been interfaced with robots in single die to three die configurations. Both forge die and hot trim die operations have been installed. Parts range in size from a few ounces to several hundred pounds.

The newer installations feature powdered metal technology in conjunction with the forging process. Thus far, the parts are small and light with the emphasis on robot control of the complete sequence. In one application, an environmentally controlled rotary hearth furnace is used to heat the parts. The robot loads and unloads the furnace thereby assuring orientation. The time the part is in the furnace is automatically interlocked with the robot.

The robot processes the part to the forge press where final compression and forming takes place. The part is automatically ejected from the forge die. Because placement of the part is critical the Unimate tooling is equipped with locators to assure proper placement of the part in the die. As in die casting, the exact repetition of the cycle through heating and forming operations assures consistent part structure and grain flow, and resultant high quality parts.

Welding - Figures 7, 8, & 9

Unimation Inc. has now constructed several automated spot welding lines for automotive frame and body assemblies. The system includes conveyorization, body buck support, process controllers, display panels, and weld guns as well as the robots. Up to 26 robots have been installed on one line.

Automatic program editing is an important feature of these systems. Sheet metal from various stamping facilities does not always match from lot to lot, and when joined, there is some variation in assembly location. The program editor is employed to modify the Unimate weld programs while the units are actually welding a body, thereby allowing a programmer to edit in changes based on observation of the weld being made. Dialing in positional changes at appropriate points in the program, the editor allows for changes in position in any axis of motion while making production.

The program so modified is transferred to cassette tape for storage off line. In case of robot memory loss or need to interchange one robot with another the program can be reestablished in the unit's permanent memory package.

This technology is now being added to by the addition of MIG welding continous path Unimates to the car and truck building lines, in combination with the spot welding units.

However, 80% of the weld wire placed in the U.S. is not amenable to welding by the standard continuous path robots because of two factors. One is that the work is too big, e.g. ship and locomotive assemblies, the other is that fixturization is not accurate because the lot size is too small to make the investment for orientation of product worthwhile. Unimation Inc. has developed the five axis Apprentice robot to meet this market. The machine is taught by leading it along the path to be welded. Welding parameters are set by a part time operator who adjusts amperage, voltage and linear speed of weld. If a vertical weld is called for the weaving parameters are also dialed into the control. These consist of frequency and amplitude of oscillation plus dwell parameters at midstroke and at the extremes. One operator can control up to 5 units and the 75# welding head can easily be transported from workplace to workplace.

The Continuous Path Unimate utilizes its Ram memory in conjunction with linear interpolation to provide 5 axis path control around tubular and irregular shapes. The machine is presently utilized for high volume families of parts such as motorcycle frames, super market baskets, and farm implement and automobile subassemblies. The programs are first generated by "leading it by the hand" and the information is recorded on backup cassette tapes for future use.

Assembly Operations - Figure 10

For several years Unimation Inc. has been working on the development of a series of assembly robots to work in conjunction with each other and with manpower. Basic characteristics of these machines are high accuracy and speed.

Two distinct concepts for their use are emerging from potential users. One concept is to utilize the substantial memory of the robot to build complete assemblies at a work station and duplicate these work stations as required. The other concept is to install robots of limited intelligence on an assembly project at each work station. The method chosen depends on the size and weight of product to be assembled, the volume in any given lot and the quality of components from which the assembly is made.

95% of an automobile's parts weigh less than 3 lbs. General Motors Corporation's Programmable Universal Machine for Assembly, nicknamed PUMA, will assemble a variety of assemblies. The robot model assigned to this task is being built at Unimation's West Coast facilities. The arm contains either 5 or 6 axes of motion, is accurate to .005" and is fast. The unit is all electric, control is by minicomputer and it has the capacity to direct future tactile and visual sensory feedback systems.

Generally, it is expected that programmable assembly robots will be utilized for annual production volumes of 200,000 to 3,000,000 pieces. The final range will be determined by the cost to the user for each installation and the flexibility the robot can provide. Rectilinear path control, even though the robot may have a polar coordinate structure, and program flexibility with regard to palletizing etc., and certain self teaching features once basic parameters are known all hold akey to the robot's future.

The Immediate Future

Not all the applications for industrial robots have been discussed herein. There are many different areas of application, some of them rather unusual, that have been installed by users, and are now providing economic rates of return. Some of these facilities resemble first generation CAM (computer aided manufacture) installations.

Japan is busily at work developing a prototype unmanned factory. This project has heavy government support. One of the benefits of this CAM

facility will be greatly reduced floor space. Also of considerable value
will be the optimization of in process component storage. Extensive robot
technology will be utilized at this facility.

Several different types of robots capable of tracking moving conveyors are
now being released for use in industry. The simplest applications require
offloading and reloading parts on those conveyors. More complicated are
those operations which require substantial work to be performed on the
parts while they are in the working sphere of the robot. These units need
additional computational abilities to work with parts that are not moving
at rigidly controlled rates.

One thing is sure. Robot workers will continue to fill onerous jobs which
are dangerous to human health or contain unacceptable working conditions.
The pace and scope of robot installations will increase as the newly
acquired skills of the machines allow easier adaptation to the work site.
The present is exciting and the future is even brighter.

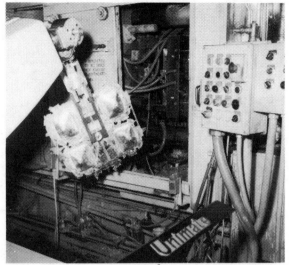

FIGURE 1.
Wrist Bend Motion Used to Increase Useful Stroke and
Manipulate Castings Through Tie Bars of a Die Casting Machine.

FIGURE 2.   Investment Casting Dipping and Stucco Operations
Using Rain Sand Equipment

FIGURE 3.   Dual Tooled Unimate at Lathe with Pallet Conveyor in
Foreground. Parts are Exchanged Utilizing Both Hands
at Lathe and at Conveyor.

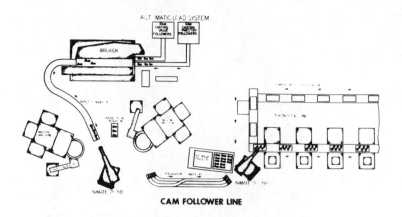

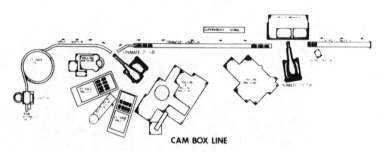

FIGURE 4. Processing Diesel Engine Parts with Two Lines
Sharing One Supervisory Controller. Robot Builder
Supplied Conveyorization, Storage Silo, Storage
Pallets and Controller.

FIGURE 5. Heavy Duty Unimate with 200# Preform on its
way from Press to Draw Bench. The Trough in
the Foreground is to Cool the Robot Fingers.

FIGURE 6. Powdered Metal Forging Operation Utilizi
Rotary Gas Furnace and 150 Ton Forge Pre
Dual Tooled Robot for Rapid Interchange
Parts at Furnace. Automatic Parts Feede
is Between Press and Furnace.

FIGURE 7.   10 Heavy Duty Robots on Automotive Framing Line.

FIGURE 8.   MIG Welding Tubular Motocycle Frames with
5 Axis Continuous Path Unimate Robot

FIGURE 9.   Apprentice 5 Axis Robot for Continuous MIG
Welding Operations on Heavy Structures.

FIGURE 10.   6 Axis Assembly Robot with Path Control
This Operation shows two Arms working in
conjunction with each other to make a
12 Part Assembly.

Presented at the Robots II Conference, October, 1977

What Can Medium Technology Robots Do?

By Walter K. Weisel
Prab Conveyors, Inc.

Over the past decade, industrial robots have played a major role in industry. The major industry usage has been automotive, in their body spot welding operations both in the U.S. and abroad.

The complexity of this type application requires that a robot remember several different body styles and when signalled, initiate the appropriate program and perform a series of complicated, six axis manipulations using a spot welding gun. Speed, dexterity and complex controls are a must.

The introduction of the same complex units into other industries has been slow by comparison. Manufacturing engineers have worked hard to use the high technology robot's capabilities to their fullest, mainly in an effort to draw out enough cost savings to justify the expenditure.

Meanwhile, the so-called simple pick and place units have gotten smarter, stronger, faster, and more reliable. In fact, they have taken on jobs, other than spot welding, by the hundreds and kept their controls and mechanical designs simple by using current, proven, and industry accepted hardware.

A major advantage for the "MEDIUM TECHNOLOGY" robot has developed in the past few years. The simpler robots have expanded their reach, speed and weight carrying capabilities to fulfill the requirement for jobs found in machine loading, die casting, plastics, forging, and material transfer. At the same time, the high technology robots have expanded their level of sophistication to meet the more complex application's requirements such as arc welding, spot welding, and continuous line-tracking.

The technologies have reached a point where today the prospective user can evaluate his application in terms of two medium technology robot arms servicing several machines in place of the one complex arm which must stand in position and try to do all of the work to justify its cost. If the idea of evaluating your application with a possibility of easily cost justifying two arms instead of one, then read on.

"What Can A Medium Technology Robot Do For You?" You'll be amazed.

After many years of agonizing over the term "robot", the thought of cat egorizing robots by technological groups may be quite distasteful to som dyed-in-the-wool roboteers. I will use the term "medium technology" to put certain capabilities of this type of robot into perspective. I also intend to show how its complex, high cost cousins can cause you to complicate your application, if you are not careful. Interesting? Read on.

The medium technology robot of today resembles the sophisticated robots of several years past. This comparison is destined to continue since the medium technology robot will continue to capitalize on today's accepted and proven hardware for its design. In most cases, the electronics and mechanics used in this type of device seem simple to your engineers and maintenance people. But more about that later, first a brief history of the industrial robot to put things into perspective.

Robots of 1977 are an accepted and proven method of automating manual operations in most industries. Usage of industrial robots has risen to the point where users and suppliers can make a good return on their investment, and this turn-around has increased the numbers of suppliers and users in the past several years.

By far the largest user of robots has been the automotive industry. There are an unlimited number of successful applications in automotive spot welding, and it is clear that the majority of the units sold, even today, are to the automotive industry both here and abroad. I believe that this trend will continue, however the simpler cousin to the complicated robot is making major inroads into the non-automotive industries where high speed machines with large memories are not the rule.

The medium technology robot exists, and literally hundreds of them have been employed by industry over the past five to eight years for applications in plastics, die casting, machine loading, forging, and material transfer jobs. They are often the least publized, so-called pick and place devices, and their role until now has been grossly underplayed.

Larger complex robots have received all of the attention in the past. Trade magazines and strong advertising campaigns have given the complex robots the spotlight, and rightfully so. A 10 or 20 robot line, welding car bodies in unison, is a spectacular sight; but how has this type of robot progressed in non-automotive applications?

By comparison in numbers to successful automotive applications, the high technology robot has problems in simple applications. The reasons are many, price, complexity, one at a time applications, lack of technical maintenance people, and the most interesting syndrome which I intend to introduce called "robot complexity dictates job complexity".

This sounds far-fetched, but in reality a complex robot applied to a too simple application will tend to increase the complexity of the job beyond the point of value. There are three points that should be discussed to clarify "robot complexity dictates job complexity".

1. The robot's complexity is dictated by the diversity of its applications base.

2. A robot's complexity dictates its cost.

3. A robot's cost can change the intent of your application.

The following explanation relates to statements 1 and 2. The basic criteria for a robot's design is to build a piece of general purpose automation, reprogrammable and non application dependent. In order to determine the base configuration of the robot, a study must be made of the various applications that you wish to apply it to before determining the proper reach, speed, weight carrying capability, memory size, and general floor layout required for each.

As soon as the robot designer takes in to account certain complex jobs, like multi-axis spot welding, arc welding, and continuous synchronous line tracking applications, the degree of complexity in control and mechanical design, as well as cost, dramatically increases. It's natural to try and design these features into your equipment since the complex application is where the action has been, and promises to be for the next several years.

It takes a very complex, high technology machine to remember three or four different body styles, each with a different pattern to be welded. These machines must also be able to wield heavy weld guns in and out of car bodies on a continuous basis with high speed and accuracy. This is a rigorous assignment for today's line rates of 60 to 80 cars per hour.

From a pure application standpoint, it is relatively easy to perform simple tasks like die casting, machine loading, plastics, or material transfer applications for the sophisticated units. In most cases, applying these complicated units to these types of applications is a gross overkill. Through the following example I will show you how the high cost of a complicated robot can complicate your application.

A manufacturing engineer employed by a die casting firm wishes to automatically extract die castings from a die cast machine. His company runs a two shift operation, and the shop labor rate plus fringes is $19,000 per year. Corporate guidelines dictate that a one year return on their investment is necessary in order for a cost savings project to be approved for purchase.

The engineer calls in a robot salesman who is selling a $40,000 robot to help analyze his job. The job is an easy one for the high technology device, and the cost of the robot plus installation and spare parts to maintain this machine come to a total of $48,000. Displacing two men (one per shift) with a single robot only gets the engineer a $38,000 per year savings, so he can't use this robot unless he incorporates more labor savings than that found in just extracting parts. However, the complex unit has the ability to unload two die cast machines and possibly introduce trimming if it is required. This gets more labor displacement into the project and helps cost justify the higher technology device.

Unfortunately, the die cast machines are located on 25' centers so it is necessary to move one of the die cast machines to bring it within the reach of the single arm. As you can see, the application starts to get more difficult based on the robot's cost and its relative displacement of manpower.

In almost all simple applications, if you were to introduce enough work to surround the robot, you would eventually get the payout required. Many times you are trying to introduce new manufacturing techniques and relocation of equipment to justify the added cost, which also reduces the chances for a successful application.

The same example, with the engineer calling a medium technology robot supplier, would have gone more like this; yes, we can extract parts from the die cast machine, and the price of our equipment is $17,000. Our robot can also quench and trim the parts for the same price. Spare parts plus installation costs for the simpler unit brought the total project price to $18,500, so you can see this project had an excellent payout when compared to a $38,000 per year labor savings.

The application remained simple, conformed to the way the plant presently operated, and gave him the expandability, if he chose, to unload and trim or unload both machines at a later date. Had the engineer gained a real benefit by unloading two die cast machines, he could have purchased two medium technology robots at less cost, avoided relocation of equipment, and avoided the disadvantages of trying to operate both machines with only one arm.

All too often an engineer trying to apply a complex unit overlooks the fact that he could easily afford two medium technology robots for the same price. The thought of two for one is a most attractive option.

Many simple loading and unloading applications to start have been pushed into complex multi-machine loading, parts palletizing, in-process inspection nightmares, just to justify a complex robot's price and utilize its full capabilities.

"What can a medium technology robot do" so far?

1. It reduces initial cost.

2. Keeps the basic job basic.

3. Allows you to think in terms of two robots for one man (usually requires a two shift operation).

   Not bad for openers!

Medium technology robots consist of a basic mechanical and electrical control package. Most units are self-contained and can easily be moved from one jobsite to another. Programming is fast and uncomplicated, and they are adaptable to most jobs since they can carry payloads of up to 100 pounds and operate with a reach of up to 10 feet across its' sphere of influence.

The basic criteria for apply industrial robots remains pretty much the same whether you use a simple or complex unit. The following is a checklist that must be satisfied:

1. Are parts oriented in a known position?

2. Does the machine to be loaded/unloaded have adequate clearance for gripper and part?

3. Does the tooling have automatic clamping and unclamping?

4. If the tooling or chucks on your table rotates or indexes, does it return to an oriented position (spindle orientation)?

5. Does location of the part into the holding device require guide pins or chamfered leads to help nest the parts?

6. Does the machine have an automatic cycle?

7. Is in-process inspection or visual inspection on a repetitive basis required?

The following are only a few applications for medium technology robots. Wherever possible, I will point out what benefits were derived by using the simple over the complex units.

Plastics: Automating the plastic injection molding machine with a robot is quite simple. The part is pre-oriented and the machine cycles are long enough so that secondary operations can normally be performed. The robot shown in the attached photo (Figure I) is seen servicing one machine, although it had a long enough stroke to unload two plastic injection molding machines from one location.

Since the robot paces the machine and works tirelessly, 20% increase in production is not uncommon. The requirements for manufacturers to comply to OSHA regulations has made this a very attractive application for the medium technology robot. Equipment manufacturers are also evaluating robots from the standpoint of their own product liability.

Machine Loading: (Figure II) Machine loading applications represent one of the biggest market areas for medium technology robots in the years to come. As stated earlier, a two for one trade-off over the more sophisticated cousins opens up new horizons for tending machine tools. Often times, one operator tends two machines and cycle rate gets to be a major consideration when trying to service both machines with one arm. By splitting the assignment between two arms, you can keep the overall operation simple and insure 50% continuing production rate should one of the arms fail. Relocation of equipment is kept to a minimum since you have the flexibility of positioning two arms instead of trying to center everything around a single unit configuration.

The attached photograph shows a robot loading and unloading a Bullard. This machine has an indexing table with dual chucks in each station, and each part must make two passes through the machine to complete the machining cycle. The part must be turned over after its first cycle and inserted into the second chuck.

To describe the operation, I'll work it backwards. The robot picks up both parts at the same time from the dual chucks and deposits the finished piece in an output conveyor. The half finished piece is then positioned in the turnover device, which rolls the part 180°, while the robot moves to the buffering and orienting station for pickup of the green part. It then returns to the turnover device, re-grasps the part, and positions both parts into the dual spindles. After signaling the jaws to close, it removes its arm and signals the machine that it is clear for the next index cycle.

In order to justify the more expensive robot for the same application, it requires that you load and unload two machines with a single robot. One of the two operators must be maintained to load parts chutes and perform gauging and tool change operations. Cycle rate is critical and the complex robot usually runs at its maximum capability to service both machines with one arm.

A major advantage in applying two arms for one is that generally you don't have to relocate either Bullard. You don't sacrifice machining time on a second Bullard while a robot is being serviced or when a man enters the work area to change tools or service the machine tool.

Another example of successful machine loading at minimal cost is to load and unload transfer line equipment where pallets are pre-oriented and automatic tooling already exists. A two robot - $40,000 package has a fantasic payout when plotted against two men per shift/two shift operation for loading and unloading transfer type equipment.

Die Casting: (Figure III) Most robots that have been on the market for any length of time have been tried in die casting applications. This application is a real test of the equipment's reliability, since the die casting environment is a very hostile one, and most maintenance skilled tradesmen haven't had much exposure to technical equipment.

The part is oriented in the die when the machine needs servicing, and through simple detection circuits, the robot can tell if it has properly extracted the part from the die cast machine. Upon sensing that it has the whole part, it recycles the die cast machine and then begins secondary operations while another part is being cast. Most robots on the market today can cope with unloading and spraying if the user requires the robot to perform die lubing as part of the extract function.

Quenching and trimming the part is usually a desired part of the operation, however the part itself usually dictates the ability to automatically trim. A careful analysis should be made of a robot's accuracy if trimming is to be undertaken. Interestingly enough, the medium technology robot with its programmable stops and simple limit switch feedback has a higher degree of accuracy than its big, sophisticated cousin. Once programmed to do the job, it can repeat to within a few thousandths of an inch rather than the $\pm$ .050" of the more complex units. Accuracy is an important consideration if trimming is to be undertaken.

The concept of unloading two die cast machines with one robot arm is foreign to the normal operating process of a die cast facility. Very rarely do you ever see one operator unloading two die cast machines, and therefore machine spacing and job scheduling is planned around the one man-one machine concept. As I pointed out earlier, it is my feeling that the main reason people unload two machines with one robot is to justify the expenditure for the high priced unit. The medium technology robot provides an economical means for automating the die cast operation on a one machine - one operator basis.

Forging: Forging applications have typically been a real problem to automate with the single arm concept. Cycle rates are very high and generally the application requires transferring a part between several positions while worrying about the load and unload time on top of it. Here again, the two arm concept is a natural way to go.

One robot stands on one side of the press to unload and load parts while the other robot takes care of intermediate die transfer points. Simple interfacing between robots, pickup and discharge points, and the forging press make this an excellent application for the medium technology robot. I might add that sufficient time exists for the unloading robot to feed a trim press. Increases in production on a job such as this are dramatic, since over a period of an eight hour shift the fatigue factor plays a big part in operator production.

Stamping Presses: Traditionally stamping presses have had too high a cycle rate for the robot to cope with die to die transfer. Presses are spaced beyond the reach of most robots' capabilities, requiring the relocation of presses and ultimately the failure of a project due to cost considerations. Two medium technology robots operating between presses can be cost justified and insure a successful operation.

Heat Treating: The manual loading and unloading of heat treating furnaces represents an environment where labor turn over is high. This in itself provides a good basis for an initial application since labor acceptance of this type of device is quite good.

Medium technology robots are capable of handling the payloads and cycle rates associated with most heat treating operations. Since cycle rates are slow, secondary operations can usually be incorporated. The increases in production are most rewarding in heat treating applications, since operator fatigue is quite high when handling hot parts.

Future Application Trends: The medium technology robot will be taking on more tasks in a variety of industries as the engineering community becomes more familiar with its capabilities. As an example, the foundry industry has been using the simpler robot to remove hot castings from sand molds. Other foundry applications which could use a device of this type are core dipping, investment casting, and core machine unloading.

The medium technology robot is also handling projectiles in the munitions industry, and as a result of new safety regulations, many hazardous jobs will be automated in the years to come. The robot will play a vital role in automating these processes.

The potential for the simpler robot's application in machine loading is unlimited, when you stop to think that at today's labor rates for multi-shift operation, the manufacturing engineer can justify two simple robot arms to load and unload machines. More companies are specifying automatic handling on their machine tools, and there has been a renewed interest on the part of machine tool manufacturers to use robots rather than go through a first time design and build for every automatic loader. The cost conscious machinery supplier is finding that the medium technology machine keeps him competitive when quoting against machine tool suppliers that quote the high technology units. You can look for the machine tool suppliers to put a heavier emphasis on the simpler robots in the years to come.

Maintenance: This is an area where the differences in robot technology affect your operation and your application. The response time required to maintain the robot in a production application is critical to the success of the job. Spare parts and training are requirements, regardless of the type of robot purchased, and you should not underestimate the importance of maintaining the equipment.

One of the major differences between the high and medium technology robots is the type of components used in their construction. Typically, you will find that the high technology robot uses many special components, both in the mechanical and electrical control systems. An inventory of spare parts is essential, and often times the cost of spare parts will range from $5 to $10,000. The high cost of spare parts could prevent you from having a cost justified project, and the natural tendency is to omit them. Don't!

Most medium technology robots are made up of standard components, and the valves, solenoids, switches, and logic components can be found locally. In some cases, your parts crib may already be stocking these components. Your shop maintenance people already understand how to troubleshoot most devices found in the simpler robots. This will have a very positive impact on robot acceptance.

Summary: The medium technology robot can be applied to many existing applications in industry today. As manufacturing engineers learn more about the medium technology robot's capabilities, its application base will expand in most industries.

I have shown how the medium technology robot can keep your operation simple and insure its success. The limits of its application are in the minds of the production minded engineers who are trying to apply them. Its future is in the hands of the supplier's designers who will always be tempted to add just one more option, or expand its capability. They must be extremely careful to weigh this additional option against the value of keeping the device simple.

Figure 1

Figure 2

Figure 3

Robots don't just handle things—they do things too

If you think robots are only good for picking up widgets at point A and moving them to point B, think again. Armed with simple tools, robots become willing workers that will boost the value added in tiresome, everyday, repetitive operations.

By JOHN MATTOX
*Application Engineering Manager
Unimation Inc., Subs., Condec Corp.
Danbury, Conn.*

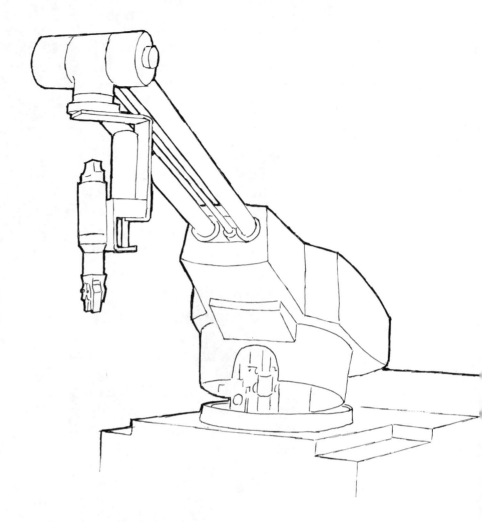

Clamp installation tool

In the assembly of formed-wire springs for automotive seats, industrial robots apply crimp-on fasteners that clamp two parallel wires together to permanently assemble the spring sections. Special fixtures are needed to preposition the components and shuttle them, stop-and-go fashion, past the robot. A tubular feed arrangement, suspended from overhead, supplies clamps without restricting movement of the application tool.

Workers are increasingly less tolerant of dirty, unpleasant jobs, regardless of the payrate. Higher levels of education and growing aspirations are leading to an insistence on employment that is challenging. Unfortunately, tool-handling jobs in production operations frequently call for a repetition of the same sequence, hour after hour. And workers assigned to these jobs soon go through the required actions almost automatically.

But when a worker's job becomes "automatic," it also becomes more difficult—more difficult for the worker to keep his attention from wandering, and also more difficult for the production engineer to control quality and production rates. What is often easier, *and* more economical, is to put a robot on a tool-handling job rather than try to obtain robot-like perfection from a worker. Industrial robots don't tire, become bored, shun unpleasant jobs, or make the errors that humans make.

Typical of the tools that can be handled by general-purpose industrial robots is the spotwelding gun. Many industrial robots with six-axis positioning control are being employed to maneuver spotwelding guns through complex welding programs.

Jobs such as welding, for which hand-eye coordination or a tactile sense might at first be thought essential, prove to be suitable for industrial robots because simple substitutes for costly-to-duplicate human abilities are often available or can be devised.

A few of the tools discussed below are not yet in use with industrial robots. But the tools will come into use as production engineers find opportunities to match existing technology to applications in their own plants.

Stud-welding head

Equipping an industrial robot with a stud-welding head is also practical. Studs are fed to the head from a tubular feeder suspended from overhead. One caution concerns accuracy with which welded studs can be located. An industrial robot can position a stud within 0.050 in., but on-the-line work positioning must be exact. The weight of the head is rarely a significant limitation. Stud-welding heads are well within the 100-lb capacity of standard robots.

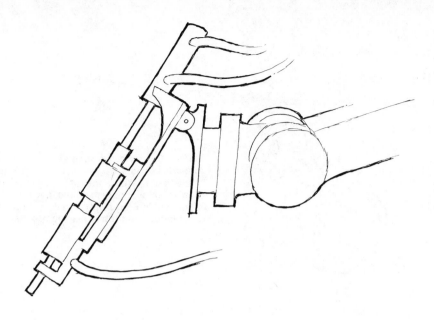

Inert gas arc welding torch

Arc welding with a robot-held torch is another application in which an industrial robot can take over from a man. The welds can be single or multiple-pass. The most effective use is for running simple-curved and compound-curved joints, as well as running multiple short welds at different angles and on various planes. Maximum workpiece size is limited by the robot's reach, unless the robot is mounted on rails. Where the angle at which the gun is held must change continuously or intermittently, the industrial robot is a good solution. But long welds on large, flat plates or sheets are best handled by a welding machine designed for that purpose. In addition to welding for fabrication purposes, wear-resistant surfaces and edges can be prepared by laying down a weld bead of tough, durable alloy. And the robot will handle a flame cutting torch with equal facility.

Heating torch

The industrial robot can also manipulate a heating torch to bake out foundry molds by playing the torch over the surface, letting the flame linger where more heat input is needed. Fuel is saved because heat is applied directly, and the bakeout is faster than it would be if the molds were conveyed through a gas-fired oven.

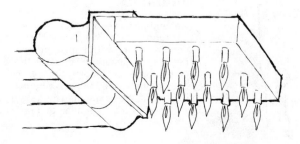

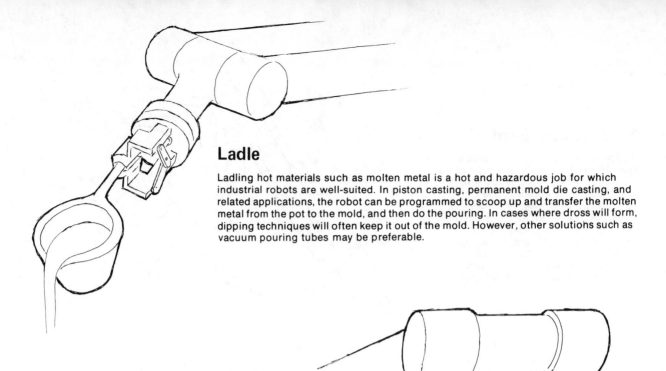

Ladle

Ladling hot materials such as molten metal is a hot and hazardous job for which industrial robots are well-suited. In piston casting, permanent mold die casting, and related applications, the robot can be programmed to scoop up and transfer the molten metal from the pot to the mold, and then do the pouring. In cases where dross will form, dipping techniques will often keep it out of the mold. However, other solutions such as vacuum pouring tubes may be preferable.

Spotwelding gun

A general purpose industrial robot can maneuver and operate a spotwelding gun to place a series of spot welds on flat, simple-curved, or compound-curved surfaces. In production line operations on appliances or auto bodies, stop-and-go rather than continuous line motion is preferred. Otherwise, weld placement accuracy suffers because the robot must track a moving target as well as place the welds. When the time available is too short for one robot to make all the welds within its reach, the number of welds can be divided among two or more robots, as is done in the automotive industry. Similarly, if all of the welds are not of the same type, there must be a different gun and so a different robot for each. The robot can position welds within 0.050 in., but the line must position the work accurately.

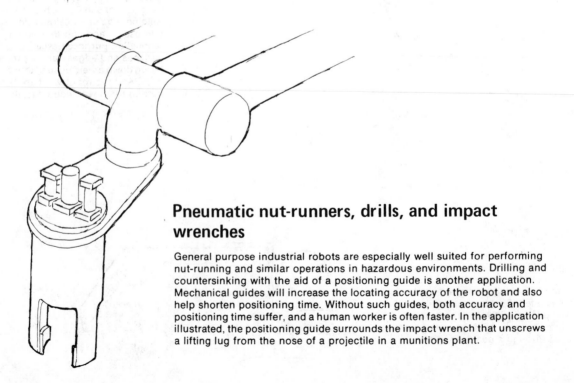

Pneumatic nut-runners, drills, and impact wrenches

General purpose industrial robots are especially well suited for performing nut-running and similar operations in hazardous environments. Drilling and countersinking with the aid of a positioning guide is another application. Mechanical guides will increase the locating accuracy of the robot and also help shorten positioning time. Without such guides, both accuracy and positioning time suffer, and a human worker is often faster. In the application illustrated, the positioning guide surrounds the impact wrench that unscrews a lifting lug from the nose of a projectile in a munitions plant.

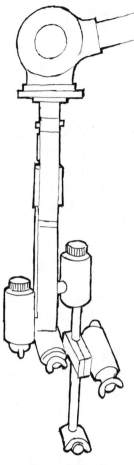

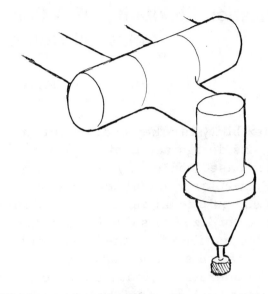

Routers, sanders and grinders

A routing head, grinder, belt sander, or disc sander can be mounted readily on the wrist of an industrial robot. Thus equipped, the robot can rout workpiece edges, remove flash from plastic parts, and do rough snagging of castings. For finer work, in which a specific path must be followed, the tool must be guided by a template. The template is a substitute for the visual—and sometimes tactile—control that a human worker would exercise. In such a case, the overall accuracy achieved depends upon how accurately the workpiece is positioned relative to the template. Usually, the part is automatically delivered to a holding fixture on which the template is mounted.

Spray gun

Ability of the industrial robot to do multipass spraying with controlled velocity fits it for automated application of primers, paints, and ceramic or glass frits, as well as application of masking agents used before plating. For short or medium-length production runs, the industrial robot would often be a better choice than a special-purpose setup requiring a lengthy changeover procedure for each different part. Also, the robot can spray parts with compound curvatures and multiple surfaces. The initial investment in an industrial robot is higher than for most conventional automatic spraying systems. When the cost of frequent changeovers is considered, the initial investment assumes less importance. Industrial robots can be furnished to meet intrinsically safe standards for installation in solvent-laden, explosive atmospheres.

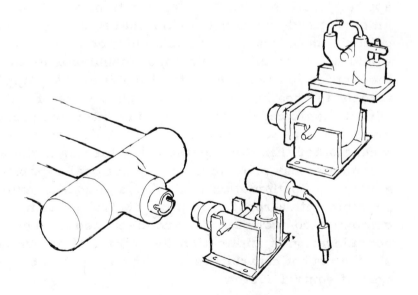

Tool changing

A single industrial robot can also handle several tools sequentially, with an automatic tool-changing operation programmed into the robot's memory. The tools can be of different types or sizes, permitting multiple operations on the same workpiece. To remove a tool, the robot lowers the tool into a cradle that retains the snap-in tool as the robot pulls its wrist away. The process is reversed to pick up another tool.

Presented at the First North American Industrial Robot Conference, October, 1976

Application Flexibility Of A Computer-Controlled Industrial Robot

By Richard E. Hohn
Cincinnati Milacron

Application flexibility is a key concept for an industrial robot. The robot configuration, its design and control system all contribute to this concept, and a computer based control system greatly enhances this application flexibility. Particularly useful features of a computer in such a system pertain to its computational and logic capabilities. The computational capability can be utilized to solve the transformation equations required to coordinate the motion of all the axes, including control of the path velocity, acceleration and deceleration. This coordinated motion allows for a number of operational features not commonly available with industrial robots, such as straight line path movement between points, ease of programming, tracking and operating on moving objects, and zero-shifting a stored program. The logic capability enables many other operational features to be realized, including branching and conditional functions. These and other features and the manner in which they relate to applications are discussed.

INTRODUCTION

An industrial robot, by its very nature, is intended to be economically applied to a broad range of applications with a minimum of special engineering and debugging. This characteristic is one of the important criteria an industrial robot should meet. Without this flexibility, it would not be feasible for a robot manufacturer to economically meet the application requirements of many users. Application flexibility of a robot is greatly enhanced through the use of a computer. The computer, with its logic capability and ease of interfacing external devices, allows many features to be easily implemented. However, while there are means other than a computer for achieving many of these features, only a computer can provide the computational capabilities required for a new type of path motion control system which expands the applications possible with a robot. Both the control system and features of a computer-controlled robot are discussed as they relate to the important consideration of application flexibility. Performance characteristics of other robot control systems are discussed and related to this new type of control system.

PRESENT CONTROL SYSTEMS

Most present industrial robots which have an easily alterable means for
storing a taught sequence of motions and functions fall into two classifications based upon the manner in which the robot motion is controlled.
The two classifications are Point-to-Point and Continuous Path.

1. <u>Point-to-Point:</u> This type of control is perhaps the simplest
and most frequently used control method. Teaching is done by
moving each axis of the robot individually until the combination
of axis positions yields the desired position of the robot end-
effector. When this desired position or point is reached, it is
programmed into memory thereby storing the individual position
of each robot axis. In replaying these stored points, each axis
runs at its maximum or limited rate until it reaches its final
position. Consequently, some axes will reach their final value
before others. Further, because there is no coordination of
motion between axes, the path and velocity of the end-effector
between points is not easily predictable. For this reason,
Point-to-Point control is used for applications where only the
final position is of interest and the path and velocity between
points are not prime considerations.

2. <u>Continuous Path:</u> This type of control is used where the path
of the end-effector is of primary importance to the application,
such as required for spray painting. The unit is generally not
required to come to rest at unique positions and perform functions as is common in applications employing a Point-to-Point
control. Typically, robots using this type of control are taught
by the operator physically grasping the unit and leading it
through the desired path in the exact manner and speed by which
he wishes the robot to repeat the motion. While the device is
moved through the desired path, the position of each axis is
recorded on a constant time base, thus generating a continuous
time history of each axis position. Every motion that the operator makes, whether intentional or not, will be recorded and
played back in the same manner. Since the operator must
physically grasp the robot, it must be designed to be essentially
counter-balanced and free under no power so that he can perform the task. Therefore, this control is generally limited to
light duty robots. Since the operator is manually leading the
robot through the desired sequence, the teaching is very
instinctive and there is no concern for the position of each axis.
The programing is direct. Another characteristic of this type
control is that considerable memory capability is required to
store all the axis positions needed to smoothly record the desired
path. For this reason, magnetic tape or disk storage means are
generally used.

These two classifications of controls, with few exceptions, have represented all the commercially available robots since their introduction. There have been many refinements to these basic control types. For example, the Continuous Path type control can make use of a master-slave system for teaching thereby removing a restriction on the physical size of the robot. For other examples of improvements, see References 1 and 2. However, the general description of each control type remains unchanged even with these refinements. Unfortunately, each type has its limitations and many applications require characteristics of both. To have the greatest application flexibility, it would be desirable to combine the smaller memory requirements and greater load capacities of the Point-to-Point control with the path control and instinctive ease of teaching found in the Continuous Path unit.

NEW CONTROL SYSTEM

The application limitations of existing control systems and the capability of the computer provided the motivation to develop a new control system, which has been termed CONTROLLED PATH. Basically, this method takes advantage of the computational capability of the computer to give the operator (i) coordinated control of the robot axes when teaching the device, and (ii) total position, velocity, and acceleration control of the robot end-effector along a desired path between programmed points in the replay or automatic mode of operation. This capability combines desirable characteristics of both the Point-to-Point and Continuous Path control systems. Also, additional capabilities are realized which would otherwise not be feasible.

With the new control system, the computer is providing two important functions. First, when teaching the robot, the axes are coordinated in a manner which allows the operator to position and orient the end-effector at desired points without having to individually command each robot axis. This feature provides the ease of instinctively teaching without having to physically grasp the robot. Secondly, when teaching, the operator is not required to generate the desired path; he only programs end-points. When in the replay or automatic mode, the computer automatically generates the controlled path at the desired velocity including acceleration and deceleration. This feature requires only the storage of path end-points and does not require "real-time" teaching of the desired path data. The importance of the coordinated axes and controlled path features of the new control system is best understood by first reviewing typical physical configurations of robots.

Robot Configurations:

Robot configurations range from the all translational configuration, similar to that used in machine tools, to all rotational or rotary-jointed mechanisms, similar to a tractor "back hoe" (Figure 1). The all translational configuration is perhaps the easiest to control, particularly to obtain coordinated axes and controlled path motion. However, other configurations are generally selected for robots since their operating range is large while requiring a smaller floor area. The all rotary configuration is the most difficult for which to develop coordinated axes control. However, it offers the possibility for a mechanically simpler design. It is possible to use rotary bearings, which are easy to seal, and rotary drives in place of exposed slide ways and linear drive systems required in the other configurations.

GEOMETRIC CONFIGURATIONS

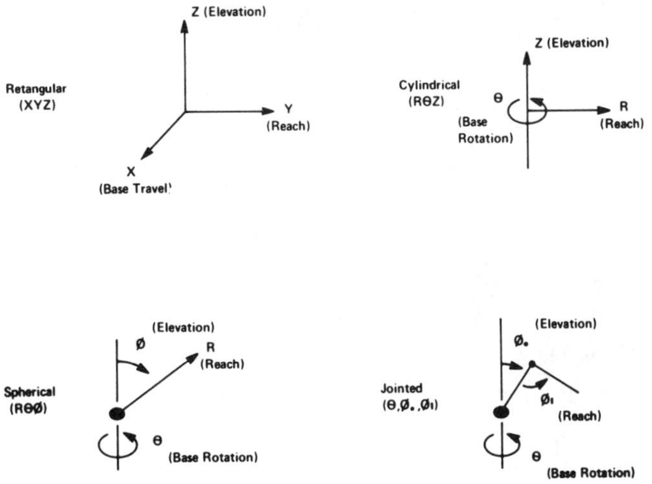

FIGURE 1

The rotary-jointed configuration can operate in a large working volume for its size, giving greater reach and thus greater application flexibility. It is possible to reach down into or onto objects as easily as if the robot were mounted overhead. The jointed arm also allows the robot to reach in close to its base at floor level which is an important consideration for many material handling and machine loading applications.

Figure 2 shows the working range of a commercially available robot which makes use of the rotary-jointed configuration. It has a reach height of 13 feet (4m) and can reach out over 8 feet (2.5m) not including tooling, for a working volume of greater than 1000 cubic feet (28m^3). Only 9 square feet (.8M^2) of floor space is required. Five of the six axes use direct drive rotary hydraulic actuators and one axis uses a hydraulic cylinder (Figure 3). While this configuration has greater application flexibility, it would be difficult to program without coordinated control of the axes.

JOINTED ARM - REACH FLEXIBILITY

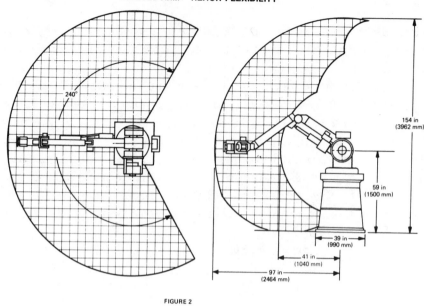

240°

154 in
(3962 mm)

59 in
(1500 mm)

39 in
(990 mm)

41 in
(1040 mm)

97 in
(2464 mm)

FIGURE 2

HYDRAULIC ROTARY ACTUATORS

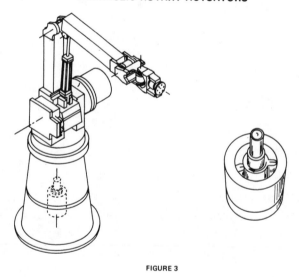

FIGURE 3

Coordinated Axes Control Motion:

An operator, when teaching a robot, is interested in the position in space of a given point on the end-effector, for instance the center of the gripper jaws, and the orientation of the end-effector relative to the task being taught. For purposes of discussion, this unique point on the end-effector will be defined as the tool center point (TCP). The teaching becomes much more instinctive if it is possible for the operator to directly cause motion of the TCP and/or orientation of the end-effector around the TCP to change relative to a defined coordinate system without having to individually command the robot axes. Thus, the operator is interested in moving the TCP in a convenient manner such as: up or down, in or out, left or right, or changing the orientation of the end-effector around the TCP.

To achieve the above using a Point-to-Point control system, the operator must move each axis until the combination of their positions properly positions the TCP. While this is no small task, he must also position the axes to give proper orientation of the end-effector. Since both the position and orientation are coupled, he must simultaneously insure these conditions. The operator's mental process involves solving the transformation equations which give pure translation of and orientation around the TCP. This is a problem for which the high speed computational capability of the computer is best suited.

Once the computer capability for translating and orienting the end-effector around the TCP exists, many options for teaching can be used. One of the most straight forward is to use a series of push buttons which allow the operator to command six coordinated but independent motions defining a teach coordinate system. Since most robot motion tasks can best be described in a rectangular coordinate system, this is an obvious choice for the teach coordinate system. Thus, the translation of the TCP is relative to X, Y, Z coordinates with orientation about the TCP defined by the three independent angular coordinates: pitch, yaw and roll. Figure 4 illustrates these coordinates.

RECTANGULAR AND ANGULAR COORDINATES

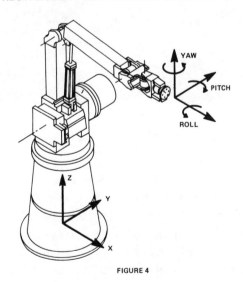

FIGURE 4

Figure 5 shows a hand held teach pendant which makes use of push
buttons. The six buttons in the lower left corner are used to position
the TCP, and the six in the lower right are for orienting the end-effector.
For the six buttons in the lower left corner, the top two cause the TCP
to move left or right (+Y, or -Y). The second two buttons cause the TCP
to move up or down (+Z, or -Z). The third two buttons cause the TCP
to move out or in (+X, or -X). In a similar fashion, the six buttons in
the lower right corner define yaw, pitch, and roll orientation respectively.

HAND HELD TEACH PENDANT

FIGURE 5

Using this teach coordinate system, the operator is directly causing
motion of the TCP as though the robot was of an X, Y, Z axis configuration.
A cylindrical coordinate system (R, θ, Z) has also been found useful in
teaching. When the operator selects this teach coordinate system, the
six buttons in the lower left hand corner of the hand held pendant are
interpreted by the computer as (+θ or -θ), (+Z or -Z) and (+R or -R)
respectively. The orientation buttons remain the same. Other teach
coordinate systems can easily be implemented, but these two have been
found to be most useful. (See Whitney, Reference 3, for a discussion
of a coordinate system related to the end-effector orientation, "hand
coordinates".)

Using either of these two teach coordinate systems, the operator can easily and instinctively position the end-effector at the points required for the application task and program these end-points into computer memory.

Controlled Path Generation:

Teaching the robot is the first operation. Once the points are taught, the computer must move the robot to each of these points in the replay or automatic mode of operation. Using the computational capabilities of the computer, a mathematically definable path between these points is generated including the control of velocity along the path and acceleration when starting, stopping, or changing velocity. Since many robot positioning tasks can be described as a series of straight line segments in space, the most obvious path of the TCP between programmed points is a straight line. By using a straight line path, heavy parts which are transferred by the robot may be smoothly accelerated and decelerated without centrifugal forces being exerted on the part. Since the acceleration and deceleration of the part are controlled, high velocities may be used, thus reducing overall cycle time. Also, the control of both acceleration and deceleration improves the robot service life by reducing jerky motions and corresponding high forces.

Figure 6 shows the resulting motion of the TCP between two programmed points. The velocity profile along this path is also shown. Note that the path velocity is constant after the acceleration span. This velocity is selected and programmed when teaching the robot. It can be

CONTROLLED PATH - STRAIGHT LINE

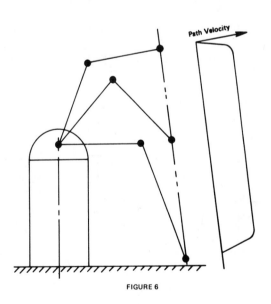

FIGURE 6

selected to be different between each pair of programmed points. If a smooth curve is desired, a number of short straight line segments may be used to approximate the curve. If necessary, a path generation algorithm, other than a straight line, can be developed to generate curved paths.

Control System Operation:

The operation of a particular Controlled Path system may be explained using the control system block diagram in Figure 7. During teaching, the operator depresses the position and orientation buttons (Figure 5) to cause the TCP to move in a coordinated manner in the teach coordinate system selected, e.g. rectangular or cylindrical. As long as the buttons are depressed, continuously changing position signals are generated. The "teach coordinates" are then transformed into "rectangular coordinates" in the first operation performed within the computer. (If the teach coordinates system used is rectangular, this operation is not necessary.) The "rectangular coordinates" are transformed in the second operation into the "robot coordinate" system. These coordinate values are then output to the axis servo systems. The six servo loops will then drive all axes simultaneously to provide the desired change in position and orientation of the end-effector. Since these operations in the computer are very fast, the operator will see an immediate motion of the TCP in the direction he commanded it to move. Once the TCP is in the desired position, the operator depresses the PROGRAM button on the pendant causing the current values of the "rectangular coordinates" to be stored in computer memory. The operator also uses other buttons

CONTROL SYSTEM DIAGRAM

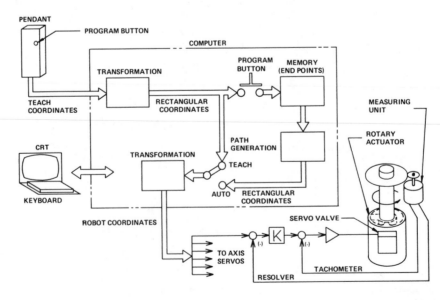

FIGURE 7

232

on the pendant and the keyboard to enter functional data, such as velocity, tool length, and the functions to be performed at the programmed point. The CRT displays the information being programmed or for editing purposes can display previously programmed data.

During the AUTO cycle of operation, the points stored in memory are recalled. Within the path generation operation, these path end-points along with the velocity information are used to incrementally compute points along the path. These intermediate points on the path are generated in a proper time sequence to cause the robot TCP to move at the programmed speed and also generate the proper acceleration and deceleration spans. The path data are in "rectangular coordinates" and must be transformed into "robot coordinates" before being output to the axis servos. The path generation is an on-line calculation done in real-time. There is no buffering of path data before its immediate use. Thus, on-line measurements of position sensor data, such as conveyor or part position, may be dynamically added to the computed path data making such features as tracking a moving part easily implemented.

While the robot may have any machine coordinate system, the control system itself makes all calculations in the basic rectangular coordinate system. Since all the end-points are stored in rectangular coordinates and all path generation calculations are also made in this same coordinate system, many new features not easily implemented with the other control systems are now practical.

PRACTICAL FEATURES ASSOCIATED WITH A CONTROLLED PATH SYSTEM

A number of features which have been developed using the Controlled Path system are discussed below:

Computed Path

The path of the TCP is computer generated between programmed points. The operator need only program transition or end-points along the path without having to traverse the desired path when teaching. The computed path used is a straight line but may be any other mathematically defined curve.

Programmable Velocity

When in the automatic mode, the velocity along the computed path is held equal to the velocity programmed. The velocity is selectable and may differ from one span to another. Some robot tasks such as seam welding require a smooth constant velocity along the desired path of the TCP.

ACC and DEC	In the computation of path velocity, the acceleration (ACC) and deceleration (DEC) along the path are automatically generated to provide smooth starting, stopping and changes in velocity between spans. This feature allows for the handling of large heavy parts at high velocities without causing high forces which could limit the life of the robot.
Coordinated Axes	The control coordinates the motion of the TCP and orientation around the TCP in a convenient selectable teach coordinate system. This greatly eases the teaching task for the operator and removes the need of master-slave teaching systems or robot design constraints which are required when the operator must physically position the robot end-effector.
Changeable Tooling Length	The TCP position relative to the robot tool mounting plate, or tool length, may take on any numerical value. If the robot is programmed using a tool of one length and it is desired to operate the robot with tooling of another length, it is not necessary to reprogram using this new tooling. It is only necessary to enter into the computer control system the new tooling length. The changes are made within the control so that the "new" TCP will move to the spatial locations defined by the existing program.
Modify Program (Offset)	After the program for a robot task is completed, this program may be entered into other robots doing identical tasks. While the repeatability between robots is good, it may not be feasible or economical to insure that the robots are identically located relative to the task. The difference may be determined and used to offset each point in the program. This feature may be used to modify a single point, all points in a branch or subroutine, or all the points making up the total program, thus, zero-shifting the programmed point locations.
Mirror Image	Many times robots are required to do identical tasks but from opposite sides of a manufacturing

line. Through the use of this feature, a robot programmed for the right side of a line may have a hard copy of this same program loaded into the robot control on the left side of the line and the program can then be mirror imaged about any of the major axes in rectangular coordinates. The modify feature described above may also be used to zero-shift the mirror imaged program.

Feed-to-Depth

Some applications, primarly material handling, require that the robot either stack or unstack objects. While in principle each point in the stack could be programmed, it may be tedious or in fact impossible for objects of varying heights. This feature enables the robot to perform the desired function, i.e. pick-up or release an object, at any point along a path between two programmed points. All that is required is a signal from a sensor mounted on the robot hand. Upon receiving this sensor signal, the robot comes to a controlled stop over a known distance and then performs the desired function before moving to the next point. Without a real-time computed path, it would be difficult to stop at an intermediate point which is determined by a sensor.

Stationary Base Tracking

Many applications or potential applications require the robot to work on a moving conveyor line. One approach is to have the robot base physically move parallel to and be synchronized with the conveyor line. This approach requires the installation of rail and carriage systems parallel to the line; however, this may not be possible or economical. Equally, it may not be economical or practical to stop the work so that a stationary base robot can perform the necessary task. In such applications, stationary base tracking is particularily desirable.

There are two types of stationary base tracking systems. In the first type, the robot is taught the task while the conveyor or part is in motion. During replay of the taught program, the speed of replay and thus speed of robot motion is synchronized with the line. While this type of tracking system is satisfactory for point-to-point tasks, continuous path tasks requiring the robot end-effector or TCP to maintain the same velocity relative to the part as when taught cannot use this synchronized tracking system unless the conveyor speed is constant without interruption in its motion. For cases where this cannot be insured, "full tracking" capability is required. This second type of tracking system is required for these velocity critical continuous path tasks. The net effect of this full tracking capability is to perform the task, relative to the part, independent of conveyor motion. For example, when spray painting, if the conveyor stops, the robot cannot stop but must continue to move at the same velocity relative to the part.

With the Controlled Path system, this "full tracking" capability is easily implemented. This system allows a continuous real-time reading of the conveyor or part position to be input to the computer to provide a "dynamic" zero-shift of the path being generated. Figure 8 shows the manner in which the conveyor position is interfaced to the computer. If the conveyor position measurement is made in a major rectangular coordinate axis of the robot, e.g. the Y axis, this signal is directly summed with the current path position coordinates generated from the non-shifted programmed points. Figure 8 also shows the manner in which the non-shifted points are programmed when teaching. The rectangular coordinate position of the TCP is differenced with the current reading from the conveyor or part position before being stored. Thus, the stored data are referenced to the start point of tracking.

Also, teaching is done with the part stationary making this task considerably easier. The conveyor may be positioned to other locations during teaching, but the same zero reference position is maintained. In the automatic cycle mode of operation, the conveyor position is then summed to the path being generated which uses the non-shifted position data. With this full tracking capability, the robot performs operations on a moving part without interruption while the conveyor is moved forwards and backwards at different speeds and also stopped.

TRACKING AND CONTROL SYSTEM DIAGRAM

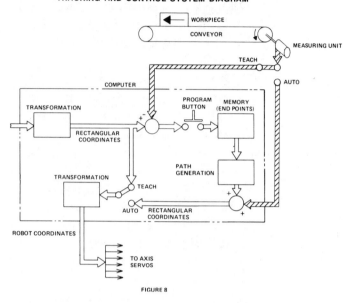

FIGURE 8

OTHER FEATURES OF A COMPUTER CONTROLLED ROBOT

Other features were developed which take advantage of the logic and interfacing capability of the computer.

Program Editing

Through the use of the computer, a keyboard and CRT, it is possible to easily implement a program editing feature. The CRT enables the operator to display the functional data stored at previously programmed points. He may delete or modify this data and restore it. He may also insert points between previously stored points by using the teach pendant.

Program and System Diagnostics

The CRT may be used to display operator teaching errors or system errors. For example, an illegal entry of data will cause a corresponding error message to appear. System error messages reduces downtime by allowing quick isolation of operating problems which might occur.

Functions

Each stored point in the computer contains the coordinate positions and orientations of the TCP, velocity, tool length, and the functions to be performed at the point. The following are descriptions of some typical non-path related functions:

DELAY - When this function is executed, robot motion is stopped for a defined period of time before moving to the next programmed point.

WAIT - Robot motion is stopped until a signal is received on a defined terminal before proceeding to the next programmed point. This function may also be made "conditional" on a set of signals received on defined terminals.

OUTPUT - This function causes robot motion to stop and a signal output on a defined terminal before proceeding to the next programmed point. This output may be defined as pulse or a level signal (on or off).

CONTINUE - With execution of this function, the robot motion is not stopped but continues to move to the next programmed point. A signal may also be output at the same time if desired.

TOOL - This is a specialized function which is used to output to one of two defined tools (end-effectors) attached to the robot hand and to verify the tool operation. Motion is stopped until the tool operation is complete before proceeding to the next programmed point.

BRANCH - The sequence of points being followed may be changed through use of this function. Using this function, robot motion remains stopped and if a signal is received on a defined terminal, another defined sequence of points, which is called BRANCH, is followed; otherwise, the previous sequence of points is followed. The BRANCH is a previously programmed sequence of points and the last point of this BRANCH may close upon another previously

programmed point. If the BRANCH is not closed
on a specified point, it becomes a SUBROUTINE.
Anytime a SUBROUTINE is entered, the operation
is always transferred back to the next point from
which the BRANCH function was used to initially
enter the SUBROUTINE. Figure 9 shows a few
possible ways a BRANCH can be used. This
function can also be made "conditional" on a set
of signals being received on defined terminals.

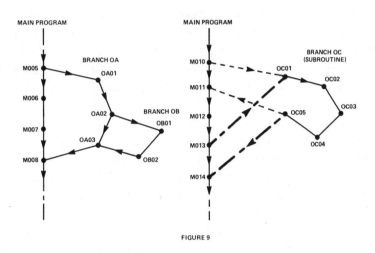

BRANCH USAGE

FIGURE 9

FUNCTION X - This function is application
dependent and is programmed to give a desired
result or combination of results. While each of
the above functions have been developed to give
broad application flexibility, some tasks require the use of a function unique to the specific
application. Since the robot is computer operated,
this specialized function may be easily added in
software.

PRACTICAL APPLICATIONS OF THESE FEATURES

The robot system described above has been used in a number of different
applications. A common characteristic of many of these applications is
the requirement of minimum cycle time. The use of the Controlled Path
system minimizes the number of programmed points that define the
sequence of motions and the controlled acceleration and deceleration
along the path between these points permits use of high velocities, thus
minimizing overall cycle time.

The jointed arm configuration, with its ability to reach into pallets near its base and reach out and into a machine work area, has made possible operations which would have required a costly overhead mounting facility for the robot or the use of more than one robot. The use of the jointed arm is therefore desirable and is easily controlled using a computer.

Since path velocity is precisely controlled, some critical process applications which are continuous path tasks are feasible. Important characteristics of the Controlled Path system for these tasks are the ability to program velocity and position data independently and modify points within the task sequence without reprogramming the total task.

Using the Controlled Path system makes many robot tasks which would be tedious to program using a Point-to-Point system much easier. Because coordinates are stored in rectangular form, the ability to transfer programs from one robot to another and easily modify or offset position data makes many potential applications much more feasible.

The greatest potential savings in using a robot are in minimizing the tooling and special fixturing required for the task. Many tasks which are performed on moving lines would meet the general characteristics which make them economically viable and feasible for a robot if the part could remain stationary. However, the mechanical hardware to make this possible is generally too costly. The ability to track a moving conveyor using a stationary base robot can make these tasks economically feasible. Also, the ability to program the task with the part stationary considerably reduces the programming time and skill required.

Descriptions of some specific applications follow:

1.   Handling of Heavy Parts -- Machine Loading
In this application, the robot was mechanically modified to handled a greater load. Part weights range up to 300 pounds (136kg), and the gripper weights 200 pounds (91kg) for a total weight of 500 pounds (227kg). The robot is required to take a workpiece from the end of a roller conveyor to one lathe, then to a second lathe, next to a drilling station and then to an output station. The parts running through this machining system are similar in geometry but vary in size. The robot services these machines on a demand basis and in the proper sequence. The Conditional Branch feature is used to enable the robot to function on this demand basis. High position repeatability is required since special fixturing to center the part in the machines is not used. Cycle time is critical and the use of controlled acceleration and deceleration allows the use of high velocities. The large reach volume of the jointed arm is also required in this application. The programming for the intricate

moves required to place the large workpiece between centers in the lathe is made easier by using the coordinated axes control.

2.   MIG (Metal Inert Gas) Seam Welding

A MIG welding gun is attached to the end of the arm.   By defining the TCP to be coincident with the weld wire tip, the operator orients the gun around the TCP making programming easier.   The ability of the control to generate a straight line between programmed points results in straighter weld lines than could be made manually and the controlled velocity results in a smooth, slow and even motion of the weld tip, all resulting in excellent weld quality and consistency.

3.   Drilling of Aircraft Wing Panel Assemblies

The wing panel assemblies are fixtured and a template is placed over the panel and used by the operator to drill all the holes in the panel.   By using the robot to position the drill to each hole location, productivity is increased by a factor of six times.   Important features required in this application are: a) Offset or Modify feature to account for positional changes of the fixtured part from set-up to set-up.   b) The ability to take alternate actions based on signals which sense tool breakage and proper alignment of the drill head to the template.   c) Orientation moves around the TCP, which are made coincident with the drill tip, allow for ease of locating the drill head normal to the panel surface when teaching.

4.   Spot Welding an Auto Body on a Moving Line

In this application, an auto body is already fixtured on a "body truck" which is moved by a floor conveyor.   A rail system is used to guide the body truck and maintain proper alignment to the robot.   The welding operation is done around the rear "wheel well" while the body is moving.   The body truck position is measured relative to the robot base. The ability to program the body when it is stationary, execute the program when the body is in motion, and a minimum of modification to the line has made this application economical.

CONCLUSIONS:

Computer control has greatly enhanced the application flexibility of a general purpose industrial robot.   The computational capability of the computer formed the basis for the development of a new CONTROLLED PATH system which allows total position, velocity, and acceleration control of the robot end-effector along a desired path.   This control system has enabled the development of many features which are very desirable and make such a robot economically viable in many new application areas.

REFERENCES:

1. Richiardi, Claudio, "Computerized Control of a Continuous Robot," Paper presented at the Sixth International Symposium on Industrial Robots, University of Nottingham, U.K., March 24 - 26, 1976.
2. "Robots: A New Force in Continuous Arc Welding," Manufacturing Engineering, September, 1975.
3. Whitney, D.E., "The Mathematics of Coordinated Control of Prosthetic Arms and Manipulators," Journal of Dynamic Systems Measurement and Control, December, 1972.

Presented at the Fifth International Symposium on Industrial Robots, September, 1975

ASEA Robot System—

Expanding The Range Of Industrial Applications

By Bjorn Weichbrodt
ASEA Corp.

A new robot system has been industrially developed for applications far beyond materials handling.

The combination of CNC-microcomputer control, all-electrical DC-drives, and a new geometry, results in a robot with unusually high programmability, accuracy and compactness.

The robot is currently installed in industry doing both materials handling, precision arc welding, grinding, and polishing.

The concept is presented, and some industrial applications are demonstrated where the robot welds and does complete steel parts finishing including sanding and polishing.

INTRODUCTION

Before discussing the design and performance of the ASEA robot system, it may be worth noting the basic reasons why robots are at all developed and marketed by ASEA.

The reasons for developing robots are about the same as for any other type of automation tool within the manufacturing industry:

First, there is an economical reason. The profitability of modern industry depends on efficient use, i.e. rapid turnover, of the invested capital. This implies a high rate of production for a given investment. Robots may help in obtaining such conditions.

Second, there is a human reason. An increasingly important demand made on industry today, is that satisfactory workplaces be provided for its employees. In many cases this is not economically possible because of the nature of the manufacturing processes involved. Robots may assist in those places which provide the most unsatisfactory working environment.

In order for a robot system to really become profitable for its user, however, it has to meet certain requirements, which are tied to the reasons above.

First, it must offer such technical features that it really may improve the capital turnover rate of a manufacturing operation. This means for example that the robot must offer all that it takes to completely automate a production group - such that the machines may be operated during several shifts a day, while the personnel only works one shift.

Or it means that the robot should be able to fully replace expensive specialty machines, such as for welding, grinding, painting etc. This puts a high demand on the robot's technical capabilities.

Second, since the robot may be called on to replace human labor in very severe environments, it must itself be able to work reliably in those environments. This also puts a high demand on the robot's design and environmental protection.

Within ASEA, Sweden's largest manufacturer of industrial electrical and electronic equipment, it was felt that the modern electronics technology and the Company's own extensive experience in production technology should provide an outstanding basis for the development of a really reliable, versatile robot system. The development work was started in 1971, and the new generation robots is now being marketed. A large number of ASEA robots are today in production use and have accumulated thousands of working hours, both within ASEA and a number of other companies.

This paper will attempt to describe some features of the robot system and its applications.

THE ASEA ROBOT SYSTEM

ASEA's all-electrical industrial robot combines the use of a simple, rugged mechanical unit with advanced elektronic control by a small microcomputer. The control system is an outgrowth of ASEA's NUCON system for numerical control of machine tools. Control of materials handling with an advanced robot has a great deal in common with the control of machining operations with an NC system.

The target for the ASEA robot development program has been to create a system that can perform both simple point-to-point handling operations and also be "trained" for welding, grinding, and assembly operations requiring precision motions along complicated curves.

The robot itself is a compact mechanical unit driven by quiet electric motors. In order to make the robot perform under really severe environmental conditions, all electronics have been removed from the mechanical unit and put into a separate cabinet. This control cabinet may be placed several yards from the robot itself, in another room if necessary.

The control system has a small microcomputer brain, a central processor which packs several thousand electronic functions into a tiny chip. Surrounding the processor is a memory system. Together the two can be programmed to guide the robot through intricate curves or in straight lines between fixed points. A large number of programs can be stored and selected on external command. Macro programs such as for pattern generation, palletizing etc may be included. All programs may be edited directly from the control panel.

The movements of the robot are programmed using a portable programming unit. Step by step, each position is stored in a memory. One programming unit may be utilized for several robots.

In addition to actual position instructions, it is also possible to store instructions for manoeuvring the gripper, switching on and off a number of outputs, testing a number of interlocking inputs, time delays and repeats.

The robot has three main degrees of freedom, including rotary movement, radial arm movement, and vertical arm movement. Furthermore, a turning and bending movement of the wrist can also be obtained, and a linear motion along a straight track. Thus, in total a maximum of six degrees of freedom.

A stable mechanical design utilizing cast aluminium for most parts in combination with an advanced servo system gives a high degree of precision. The repeatability is better than 0.2 mm for the smaller robot and 0.4 mm for the bigger robot.

The basic system includes robots for 6 to 60 kg handling capacity, control equipment and equipment for programming the robot. Furthermore, a battery is included as a stand-by source of supply should the main supply be interrupted.

As compliments to the basic system, a number of accessories may be added, such as

o tape recorder for transmitting and receiving of programs and for long term program storage

o additional memory cards increase the capacity of the memory up to 700 program points plus a large number of input and output commands

o pneumatic unit for installation in the robot arm when a pneumatic gripper is used

o pneumatic linear track for controlling the travel of the robot between two end positions

o diagnostic test panel for fault tracing and servicing purposes

The small, 6 kg robot, with its control system is shown in Figure 1.

The big, 60 kg robot, is shown being programmed in Figure 2.

Both these mechanical units use the same control system, shown in Figure 1.

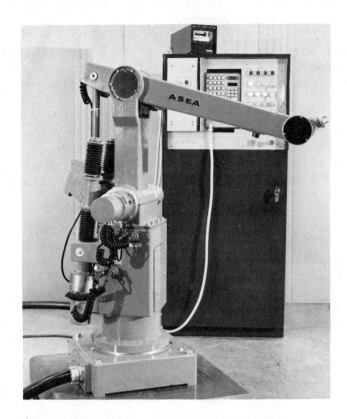

Figure 1: 6 kg ASEA robot with control system

Figure 2: 60 kg ASEA robot being programmed

The control system panel with its programming unit is shown in Figure 3.

Figure 3: ASEA robot control system with programming unit

The portable programming unit in the middle of the panel has controls for manual drive of the robot's movements (lower part), controls for programming (upper left), and a block number indicator (upper right). The key board is used for block search and insertion of data such as waiting times, jumps between programs etc.

Four main programs and any number of subprogram may be simultaneously stored in the memory of the control system.

SOME APPLICATIONS

The ASEA robots have been used in several industrial application areas, including materialshandling, continuous arc welding, sanding, polishing and deburring of castings.

Figure 4 shows the 6 kg robot at work in a Swedish die-casting manufacturing operation, which today employs a total of 6 robots. Generally, two robots work in each production group. One robot serves the die-casting machine by taking out the castings (point-to-point operation), lubricating the dies (continuous path operation), checking that a complete detail has been cast (robot working with sensors), and delivering the casting to the trimming press. The second robot (shown in Figure 4) handles the casting between various special machines for boring, tapping, and sanding. (The ASEA robot, with its continuous path control, actually itself deburrs the edges through manoeuvering the casting over a sanding disc). The result is an automatically produced, completely finished part, ready for assembly.

Figure 4: ASEA robot with double grippers, working in die-casting shop

Figures 5 and 6 show the 6 kg robot handling tablets from a press, one every 5 seconds, and palletizing them on a pallet. The robot has a special program which permits easy programming of repeating patterns.

Figure 5: ASEA robot picking tablets from press

Figure 6: ASEA robot palletizing pressed tablets

Figures 7 and 8 show the robot arc welding using welding equipment developed in cooperation with the ESAB Corp., Laxå Sweden. In Figure 7 a transformer casing is welded. In Figure 8 the robot automatically welds a frame for DC-motors. In this production installation the robot performs 64 arc welds about 4 in. each, all of these stored in one program.

Figure 7: ASEA robot welding transformer casing

Figure 8: ASEA robot welding DC-motor frame

A robot installation at the Magnusson Corp., Genarp, Sweden, employs 2 robots which are being run in 3-shift operation, while the personnel only works 1-shift. The two robots do a complete finishing job on stainless steel tubing bends, including sanding and polishing. The tubing bends are stored in a magazine (Figure 9) where they are picked up by a robot. The magazine is sufficiently large for 1-shift of unattended operation. Each tubing bend is then sanded (Figure 10), using a band sander. In some cases the robot must change its grip during the operation, in order to permit sanding on all sides. This is done in a special fixture (Figure 11). After sanding, the tubing bend is polished, still held by the same robot (Figure 12).

In this operation, the robots use their combination of high accuracy, compactness, and contouring capabilities. The cycle time is about 30 % shorter than for manual operation. (Operators for this job have been very hard to get since the job is difficult and requires precision at the same time as it is repetitive). But, more important, the production may be extended to 3-shift operation, without the need for increasing the capital investment in sanding and polishing machines, or increasing the floor space. Thus, highly increased production capacity at a very small capital cost.

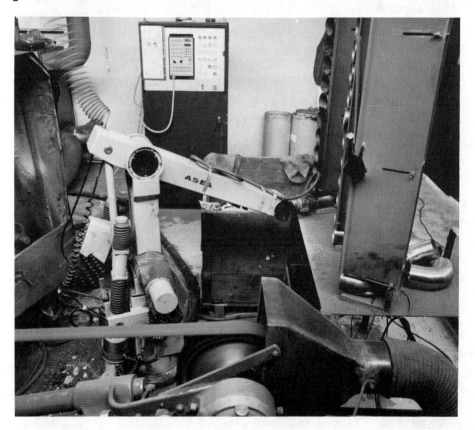

Figure 9: ASEA robot picking stainless steel tubing bends from magazine

Figure 10: ASEA robot manipulating tubing bend for sanding

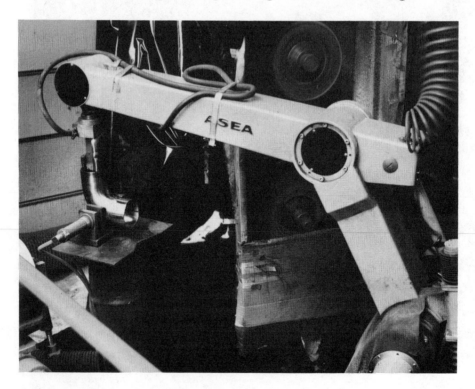

Figure 11: ASEA robot changing grip of tubing bend during sanding
operation

Figure 12: ASEA robot polishing tubing bends

The same type of automated sanding and polishing jobs apply to many other products, including stainless steel houshold goods, food processing equipment, silverware, and buckets for turbines and jet engines.

In conclusion, all the applications which have been described above require a robot like ASEA's which has the capabilities of easy programming, contouring, high precision, 3-6 degrees of freedom, and compactness. Most of the jobs, i.e. welding, sanding, deburring, are unattractive to do manually since they usually have to be done in poor environments.

When a robot has all the needed capabilities, it can be used for complete automation of a defined production group. This is important, since it opens up the possibilities of unattended operation, and thereby more efficient use of available machines through multi-shift operation.

In most of the ASEA robot installations which have been described above, the total capital investment, including machines, buildings, work in progress, etc, for a given production capacity, has actually been decreased through the robot installation. This investment reduction in relation to produced output, which indicates that the available capital has been better utilized, is a good criterion of a profitable robot installation.

The ASEA robot system is marketed in the United Stated by ASEA Inc., Four New King Street, White Plains, New York 10604, Telephone No. (914) 428-6000.

DEVELOPMENTS IN PROGRAMMABLE AUTOMATION

Reprinted from Manufacturing Engineering, September, 1975

Previously limited to parts-handling assignments, the modern robot is growing in sophistication. The reason: the advent of artificial intelligence

C. A. ROSEN *and* **D. NITZAN**
Stanford Research Institute

NEEDS FOR ADVANCED AUTOMATION

Previous studies and surveys have shown that there is a national need to develop advanced automation tools, techniques, and systems that would be applicable to a wide variety of manufacturing processes that are still highly labor intensive. In particular, there are demonstrable needs to increase worker productivity and finished goods quality while simultaneously reducing worker discontent on dull, repetitive, arduous, dangerous and otherwise undesirable jobs.

These major factors were deemed important enough several years ago to initiate a series of workshops and study groups sponsored by the National Science Foundation and carried out by the Automation Research Council. A series of reports and recommendations have emerged from these studies that bolster the early conclusions. More recently, two important concerns of the U.S. — inflation and the foreign exchange deficit — have provided additional justification for this work.

Economists believe that in the long run these factors can be brought under control only by effectively increasing the real productivity of our workers, that is, by decreasing the cost of goods produced and increasing our competitive stance in world markets.

Labor Intensive Areas. Several major areas that are still labor intensive and appear to offer good opportunities for mechanization are material handling, inspection, and assembly operations in discrete-part manufacturing industries for which batch or multimodel production is the normal mode of operation. These areas were chosen as the research domain.

To facilitate early technological transfer of the results of this research and to focus the research on real production problems, a number of industrial firms participate actively in this program. Their principal roles are to acquaint our staff with typical factory problems to be solved, to help choose economically feasible constraints to simplify machine requirements, and to carry back the research results for their own factory use or further development. Other important contributions by industrial participants include:

► Assisting in the evaluation of the economic consequences (such as cost effectiveness and worker acceptability) of the proposed techniques.

► Considering which design changes are possible and which economically permit material handling, assembly, and inspection tasks to be performed more readily or simply by the new mechanized equipment.

► Determining the system consequences (such as matching throughput, the effect of breakdowns, and back-up strategy) of introducing new mechanization into existing production lines.

PROGRAM RATIONALE

We expect our research to lead to cost-effective, programmable systems for automated material handling, inspection, and assembly. This will require modular, general-purpose software packages, programming techniques, and appropriate hardware that can be assembled into complete, integrated systems. We believe that such systems will be evolutionary for the next few years. Thereafter, they may be revolutionary as various industries begin to apply their developmental resources to their own specific problems.

We visualize factories in which many repetitive jobs are done by computer-controlled machines supervised by a smaller but more highly trained work force than has previously been employed. The workers will be capable of programming (*i.e.*, setting up) each job, modifying procedures as necessary, changing over for new models or batches, maintaining the equipment, and coping with stoppages and breakdowns. Thus, we will timeshare the workers, multiplying their capabilities by relieving them of the relatively low-level jobs that can best be done by machines. Workers will perform more challenging tasks that, at present, either cannot be done by machines or can be done only with expensive computer hardware and software.

Development of such computer-controlled systems is based on the confluence of two technologies:

1 *Programmable manipulators, commonly known as industrial robots, which have been developed to the stage of being acceptable to industry.*
2 *Artificial intelligence research in visual and tactile sensing, pattern-recognition techniques, robot manipulation, and robot systems.*

1. Block diagram of the computer-controlled system.

2. *Laboratory test bed used in AI research.*

3. *Forgings identified by robot.*

The first technology provides the "muscle"; the second provides the "sensors and brain"

LIMITATIONS OF EXISTING HARDWARE AND SOFTWARE

The industrial robots in use today are pick-and-put devices that are used for such jobs as loading and unloading presses, stacking parts, spot welding, and paint spraying. A key to their attractiveness as alternatives to specially designed equipment (fixed, or hard automation) is their ability to be easily reprogrammed to handle different tasks. This is particularly important in factories which have small production runs, or where models may change frequently. Such factories are estimated to produce the majority of manufactured goods.

A major limitation of these robots is their lack of anything but the most primitive sensory feedback. In particular, no commercially available industrial robot has visual, range, or tactile and force sensors; hence, the present machines are limited to highly constrained situations. They cannot determine the position and orientation of parts, nor can they manipulate objects that are not precisely located, or perform any kind of inspection task.

The extensive research efforts in computer science in general and in artificial intelligence in particular have provided many tools and concepts for solving this problem. Robot and hand-eye research has demonstrated the possibility of guiding a vehicle or a manipulator by computer-processed visual information. Many of the picture-processing techniques are directly applicable to industrial automation, and the associated experience with programming languages is certain to be helpful.

However, artificial intelligence research has been primarily directed toward finding very general methods that demonstrate principles. Thus, relatively little attention has been paid to questions of computational cost and

program complexity — questions that are of major importance to industrial applications.

For example, the goal of most visual scene-analysis programs has been to be able to identify all the objects in a scene regardless of orientation, with parts of some objects occluded by other objects, and under varying and rather difficult lighting conditions. Computer programs that can barely cope with fairly simple scenes under these conditions are huge, difficult to code or modify, time consuming, and require very large computers.

In a factory situation, recognition of a relatively small number of parts — one at a time against a controlled background and under controlled lighting — can be accomplished by using a number of algorithms that will run quickly on a small computer. The key point is that the majority of experienced and competent artificial intelligence scientists have barely begun to apply their knowledge to the constrained and, therefore, simplified problems of the factory. Furthermore, cost effectiveness and efficiency, which are usually set aside in long-range artificial intelligence research, must now be considered.

TRAINING AND COMPUTER CONSTRAINTS

In addition to the preceding considerations, our desire to develop practical, programmable systems leads to two additional overriding constraints which characterize our approach: training by factory personnel and minicomputer control.

Training by Factory Personnel. It is probably completely impractical to replace factory-trained people experienced in assembly, inspection, and production skills with trained computer scientists who are expert in computer programming but know little of factory work. Therefore, our system is based on the assumption that programming or training the sensory-manipulation and inspection systems must be done by factory trained people who are not high-level computer system programmers.

Manipulative actions are to be *programmed by doing*; that is, a factory worker will train each manipulator by leading it through a sequence of desired actions. Use will also be made of training aids, which include word or phrase recognition devices for natural language communication, joysticks, and a graphic terminal supported by a

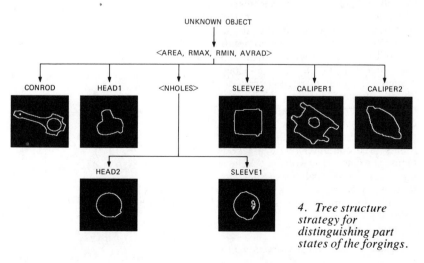

4. *Tree structure strategy for distinguishing part states of the forgings.*

5. *Lamp bases inspected through interactive vision.*

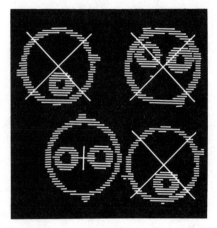

6. *Computer display of lamp base inspection.*

library of programs that can perform many useful functions.

Visual sensing routines will be programmed with an interactive graphic terminal. After a brief period of training, a factory worker will be able to select, modify, and assemble appropriate programs, using a light pen or equivalent device with the keyboard input of the terminal. In this way, reprogramming for short production runs will be facilitated.

Minicomputer Control. The computer programs that form the backbone of the system, including the various inspection, sensing, and manipulative routines, must be small and simple enough to require modest computer capabilities with respect to both the central processor and the random-access memory. Thus, an inspection or assembly operation will require a minicomputer backed up by an inexpensive disc bulk memory. It is possible to develop a more ambitious system with a very large time-shared computer that can service many stations. However, we believe that the present reliability of such systems is not acceptable for production; considerably more sophistication in software will be required before they are practicable.

7. *End-effector and sensory elements.*

OBJECTIVES AND RESEARCH PERFORMED

Our primary objective has been the development of computer programs that control a system of manipulators; visual, tactile, and force sensors; endeffectors; and auxiliary devices in performing a wide range of material handling, inspection, and assembly tasks. Our second is the development of a user's language so that these programs can be operated by factory personnel with no computer science expertise. Third, these programs will demonstrate economic feasibility suitable for basing future factory-installed systems upon our techniques and software designs.

Since the research reported here, covering the period from April, 1973 to June 30, 1974, two additional progress reports have been written and the program is continuing. The following is a summary of the major accomplishments and a brief description of each task performed by our research team during the first fifteen months of this project.

Test-Bed Hardware. Extensive work has been done in assembling a suitable test bed. Its components can now be conveniently assembled into many different systems, using software rather that hardware control and connections. This has increased our capabilities for conducting new experiments.

The test bed, previously controlled by a time-shared computer system (Digital Equipment Corporation PDP-10 and PDP-15 computers), is now controlled by a standalone PDP-11/40 minicomputer when operating in execution mode. Training or programming, however, still requires the use of the larger, time-shared PDP-10 computer in addition to the PDP-11/40. A block diagram of the computer-controlled system is shown in *Figure* 1. This computer system provides an excellent facility for initial software development through

the use of high-level language and debugging tools. The test bed consists of the following devices:

► Unimate Industrial Manipulator, Series 2000A.

► Stanford arm, designed by V. Scheinman and built by Dest Data Corporation.

► Television cameras with silicon target vidicon.

► Two linear diode array optical sensors, built by Dest Data Corporation.

► Several end-effectors or "hands" (built at SRI) equipped with tactile, force, and torque sensors; one hand is equipped with a proximity sensor.

► Servosystem for the Stanford arm to provide position, velocity, and force feedback servocontrol (still under construction at SRI).

► Strain-gage force and torque sensor for the wrist of the Stanford arm (developed and built at SRI; undergoing test and evaluation).

► Word or phrase recognizer, Threshold Technology, Model VIP-100, a self-contained system with its own minicomputer. This system is now providing voice control of all the other devices.

► Joystick control, to facilitate training of the Unimate and the Stanford arm (designed and built at SRI).

► Lighting systems that can be adjusted to enhance contrast and deliberately produce or eliminate highlights (built at SRI).

► Auxiliary devices, including a conveyor belt with positional sensing, and a minicomputer-controlled turntable for positioning and holding subassemblies and parts (built at SRI).

A few of these devices (the Unimate, a TV camera, an end effector, a lighting system, the conveyor belt, and the turntable) are shown in *Figure* 2.

Interactive Vision for Sensing and Inspection. Software and hardware for an interactive vision system have been

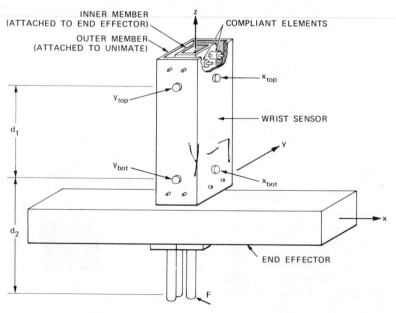

developed for two major applications. These are: the determination of identity, position, and orientation of randomly placed workpieces for material handling; and the visual inspection of workpieces.

We have written programs in the LISP and FORTRAN languages for the decision-making and mathematical analysis processes needed to solve problems in these areas. Using an interactive graphics terminal, a programmer can conveniently write picture-processing routines and apply them to images of real objects produced by a television camera. Successful routines are then stored on a disc file. Later, selected routines can be retrieved and incorporated into complete material handling or inspection programs.

A substantial, and still expanding, library of routines has been developed, debugged, and applied experimentally to real television images. These include:

DATA REDUCTION ROUTINES — e.g., for extracting the two-dimensional outline of the image of an object.

DISTINGUISHED FEATURE EXTRACTION ROUTINES — such as those that compute area, the center of area, second moments, minimum and maximum radius vectors, and perimeter. These features are to determine the identity, orientation, and position of an object.

HIGHER-LEVEL FEATURE EXTRACTION ROUTINES — useful for qualitative or semiquantitative inspection. These include routines for finding corners, separating multiple objects, finding holes, and measuring angles between specified radii.

RECOGNITION ROUTINES — to permit identification of a workpiece on the basis of its distinguishing features. The presently preferred method makes use of a tree structure strategy in which recognition tests are applied sequentially and depend on the outcome of each preceding test.

TRAINING AND EXECUTION ROUTINES — with which the trainer/programmer can specify the grip points (where the manipulator is to grip the workpiece), the height at which such acquisition is to be made, and the orientation of the workpiece.

Experimental Results. In a laboratory experiment, this system was used to identify each of the four different foundry castings shown in *Figure 3* — three of which have two stable positions each — and to determine the position and orientation of each so that a manipulator can acquire a workpiece randomly placed on a moving conveyor belt. The outlines, the names, the distinguishing features, and the tree-structure strategy for distinguishing the seven possible part states are illustrated in *Figure 4*, where <AREA> de-

notes the value of the image area; <RMAX, RMIN, AVRAD> denote the maximum, minimum, and average radius-vector magnitudes, respectively; and <NHOLES> denotes the number of holes.

The recognition and acquisition capabilities are important because, in many instances, it is too expensive to position and orient a workpiece in a precise and known manner (especially if the part is semirigid and may be deformed). In other instances, it is preferable not to segregate different parts, and this visual machine recognition is a useful technique for sorting.

In another series of experiments the interactive vision system was programmed to inspect washing machine water pumps and incandescent lamp bases. The pumps were inspected to confirm that the handle of each pump was present and to determine which of two possible positions it was in. The lamp bases, *Figure 5*, were inspected to verify that the correct number of electrical contact grommets were present on each base and located in correct positions within acceptable tolerances. An example where three lamp bases were rejected and one was accepted is shown in *Figure 6*. These programs illustrated the versatility of the interactive programming system, with which new inspection jobs could be programmed.

SENSOR-CONTROLLED MANIPULATION

A Unimate (Series 2000A) industrial robot was modified and brought under control of our large computer system. It was then possible to extend its performance substantially in several different ways. With a computer in the control loop, it was also possible to augment the performance of the system by adding sensors. These, together with the means to interpret the sensory data in a nontrivial way, permitted adaptive or corrective control of manipulation. Several examples will illustrate the power of this type of control.

Without the ability to "feel" or "see" objects in their work environment, present manipulators (such as the Unimate) must rely on their internal positioning accuracy for acquisition and placement of parts, and on prepositioning and orientation of the parts themselves within rather narrow known limits.

Beyond these limits (about 1 millimeter for the Series 2000A Unimate), compliant pegs and fixtures, or a vibrating end-effector or workpiece, could be used to take up the positioning error that varies from one move to the next. Such requirements impose costly relocation and reprogramming for different jobs or modifications. Intrinsic performance could be improved by increasing the inherent accuracy of the manipulators, an improvement associated with rapidly increasing cost,

8. Water pumps in a tote box.

9. Force-sensor control of end-effector in pump packing.

especially if high throughput must be maintained.

An alternative method to control manipulation is to make use of sensors that provide feedback at the workpiece. To accomplish this control, we have developed several end-effectors, one of which is shown in *Figure 7*. The end-effector is decoupled from the rest of the manipulator by compliant suspension at the wrist, and sensors are incorporated to measure forces and torque applied to the wrist. In *Figure 7*, x_{top}, x_{bot}, y_{top}, and y_{bot} are linear potentiometers whose output voltages are used to determine the x and y components of the force and torque acting on the wrist. Potentiometers for the z components of force and torque, not shown in *Figure 7*, are inside the inner member of the wrist sensor. A small minicomputer (to be replaced by a microprocessor) sequentially polls all sensor signals and processes the data, passing on commands to the main minicomputer system. This processor has the ability to stop the Unimate quickly when selected forces or torques at the wrist reach specified threshold values.

In the future, it will be possible to steer the manipulator adaptively to null out forces or torques so as to bring mating parts together with precision. Such control of inserting a peg in a hole has been demonstrated by others. Using the existing system, we have shown that the Unimate arm can be stopped, when operating at 10 cm/s, within a maximum deviation of ±0.05 mm — a 20-fold increase in the nominal accuracy of the Unimate. This arm speed, however, is only 10 percent of the maximum speed

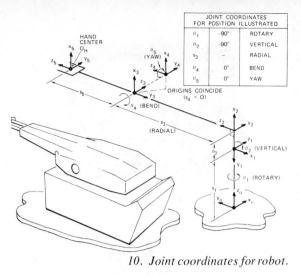

JOINT COORDINATES FOR POSITION ILLUSTRATED		
θ_1	-90°	ROTARY
θ_2	-90°	VERTICAL
s_3	-	RADIAL
θ_4	0°	BEND
θ_5	0°	YAW
ORIGINS COINCIDE $(s_4 = 0)$		

10. *Joint coordinates for robot.*

CONTROL COMMANDS (a)	MOTION COMMANDS (b)	QUANTIFIERS (c)
START RECORD	MOVE PLUS X	
EXIT	MOVE MINUS X	ONE INCH
REPLAY	MOVE PLUS Y	TWO INCHES
AGAIN	MOVE MINUS Y	THREE INCHES
TOUCHDOWN	MOVE UPWARDS	FOUR INCHES
FASTEN	MOVE DOWNWARDS	FIVE INCHES
RIVET		SIX INCHES
		A LITTLE
		A LOT

11. *Voice commands for robot control.*

attainable. Better servocontrol applied to the Unimate would probably extend this performance with some degradation at higher speeds.

In material handling applications, it is often necessary to acquire workpieces whose position and orientation are not known precisely and which would be too costly (in terms of special equipment or process time) to position or orient prior to automatic acquisition. We have, therefore, used the previously described visual sensor (television camera) and interpretive software to determine the position and orientation of each workpiece. Other control software then makes use of this information to adaptively position the manipulator so that it can acquire the workpiece at the desired grip points.

Experimental Results. Laboratory experiments have demonstrated the augmented capabilities of the Unimate industrial robot equipped with visual and tactile sensors. In one experiment, an automotive casting randomly placed on a moving conveyor could be acquired by the Unimate, transported, and deposited with specified position and orientation. In another experiment, a number of washing-machine pumps were packed neatly in a tote box, *Figure* 8. This experiment involved the use of force sensors to control the motions of the end-effector of *Figure* 7 in packing each water pump. For example, completed packing of the first pump is

shown in *Figure* 9. A proximity sensor mounted on one finger of the end-effector controlled the initial placement of the end-effector when acquiring the next water pump.

PATH CONTROL BY COORDINATE TRANSFORMATIONS

In programming the actions of an industrial manipulator, we are mainly concerned with directing the position and orientation of the hand (*i.e.*, end-effector) at rest or in motion. These actions can be described in terms of rotary and linear motions of the manipulator's joints (joint coordinates) or in terms of the hand's position and orientation in relation to the room or workpiece (world coordinates). For example, the joint coordinates of the five-joint Unimate and their designations are shown in *Figure* 10. The five nodes in the schematic diagram represent four rotary joints, whose angles of rotation are Θ_1, Θ_2, Θ_4, and Θ_5, and one linear (sliding) joint, whose variable length is s_3.

It is difficult for humans to use joint coordinates, which depend on the kinematics of the arm and hand. It is far more natural to specify the position and orientation of the hand in world coordinates, which are independent of the machine design. By making use of a coordinate transformation appropriate to the specific manipulator, the computer can carry out the needed computations automatically, determining the correct sequence of joint rotations and extensions. For example, it is possible to choose arbitrary x, y, and z axes as our world coordinates, and control the end-effector to move smoothly along any curve in this system.

The reverse computation — that is, from known joint coordinates to world coordinates — is useful in determining where the end-effector is. This capability can also be used to determine the distance between objects, the height of an object or edge, the dimensions, position, and orientation of a tote box (by moving the hand and reading joint coordinates when its tactile sensors indicate contact), and so on.

We have programmed the transformation computations for a five-joint Unimate, a six-joint Unimate, and the Stanford arm. The five-joint Unimate programs, which have run on the time-shared PDP-10/PDP-15 system, are now operational on the PDP-11/40 minicomputer.

TRAINING THE MANIPULATOR SYSTEM

The use of computer coordinate transformations, as described in the previous section, aids the programmer in training the manipulator. We are exploring and devising several other programming aids, such as control of manipulator motion by joysticks and by spoken words and phrases. Voice con-

trol is especially important when the trainer's hands are busy at other tasks — manipulating the joysticks, for example.

We have built and interfaced a dual joystick control that replaces the Unimate training gun and its individual binary joint-control switches. With the aid of the aforementioned on-line coordinate transformation program, these joysticks can control the motion of the Unimate along any path specified in world coordinates, considerably simplifying training by doing.

We have interfaced a word (or phrase) recognition device (Threshold Technology Model VIP-100) to our computer system. This interface provides a separate voice communication channel primarily during training, but also during task execution. In a limited fashion, it is now possible to control the position and velocity of the end-effector in world coordinates, and to train the Unimate to perform an elementary task composed of a sequence of actions using a small vocabulary of spoken words and phrases. We plan to expand this limited language to cover many more manipulative operations. including the incorporation of sensors, and implementing an editing system for program modification and debugging.

An experiment simulating a one-sided fastening operation by the Unimate made full use of these capabilities. The Unimate was controlled during training by the word/phrase recognition device. Using only spoken words and phrases for control, the operator trained the end-effector to go through a sequence of positions in a desired pattern and simulate the fastening action. The voice commands for this experiment are shown in *Figure* 11. The coordinate transformation program aided the trainer by enabling him to specify straight or circular paths for the end-effector to follow. After training the system in this manner, a spoken command caused the system to retrace its sequence of stored actions, illustrating both training and execution. This experiment clearly showed promise in the use of training aids for both simplifying the programming of a manipulator and improving its performance by having far greater control of its end-effector.

In the next phase of our work, we plan to develop computer software applicable to a wide range of real factory tasks in material handling and assembly operations, using one or two manipulators. Further development in the vision system will include automatic feature selection ("training by showing") and more extended part inspection. ■

This work is funded by Grant GI-38100X1, the National Science Foundation, and supplemented by twelve industrial affiliates.

CHAPTER 6
APPLICATION

Commentary

Beginning with an overview of applications of industrial robots, this chapter moves on to consideration of specific robot-application relationships.

Based upon the capabilities of certain robots, a number of actual and potential applications are enumerated. These examples should prove especially helpful to the novice user by, first, providing suggested robotic operations and, secondly, by assuring the reader that such robot applications are feasible.

This chapter brings into focus the relationship between a robot's capabilities and its applicability to certain kinds of manufacturing operations.

Reprinted from Machine and Tool Blue Book, March, 1980

Industrial Robots Today

Robots can be a most efficient and economical alternative to human labor in material handling, welding, assembly, machining, forming and finishing; especially in hazardous or unpleasant environments. Here is a detailed listing of most industrial robots currently available in this country.

By WILLIAM R. TANNER, President, Tanner Associates, Farmington, Mich.
and RAYMOND H. SPIOTTA, Executive Editor

In the fourth century B.C., Aristotle wrote, "If every instrument could accomplish its own work, obeying or anticipating the will of others ... if the shuttle could weave and the pick touch the lyre without a hand to guide them, chief workmen would not need servants nor masters slaves." Twenty-four centuries later, in factories around the world, his vision is finally becoming reality.

Some of this reality is embodied in the industrial robot, which is defined by the Robot Institute of America as, "a programmable, multifunction manipulator designed to move material, parts, tools or specialized devices through variable programmed motions for the performance of a variety of tasks." Development of industrial robots began in the 1950's and the first units were put into production by Unimation, Inc. and (then) AMF Versatran in the early 1960's.

Today, more than 60 companies around the world produce industrial robots and several thousand robots have been installed in factories in the United States. There are also more than 2000 robots in use in Japan and nearly that many operating in Europe. Annual shipment of robots by U.S. manufacturers exceeded 1000 units in 1979 and is projected to grow to around 5000 units by 1985. Overseas, the production and installation of robots is expected to show similar growth.

What Is a Robot?

Just what are industrial robots and how do they work? Although they vary widely in shape, size and capability, industrial robots all generally are made up of several basic components: the manipulator, the control and the power supply.

The manipulator is the mechanical device which actually performs the useful functions of the robot. They are hydraulically, pneumatically or electrically driven jointed mechanisms capable of as many as seven independent, coordinated motions. Feedback devices on the manipulator's joints or actuators provide information regarding its motions and positions to the robot control. A gripping device or tool, designed for the specific tasks to be done by the robot, is mounted at the outermost joint of the manipulator. Its function is directed by the robot's control system.

The control stores the desired motions of the robot and their sequence in its memory; directs the manipulator through this sequence or "program" upon command; and interacts with the machines, conveyors and tools with which the robot works. Controls range in complexity from simple stepping switches to minicomputers. Memory devices include drum switches, magnetic plated-wire, ferrite cores, floppy discs, cassette tapes and solid-state electronics.

Hydraulically actuated robots also in-

Cincinnati Milacron T^3 robot tends to loading and unloading Cinturn NC universal turning center.

clude an electrically driven pump, control valves, reservoir and heat exchanger in a power supply unit which provides fluid flow and pressure to drive the manipulator. Pneumatically driven robots are usually connected (through a filter-regulator) to the factory compressed air system. Cooling water is also required by some robots to maintain hydraulic fluid temperature or for control cabinet temperature regulation.

Basic Robot Types

There are two basic types of industrial robots: nonservo and servo controlled.

Motions of the nonservo robot are limited by fixed stops on each mechanical joint, or axis. Thus, each axis of this robot can generally move to only two positions. These stops are adjustable so that the end positions of each axis can be set up as required for the task to be done. This limits the capabilities of a nonservo robot to the performance of basically simple tasks. With some nonservo robots, such as the Prab and the Mobot, additional stops are provided on some axes. These indexable stops are inserted or withdrawn automatically on the appropriate steps of the program to provide more than two positions for an axis.

Nonservo robots may use simple sequencing controls such as stepping switches or pneumatic logic sequencers capable of executing single programs of about 24 consecutive steps, or electronic programmable controllers of greater program capacity. These controls initiate the motions of the manipulator, actuate tools or grippers and transmit and receive signals to and from the other equipment with which the robot operates.

Positioning repeatability of nonservo robots is typically in the range of ±0.010 inch. Payload capacities range from 2 to 100 pounds. Horizontal reach ranges from 1 foot or less to as much as 8 feet; vertical reach from 6 inches to several feet. From two to five independent axes of motion are available. They are relatively inexpensive (from $5000 to $35,000), are simple to program or set up for the task to be done and require little maintenance.

Typical nonservo robots available in the U.S. include the Auto-Place and Seiko, which are small pneumatically actuated units available in several models, and the slightly larger Auto Mate. Modular Machine Co. produces a series of linear and rotary motion devices, in a range of sizes, which are combined into Mobots for specific applications. Prab robots are larger, hydraulically driven nonservo units with payload capacities ranging over 100 pounds and horizontal reach out to 98 inches.

While outwardly similar to their nonservo counterparts, the servo-controlled robots incorporate feedback devices on the joints or actuators of the manipulator which continuously measure the position of each axis. This permits the control to stop each axis of the manipulator at any point within its total range, rather than at only two, or a few, points. Servo-controlled robots thus have much more manipulative capability than nonservo robots by being able to position a tool or gripper anywhere within the total work envelope.

Servo-controlled robots use electronic sequencers, minicomputer or microprocessor-based control systems and magnetic or solid-state electronic memory devices. They are often capable of executing more than one program containing several hundred sequential steps. Simple point-to-point positioning and trajectory control are both available, in some cases in the same unit. Other control capabilities available include branching, subroutining, speed control and multiple-axis tracking of moving objects.

Positioning repeatability of servo-controlled robots is in the range of ±0.060 inch. Payload capacities of the various units range from less than 10 pounds to more than 500 pounds. One robot, the Versatran Model FC, can handle end-of-arm loads up to 2000 pounds. Vertical and horizontal reach range from about 18 inches to 10 feet or more. One of the largest robots, the Cincinnati Milacron T³, has a working volume of 1000 cubic feet and can reach 154 inches vertically, while the smallest servo-controlled robot available, the Unimate Puma 250, has a working volume of about 9 cubic feet.

Prices of servo-controlled robots start around $35,000 and go up to about $125,000. All are either hydraulically or electrically driven. Hydraulic-drive units include Versatran, Cincinnati Milacron T³, Unimate and P.A.C. Pacer—which are basically point-to-point robots—and DeVilbiss/Trallfa, Nordson, Binks and Thermwood, which are continuous-path units. Electric servodrives are used on the Asea, General Numeric and Unimate Puma and Apprentice robots.

Where Are They Used?

Industrial robots are being used for a wide variety of tasks in factories, shops and foundries around the world. Robots unload parts from die casting machines and plastics injection molding machines. They load and unload parts at machine tools and stamping presses. They handle cores and castings in foundries and make shell molds for investment castings. They handle hot billets and parts in forging operations.

Robots are used for resistance (spot) and continuous fusion (arc) welding. They spray paint, stain, porcelain frit and plastic resins. They perform many material handling operations, stacking and unstacking parts and handling hot, cold, fragile, large, small, light and heavy parts quickly, safely and reliably; often handling several parts at a time. Robots are used for drilling, grinding, polishing and deburring, handling either the parts or a power tool.

In die casting and plastics injection molding operations, a robot may unload a single machine or as many as three machines. Lubrication of dies and quenching and trimming of die castings, or degating of plastic mold-

Asea robot handling MIG welding torch.

ings, may also be done by the robot. Users cite increases in productivity and reduction of scrap as significant advantages of robots in these operations.

In machine tool loading and unloading, the robot may also tend more than one machine—loading and unloading each in turn, or on demand, and transferring parts from machine to machine, as well as placing parts in gages for dimensional checking. Two-handed tooling is often installed on the robot so that it can handle both the rough and the finished part with a minimum number of movements. Productivity improvement is a major factor in robot machine tool tending.

Robots load and unload stamping presses and transfer parts from die to die or from press to press. Two-handed tooling is often used on these operations also. Because robots are relatively slow compared to average press cycles, they are generally used to handle large, heavy parts and for press-to-press transfer involving longer distances or where turn-over of the part is required.

Both nonservo and servo-controlled robots such as the Seiko, Auto-Place, Prab, Versatran, Unimate and Cincinnati Milacron T[3] are being used for all of these operations.

Applications in foundries include transfer of hot castings from pouring lines to shakeout tables, installing and removing cope-and-drag clamps on molds on pouring lines, setting of cores and venting and drying of molds. Robots are also used in casting cleanup operations, handling cutting torches or abrasive cutoff wheels to remove gates and risers, and for grinding flash from parting lines.

In cleanup operations, either the casting is handled over a stationary grinder or cutoff wheel or the torch or power tool is held by the robot and moved over a stationary casting. Increased productivity, improved quality and removal of workers from unpleasant or hazardous environments are considerations in these robot applications. Because of the complex motions required, servo-controlled robots such as the Unimate, Asea and Cincinnati Milacron T[3] are used for casting cleanup operations.

In forging operations, robots are used to transfer hot billets from furnaces to forging presses, to transfer parts from die to die in successive forming operations and to handle hot and cold parts in trimming operations.

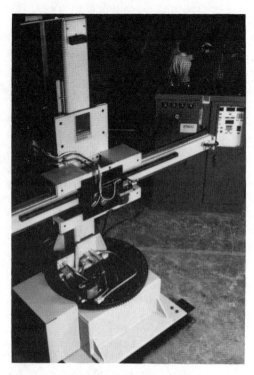

Versatran Model FC robot has 2000-pound load capacity.

Advantages are increased productivity, especially where heavy parts are involved, safety and working condition considerations. Here also, servo-controlled robots such as the Pacer, Versatran and Unimate are generally used.

Continuous-path servo-controlled robots such as the Binks, Nordson, De-Vilbiss/Trallfa and Thermwood are used for spraying a wide variety of parts and materials. Paint, plastic powder, porcelain frit, stain and asphalt insulating materials are applied using air, airless and electrostatic spraying equipment. Metal and wood furniture, automotive panels, plastic parts, sanitary ware and domestic appliances are among the wide variety of parts sprayed by robots.

Most of these operations are done on moving conveyor lines, with the robot's motions synchronized with the movement of the conveyor. A random arrangement of different parts may be sprayed in succession, with the robot's control selecting and executing the proper program at the appropriate time, triggered by signals from a part identification system. A major factor in the use of robots for spraying is a significant reduction in material usage. Quality improvement, as measured by appearance and uniform coating thickness, is also significant. Another consideration which may be important in the future is the prevention of human exposure to toxic materials.

Another application for continuous-path servo-controlled robots is spraying plastic resin and chopped glass fiber into molds in the manufacture of glass-reinforced plastic products. Yet another is the application of adhesives and sealers in complex patterns. In these operations, uniformity of materials application is critical to product quality: avoidance of worker exposure to irritating or toxic materials is also a consideration.

In material handling applications, both nonservo and servo-controlled robots are being used. They may simply transfer parts from one location to another, handling one or more at a time. With simple sensors installed on the end-of-arm tooling, a robot can remove parts and materials from stacks, one piece at a time. It can place parts in precise, complex patterns on pallets and in containers. Many of the servo-controlled robots can remove parts from, or place parts on, continuously moving conveyors or work fixtures.

End-of-arm tooling for material handling operations includes vacuum and magnetic devices and a wide variety of pneumatically or hydraulically actuated mechanical grippers, usually designed especially for the parts being handled. Multiple grippers or holding devices are often used to handle more than one part at a time.

Material handling robots can reduce damage in handling fragile parts and can remove operators from hazardous or unpleasant environments or relieve them from monotonous or fatiguing tasks. They can increase productivity where heavy parts, short cycle times or long transfer distances are involved. Interlocked with production equipment, robots can eliminate delays and interruptions, thus increasing system efficiency.

Some Basic Considerations

Use of industrial robots in production operations is, for the most part, a relatively new experience. The development and implementation of robot applications generally follows the same basic approach as for any other manufacturing process. However, the robot's flexibility and limitations are a somewhat unique combination requiring some special considerations if success is to be achieved.

Except for a few small nonservo units, robots are not particularly fast-acting devices. In most operations, except where weight or size of parts or tools is significant, a human can operate as fast as, or faster than, a robot. On the other hand, a robot has a decided advantage in handling heavy objects and in its effective work envelope. Also, the robot is not subject to fatigue and will operate at a steady rate under conditions where a human's pace tends to decrease with time.

In high-speed or high-volume manufacturing operations, custom-designed, special-purpose automation devices are generally more practical than industrial robots. At more moderate production volumes, the cost to design and build special-purpose devices often cannot be justified. Here, "off-the-shelf" automation in the form of robots is a logical alternative. The relative ease of programming and the capability of many units to store several programs in memory and execute, at random, any one of them also makes robots attractive for batch manufacturing operations.

As with any automated equipment, a robot requires an ordered environment. It is a device which is programmed to move through, and stop at, various positions in space. It performs this task with a fair degree of accuracy and repeatability; however, if the object it is intended to handle or work on is not consistently oriented and positioned with at least a similar degree of accuracy, the robot will fail in its task.

Likewise, a robot is not capable of exercising judgment, except when pre-programmed to act upon externally generated data. Thus, it cannot intuitively determine whether a part being handled is "good" or "bad" or if the tool it handles is functioning properly. Preinspection of parts or monitoring of tool performance may be necessary. Sometimes such inspection or monitoring cannot be practically accomplished except by human observation; in such cases, application of a robot may be difficult to justify.

An Image Problem

"Robot" is a relatively recent addition to our language. The term was first used by the Czech writer Karel Capek in 1921 in his play, *R.U.R.* (Rossum's Universal Robots), to describe artificial beings manufactured to replace workers. Eventually the robots revolted, turned against their creators and eradicated them.

Robots were further portrayed as a menace to humanity in Fritz Lang's 1926 movie, *Metropolis,* and in early science fiction. To some extent, the dislike and mistrust of robots fostered by these early dramatizations still exists. Certainly, a worker in a factory today, faced with displacement by a robot, must feel some concern and perhaps resentment.

Much of this concern and resentment can be avoided or eliminated by an open, humanistic approach to applying robots. Workers should be aware that they are being *displaced,* not *replaced,* and job reassignments for the displaced should avoid any loss of position or income; where practical, these workers should be given an opportunity for retraining or upgrading of skills. Emphasis should be placed on applying robots to monotonous, fatiguing or hazardous operations.

As George Mechlin, vice president of research at Westinghouse, said at the 9th International Symposium on Industrial Robots (ISR) last year, "It's ironic that many people see robots as mechanical workers in a factory threatening human employment and attempting to dehumanize them. That very dehumanization is a problem that robots can eliminate." A concerned approach to robot applications can go a long way toward overcoming resistance to their use and toward improving conditions in the workplace.

Future Developments

As with numerical control and other manufacturing equipment, microprocessors are moving into the area of industrial robotics. Microprocessor-based robot control systems are becoming more and more common. As a result, robot control capabilities are enhanced and new techniques, such as sensory feedback, are becoming practical. In the near future the force, tactile and vision feedback systems now under development will move from the laboratory to the factory floor.

The immediate impact of sensory feedback capabilities will be significant. No longer will a robot require the highly ordered environment of today. Sensor-equipped robots will be able to work with mixed, randomly positioned, unoriented parts. Sorting and visual inspection operations will be feasible as robots attain the ability to apply sensors and rudimentary judgment to their programmed tasks. Sensory feedback will enhance the use of robots for assembly operations also.

The advent of high-capability microprocessor-based controls will also have an impact upon robot programming techniques. Off-line, high-level language programming will be employed, particularly on complex, long or high-precision operations. Control of groups of robots in manufacturing systems by a central computer, already done in a few isolated installations, will become more common. Off-line computers will be used more extensively for program storage, with programs

Three Unimate robots spot weld automobile body.

Auto-Place robot transfers glass plates from rack to processing conveyor.

down-loaded to the robots as required to enhance batch manufacturing applications. Off-line simulation with interactive graphics systems will permit development and prove-out of new programs without interruption of production.

Of equal importance with the development of higher-capability control systems and sensory feedback is the development of more specialized robots. Without sacrificing the flexibility of programmable automation, the trend toward special robot configurations for certain applications will continue. Some specialization has already been accomplished, as evidenced by the "painting robots" of DeVilbiss/Trallfa, Nordson, Binks and Thermwood, and by Unimation's Puma assembly robot and Apprentice portable fusion-welding robot.

Such areas of specialty will include robot fusion (arc) welding systems such as those now offered by Cincinnati Milacron, Asea-Esab and Unimation. New robot welding systems will incorporate sensory feedback for joint tracking and gap filling; programmable, coordinated-motion part positioners; and servoed welding controls. New mechanical configurations may also be developed, as well as new programming methods.

To answer the needs of metal fabricators, a family of new robots for press operations is needed. Except for the very small nonservo robots such as Seiko and Auto-Place, existing robots

lack the speed and positioning repeatability for effectively handling parts in high-speed stamping operations. A series of larger nonservo robots capable of high-speed motions and close repeatability, with multiple program storage or fast program exchange, is likely to be developed for these applications.

In assembly operations, the microprocessor-controlled robot with sensory feedback capability will perform the complex part and tool-handling tasks. For the most effective utilization of these high-capability devices, simple high-speed nonservo robots should be used to transfer parts from magazines

DeVilbiss/Trallfa robot carries dual electrostatic spray guns while applying paint to bicycle frames.

Industrial Robots Today

and part feeders to the assembly robots. These simple robots will be of modular construction and will have control systems which can interact with a supervisory computer for coordination of motions and fast program interchange.

Although these future developments will open up new areas of opportunity for industrial robots, present capabilities and applications have by no means been fully exploited. At the 9th ISIR, Jack Wallace, president of Prab Conveyors, Inc. and past president of the Robot Institute of America, stated, "Robots are underused in the U.S. today. For each one in place, there are an estimated 10 other possible applications."

Industrial robots are powerful tools for increasing productivity and for solving problems of worker safety and poor working conditions, and they are available now in a wide variety of shapes, sizes and capabilities. Arranged alphabetically by manufacturer or supplier in the listings that follow are the majority of the industrial robots available in this country today.

For a single copy of this article, circle No.4 on reader information card.

● ● ●

Manufacturer	Name or Model	Machine Description	Drive System	Work Envelope	Load Capacity	Program Capacity	Repeatability (± in.)	Remarks
Asea Inc. For more information, circle 25	IRb-6	3 to 6 axes; servo PTP; servo cont. path	Electric d-c servo	Max. radial stroke 22.7''; max. radial reach 41.4''; max. vertical stroke 31.7''; max. vertical reach 340°	13 lbs	350 steps (250 points + 100 other functions)	0.008	Options: additional memory of 350 steps; tape recorder; mass storage; computer link; adaptive control; servo-controlled linear track
	IRb-60	''	''	Max. radial stroke 37.7''; max. radial reach 82.2''; max. vertical stroke 54.3''; max. vertical reach 76.7''; base rotation 330°	132 lbs	''	0.016	Options: additional memory of 350 steps; tape recorder; mass storage; computer link; force sensor
Automation Devices, Inc. For more information, circle 26	Model 6100, 6200, 6300 Placement Device	Air-powered lift-and-transfer device incorporating a special Tangen Drives gear and cam system; mechanism enclosed in oil-filled housing	Double-acting electric solenoid valve activates an air cylinder	90° transfer (left and right); 1/4 to 1 1/2 lift	1 to 3 lbs	Adjustable to 100 cycles	0.001	Small size excellent for limited spacing; low maintenance
Auto-Place, Inc. For more information, circle 27	Series 10	5 axes; PTP	Pneumatic	Lift 0-2''; reach 0-12''; rotate 0-270°	5 lbs	24 steps	0.003	7 axes of motion with additional equipment; may be furnished and interfaced with standard programmable controllers
	Series 50	''	''	Lift 0-5''; reach 0-24''; rotate 0-200°; turnover 0-270°	30 lbs	''	0.015	''
Binks Manufacturing Co. For more information, circle 28	Model MKII-6-90	6 axes; cont. path; System "90" solid-state controller	Hydraulic	Nominal 7' high x 9' wide x 3' deep	30 lbs	4-minute min. per solid-state memory module; 2 modules standard; can be expanded to 15 modules	0.125	Designed for industrial spray painting applications
Cincinnati Milacron, Inc. Robot Division For more information, circle 29	T-3	Computer-controlled; 6-axis servo-controlled path	Hydraulic	97'' x 240°; 1000 ft^3	100 lbs 10'' from tool mounting plate	700 points (1500 points optional)	0.050	Maximum velocity 50 ips
	HT-3	Computer-controlled; 6 axis servo-controlled path; (heavy-duty version of the T-3)	''	102'' x 240°; in excess of 1000 ft^3	225 lbs 10'' from tool mounting plate	''	''	Max velocity 35 ips

Industrial Robots Today

Manufacturer	Name or Model	Machine Description	Drive System	Work Envelope	Load Capacity	Program Capacity	Repeatability (±in.)	Remarks
DeVilbiss Co., The For more information, circle 30	DeVilbiss Tralfa Series TR-3003 S or W	6 axes; servo controlled cont. path with built-in PTP capability; lead-through teach; microprocessor control with dual floppy disc memory	Hydraulic	Horiz. motion 10.4" (93° or 135°); vert. reach 6.8"; reach (in & out) 3.2"; wrist motions—pitch 176° or 210°; yaw 176° or 210°; roll 210° or 420°	Arm equipped with rotary actuators. 25 lbs at end effector. Flexiarm 10 lbs at end effector	Computer Robot Control (CRC). 64 programs. 128 minutes (depending on sampling rate). Single Cassette Tape (SCT). single program. 85 or 105 seconds	Flexiarm 0.090; rotary actuator 0.020	Designed specifically for spray finishing. opt. PFS (path, function, speed) for programming gun path on a stationary part
General Numeric Corp. For more information, circle 31	Model GN0	6 axes; 1 or 2 hands; servo PTP	d-c servo	33" radius hemisphere	11 lbs each hand	9999 part program numbers; 1050 of punched tape equiv.	0.020	If used with built-in CNC control, power source depends on specs of CNC power supply
	Model GN1	4 axes; servo PTP	..	Stroke 31.5" or 43.5"; 20" vert. travel; pivot through 210° horiz. arc; 180° gripper wrist rotation	45 lbs	..	0.039	Double-hand avail.; can be had with CNC either attached to robot or remote
	Model GN2	5 axes with simultaneous control of 3 axes; servo cont. path	..	Vert. arm travel 31.5"; arm pivots 300° horiz & extends 43"; gripper hand rotates 270° & can pivot 120° vert	..	50-point memory; 704 points opt.	0.039	
I.S.I. Manufacturing, Inc. For more information, circle 32	Variable	Standard products of modular design providing the degrees of freedom required for specific applications	Pneumatic, hydraulic or electric per customer request	Variable	150 lbs	As per customer specification	0.002	Modular design allows user to select only movements required
Industrial Automates, Inc. For more information, circle 33	Model 9500	Nonservo PTP, 5 axes	Air/oil	24" extend; 7" lift; 7" elbow; 90° or 120° swing; 0°-180° wrist	10 lbs	32 I/O. 1000 memory steps	0.015	Optional equip.: end-of-arm tooling; base subsides (X & Y axes); extract function module
Manca, Inc. For more information, circle 34	Fibro/Manca Universal Modular Parts Handling System	No. of axes depends on application; nonservo PTP; 3 basic modules—linear translator, rotary unit, gripper	Pneumatic, hydraulic, electric	Arm move – horiz (arc). up to 360° (100° 200° 360°). vert. 10 to 1250 mm stand. radial. 1250 mm. Wrist move – pitch (arc). up to 360°; yaw (arc). up to 360°	Max. at low speed. up to 1700 kg depending on size of modules	Memory, mechanical step sequencer, pegboard control; accepts any programmable or hardwired controller	Up to 0.01 mm (0.0004") depending on speed & load	Off-the-shelf parts-handling system is tailor-made to suit specific need
Modular Machine Co. For more information, circle 35	Mobot	1 to 7 axes linear & rotary; servo & nonservo. PTP & continuous path	Pneumatic, hydraulic, electric options	Up to 6 x 10 x 50'	Up to 600 lbs	Unlimited	0.005	4-stage programming (configuration, position, power option, & sequence); standard modules combined to suit each task. 1-year payout
Nordson Corp. For more information, circle 36	Industrial Robot	Driven by rotary actuators through servo valves; electronic control through 6 axes of motion; 7th axis is manually adjustable	Hydraulic	Arm (swivel base) – horiz. circum. 116.25"; in & out 39.37"; Wrist—horiz. 240°; vert. 240°. Arm—vert. 107.8". manual pivotal adjust. up 60°. down 60° in 15° increments	33 lbs	30 minutes of operation in mass memory		

Industrial Robots Today

Manufacturer	Name or Model	Machine Description	Drive System	Work Envelope	Load Capacity	Program Capacity	Repeatability (± in.)	Remarks
PickOmatic Systems, Inc. For more information, circle 37	Robotic Parts Handlers	Consists of linear motion units & radial swing units; 2-axis parts handlers are cam-controlled rotary indexing devices with cam rollers following a precharted motion path; no programming is required	Electric	9 models available	Up to 35 lbs	Preprogrammed cam	0.005	
Prab Conveyors, Inc. Robot Division For more information, circle 38	Prab 4200	5 axes, nonservo PTP	Hydraulic	Extend 42"; rotary 250°; vert. 20°; end-of-arm pitch 90°; yaw 90°; roll 180°	75 lbs	24-100 steps depending on control option	0.008	
	Prab 4200HD	"	"	Extend 58"; rotary 250°; vert. 20°; arm pitch 90°; yaw 90°; roll 180°	125 lbs	"	"	
	Prab 5800	"	"	"	50 lbs	"	"	
	Prab 5800HD	"	"	Rotary 300°; vert. 0-6'; extend 42"; pitch 220°; yaw 220°; roll 359°	100 lbs	"	"	
	Versatran "E" Series	7 axes, servo PTP	"	Rotary 300°; vert. 0-10'; extend 60"; pitch 220°; yaw 220°; roll 359°	100 lbs	4000 steps	0.030	
	Versatran FA	"	"	"	300 lbs	"	"	
	Versatran FB	"	"	"	600 lbs	"	"	
	Versatran FC	"	"	"	2000 lbs	"	"	
Production Automation Corp. For more information, circle 39	Pacer I	1-10 axes, servo PTP	Hydraulic	Max. horiz. 20", vert. 12"; az. 180°	25 lbs	Up to 1200 command functions; to 120 program steps	0.005	Uses same space as one workman
	Pacer II	"	"	Horiz.30", vert. 24", az. 200°	50 lbs	"	0.010	
	Pacer III	"	"	Horiz. 48"; vert. 36"; az. 240°	150 lbs	"	0.025	
	Pacer IV	"	"	Horiz. 60"; vert. 48"; az. 240°	400 lbs	"	0.050	
	Pacer V	"	"	Horiz. 72"; vert. 48"; az. 270°	800 lbs	"	0.075	
	Pacer VI	"	"	Horiz. 84"; vert. 48"; az. 270°	1200 lbs	"	0.100	
	Pacer VII	"	"	Horiz. 96"; vert. 60"; az. 270°	1800 lbs	"	"	
Reis Machines For more information, circle 40	1215.5	5 axes; nonservo PTP & cont. path	Electric	118" dia., 47" vert.	110 lbs	200,000 steps	0.0235	Modular mechanical grips permit interchange of "fingers" with same grip body
	1215.6	6 axes; nonservo PTP & cont. path	"	"	"	"	"	
Seiko Instruments, Inc. For more information, circle 41	Applies to all Seiko robots in addition to following specific model data	Nonservo PTP; modular; multiplane mounting; internal damping; adjustable speed; internal sensor/limit switches for feedback (except Model 700)	"	V = vertical; H = horiz.; P-R = plane rotation; G-R = grip rotation	"	Equal to controller program capacity; Types of applicable controllers; mechanical, electrical, electronic, computer, air logic, custom	"	

Industrial Robots Today

Manufacturer	Name or Model	Machine Description	Drive System	Work Envelope	Load Capacity	Program Capacity	Repeatability (± in.)	Remarks
Seiko Instruments, Inc. (Cont.)	Model 700	4 axes; interchangeable arms (6", 8", 10")	Pneumatic	V=1.6"; H=6", 8", 10"; P-R=90°, 120°, custom; G-R=180°	2.2 lbs	"	0.001	25 cpm max. cycle speed; no accessories necessary
	Model 200	2 axes/std., 1-2 add'l axes with accessories	"	V=0.4-0.79"; P-R=90° std., 120° max. plus custom	1.65 lbs	"	0.0004	60 cpm max. cycle speed; avail. accessories—H, V-extension. G-R; intermediate stop, dual arm
	Model 400 Series	"	"	Model 400—V=0.79-3.94"; H=15.75"	8.82 lbs	"	0.001	Avail. accessories—G-R, grip slide sideways action; 25 cpm max. cycle speed
		"	"	Model 400(L)—V=0.79-3.94"; H=27.56"	6.61 lbs		0.002	
	Model 100	"	"	V=1.97"; H=7.8"	3.31 lbs	"	0.0004	30 cpm max. cycle speed; avail. accessories—G-R, grip perpendicular
Sterling-Detroit Co. For more information, circle 42	Model U Robotarm	3 axes; PTP	Hydraulic	Designed to suit	Up to 400 lbs	Relay	0.005	Modules assembled as reqd.; std. units with custom features
	Model QT Robotarm	4 axes; PTP	"	"	"	"	"	
	Model DQT Robotarm	4 axes; PTP; dual arm	"	"	"	"	"	..
	System II Robotarm	6 axes; PTP; modular	"	"	"	EPROM memory; 1K commands; indefinite memory storage; keypunch entry; circuitry for transf. RAM into EPROM		..
Thermwood Machinery Manufacturing Co., Inc. For more information, circle 43	Series Six	Servo cont. path/PTP; 6 axes (add'l. optional), floor/ceiling mount	Hydraulic/mechanical	Horiz. 48", vert. 84", rotary 135°, wrist roll 450°, wrist move. 180° cone	18 lbs	5 minutes cont. path; 8 programs; lead-through teach	0.125	Microcomputer-based control; analog feedback; semicond./hard disc memory opt.; program edit std
Unimation, Inc. For more information, circle 44	Unimate 1000	5 axes; servo PTP	Hydraulic	Rotation 208°, vert. traverse 46°-56°, boom ext. 7.5'-10'	50 lbs	32; plated-wire memory; lead-through programming	0.050	
	Unimate 2000B	6 axes; servo PTP or 500" cont. path	"	"	300 lbs	2048; plated-wire memory; lead-through programming	::	
	Unimate 2100B	"	"	"	"	"	0.080	
	Unimate 2000C	"	"	"	"	"	0.050	
	Unimate 2100C	"	"	"	270 lbs	"	0.080	
	Unimate 4000B	"	"	"	450 lbs	"	::	
	Puma	6 axes; servo PTP or cont. path	Electric	Vert. 60", horiz. 34", rotation 320°	5 lbs	As reqd.; LSI computer memory; lead-through programming	0.004	
	Apprentice	5 axes; servo cont. path	"	Vert. 35", horiz. reach 64", rotation 180°	10 lbs	8K; semiconductor memory; lead-through programming	0.040	

Robots: New Faces On The Production Line

CARL REMICK

"Robots of the world! The power of man has fallen! A new world has arisen: the Rule of the Robots."
> —RADIUS, a revolutionary leader
> *R.U.R.*

ON OCTOBER 9, 1922, A NEW PLAY OPENED at the Garrick Theater in New York; a new word entered the English language; and a new statement of technological fear found its way into the public consciousness.

The futuristic play was *R.U.R.* by Czechoslovakian dramatist Karel Capek, and the new word, "robot," was taken from the Czech word for work, *"robota."* Action was set at the headquarters of Rossum's Universal Robots (R.U.R.), a firm that developed and manufactured a line of artificial humans. As Rossum's robots increased in sophistication, they developed a lust for power and eventually seized control of the world. Critic Alexander Woollcott perhaps best summed up popular reaction when he termed the play "hair-raising."

Technology has enjoyed more than a half century of advance since that production, and things called robots have moved off the stage, out of the laboratory, and into industry. There are now an estimated 5,000 of them toiling away in factories throughout the world; but seeing them in motion does not—at least at this time—cause the scalp to prickle. Essentially mechanical arms, today's robots are humble machines performing humble tasks in such areas as materials handling, machine loading, die casting, investment casting, forging and heat treating, foundry operations, plastic molding, welding, machining, finishing, and simple assembly.

Robots have a practical simplicity their human counterparts lack and are designed with considerably greater creative flair. They can be driven by hydraulic, pneu-

matic, or electrical mechanisms. Their control may be by mechanical stops, magnetic tape, or computers. Their hands can be of any configuration. "A robot doesn't have to go home at night, so if we give it vision, we can leave its eye on a pedestal," notes Joseph F. Engelberger, president and founder of Unimation Inc., which pioneered in industrial robots. "Another good place for an eye is in its palm."

Robot eyesight is still on the frontier as far as industry is concerned. But General Motors's new PUMA (Programmable Universal Machine for Assembly) system—which uses robots built by Unimation—will eventually be fitted out with vision.

The auto industry has long held the lead in practical robotics, and PUMA is a state-of-the-art system. "It is designed to fit right where a human being would fit in assembling automobile dashboards, brakes, tail lights, carburetors—all those operations where people stand cheek-by-jowl and jam middle-sized pieces together," Engelberger says. Such assemblies represent 90 percent of the parts in a GM passenger car.

"Anyone who has looked into human anatomy will have seen at once that man is too complicated and that a good engineer could make him more simply."—HARRY DOMIN, *General Manager, Rossum's Universal Robots.*

A PUMA robot is being sent to Delco Products in Rochester, New York, to begin its first job in late spring. The robot will grab an electrical motor armature as it comes from a 450-degree furnace, attach a commutator ring, and run it back for heat curing. It has been a gloves-and-tongs operation and is "kind of a nasty job," says Frank Daley, director of GM Manufacturing Development. In all, ten PUMA installations will go on line this year.

Though sophisticated, the present PUMA is a first-generation robot system. "These first-generation robots are fixed sequence devices," explains Donald E. Hart, head of the Computer Science Department at GM Research Laboratories, and he points out that the robots' consequent stubbornness can cause problems. "Since the sequence of operations is preprogrammed and fixed, the part to be operated on must be in precisely the right place at the right time," Hart explains. "This means that the parts must be accurately fixtured and stationary. If a body

Carl Remick is an associate editor of Management Review.

SEEING IS RETRIEVING: *A vision-equipped robot lifts a part from a moving conveyor belt in General Motors' experimental system CONSIGHT.*

panel isn't where it's supposed to be, the welding robot will happily weld thin air."

Which is no way to boost ROI. So GM researchers are pressing the effort to make PUMA a second-generation, or "adaptive," robot system through sensory feedback.

"Tomorrow's robots will have two senses: vision and touch," says Lothar Rossol, a staff research engineer in GM's Computer Science Department. "In a limited way, these second-generation robots will be able to sense and react to their environment—that is, their path will change on each cycle as the position of parts changes or as the requirements of the job change."

GM has already developed a "seeing" computer that it uses to inspect circuit chips. But checking components in the relatively tidy world of electronics is far less demanding for a computer than figuring out the visual chaos of heavy industry. States the chief of GM Computer Science: "In general, the real world of GM's plants does not consist of stationary white parts on black backgrounds.

Parts are grey, shiny, dull, dirty, and greasy. They move on conveyor belts, hang from hooks, and are jumbled in bins."

GM has teamed computer vision with a Cincinnati Milacron robot in a system called CONSIGHT, which permits the robot to pluck a part from a moving conveyor belt—*if* the part is clearly separated from other parts. Massed bits and pieces continue to bewilder the computer, and work goes on.

Cautious acceptance

Talk about robots that see and feel may quicken a layman's pulse, but robot authorities are rather casual about the whole thing. Technical feasibility isn't a problem so much as is lack of commercial demand, indicates John A. Fulmer, manager of Cincinnati Milacron's Industrial Robot Division.

In fact, lack of demand is something that hampered

Automatons are all around

The future of industrial robots looks black—in terms of ink, that is. It took the robot business years to pass from prototype to profit, but the industrial robot's ability to, quite literally, lend a hand around the shop is gaining recognition. Here are just a few specific applications robots have received:

- International Harvester/Canada has applied robots to a heat-treating and forming line used in producing harrow disks. The robots have resulted in improved quality, increased production volume, and better production control. And the workers who were replaced are delighted, having been reassigned to work that is not so hot and heavy.

- Xerox uses robots in manufacturing a family of duplicator parts. Robots were chosen over single-purpose hard automation because their flexibility can accommodate the model changes and variable demand of this operation. The robots' manufacturer also notes that this soft automation "resulted in a high degree of unit machine independence, minimized control system complexity, and included the ability to support and maintain the equipment by plant maintenance personnel."

- Robot welding systems at several Chrysler assembly plants have reduced the time and cost required by model changeovers. "This, along with the flexibility of the system during operation, is important to us because we operate with fewer assembly lines than most of our competitors," said Richard A. Vining, vice-president of Chrysler's Stamping and Assembly Division, where the first system was planned. "When you've got fewer lines, you need more flexibility to handle your product mix."

- Labor costs are rising, but small-batch manufacturers may find hard automation too expensive for their low-volume needs. Do-ALL Co., a leading supplier of machine tools, cutting tools, and precision measuring equipment, increased productivity by incorporating robots in a small-batch system using conventional and NC equipment. The robot operation combines the programming convenience of computer numerically controlled equipment (CNC) with a simplified approach to parts classification, thus boosting efficiency without the expense of group technology using computer-aided design/computer aided manufacturing (CAD/CAM) techniques.

---

the robot industry in general until recently. John J. Wallace, president of Prab Conveyors, which manufactures robots, points out that the industrial robot business has been around for 20 years. "There was a feeling that the minute robots were cleaned up so they were saleable, people would knock manufacturers down trying to buy them," says Wallace, who was also 1978 president of the Robot Institute of America (RIA). "It hasn't been that way—only in the last couple of years have we seen a real, serious interest in the product and what it will do."

Unimation's Engelberger reminisces about the early days, in 1956, when he and his associates were "starry-eyed"—certain that sufficient technology was on hand to make a commercially successful robot. "We were absolutely right!" he says. "Just 16 years and 12 million dollars later, Unimation Inc. turned its first profit in its sole business, robotics."

Lack of R&D may not be what afflicts American industry today. "You don't need research and development to use something you already have. What you need are incentives, and certainly there's nobody in government giving people extra incentives to use robots. This is an under-used technology, and it's not the only one," says Wallace of Prab Conveyor.

What incentives would he suggest? "An extra tax deduction for under-used technologies that would increase productivity—that's such an obvious one," Wallace states. "Here's a funny thing. We're in the conveyor business, and scrap recycling is our specialty; we're also in the robot business. Every time we sell a scrap conveyor, the buyer can get a 20 percent tax deduction on it, and when he buys a robot, he gets 10 percent. Now that's cockeyed—it should be the other way around."

Industry's caution about technological innovation is understandable, says Wallace: "We're talking about production managers here. They have to accept a technology and commit themselves to its success before it *will* be a success. Now, these people are production-minded rather than efficiency-minded. They say, 'You want 200 widgets a day instead of 100 widgets? Fine, give me twice as many machine tools, twice as much floor space, and you'll get your 200 widgets.' And they know for a fact they can deliver the widgets that way. They *don't* know for a fact that they could retrofit their existing facilities with robots and get, say, 150 widgets just for openers. And they're not particularly interested in betting their status on that possibility because it's risky. There's no particular incentive for them to stick their necks out unless the boss says, 'Hey, we're going to get an extra tax benefit if we use these things—get busy and figure out how to use them.'"

Yet even without extra tax incentive, the word on robots is slowly beginning to get around, Wallace says. And Donald A. Vincent, manager of RIA, reports that most robot manufacturers now have substantial back orders.

Effect on human jobs

For every robot order on the books, there is a job that won't be done by human labor anymore. But whatever the future may hold, it's unlikely anyone would claim that today's robots rob workers of satisfying service. "A job can be done by a robot only because the task was subhuman in the first place," says Unimation's president. "If a job requires any judgment at all, there is no robot that can do it, and there won't be for a long time."

"When people have a job that a robot can do, they know it, and they really don't like what they're doing," says Jerry Kirsch, president of the robot manufacturer Auto-Place, whose firm has done extensive work in robot vision. "I was out at a Ford plant one morning recently and saw a bright-looking kid who spends all day working on the line, picking up a part and putting it into an assembly machine—it's really demeaning."

Former RIA President Wallace makes a similar point: "We see this attitude, 'Don't take my job away from me!' And our response is, 'Well, if your job is so dumb and so repetitive and so monotonous that a robot can do it, why, it's not much of a job, and you'd better look for another one.' And the union people are the ones who put those words in our mouths!"

If that is true, and workers and their unions favor robots, then what about the troubles at Lordstown? Really the last word in automation today, the GM assembly plant at Lordstown, Ohio, has the most robots of any GM operation and had severe and widely publicized labor problems during the early '70s.

"Lordstown had labor problems, and Lordstown had robots, but it was strictly journalism that connected the two," says the head of Prab Conveyor. "The labor problems had nothing to do with the robots."

Joseph Engelberger's firm supplied the robots, called Unimates, that were put in service at Lordstown. "They set up the line there to run at 120 jobs an hour," says Engelberger. "A classical automotive line runs at 60 jobs an hour, and there was a lot of complaining that there was a speed-up. In fact, there was not—the job content was cut in half as well. But that didn't change the attitude of the workers on the line."

And there was another problem, he notes: "In addition to the robots that got all the publicity, there was a vast amount of hard automation at Lordstown that did not work when the plant opened." Soft automation (robots) can be moved from place to place and reprogrammed to perform a variety of tasks; hard automation is machinery fixed in one location and designed to perform only a single function.

"Without fanning up the Unimate, a[dds] Engelberger, "the fact is that the Unimate h[ad] hours of field experience and millions of do[llars de]velopment behind it when it went into Lordst[own. It] was a reasonably reliable device. But when you bu[y h]ard automation, there tends to be a lot of debugging involved. The GM people had a hot car [the Vega], and they wanted it out. They had 750 more workers in that plant than they needed, because one of the things you do when automation doesn't work is to throw bodies at it."

"Minding" robots

The job of robot "attendant," or supervisor, may not require much skill, but it keeps a person busy and is not tedious if designed properly, says David Nitzan, program manager, Industrial Automation Group, SRI International. "One attendant can attend several machines; so if he is bored, that means somebody goofed in deciding how many machines he would be assigned to attend," says Nitzan, whose research group advises 18 corporations that make or use robots. "If somebody is assigned to look at just one machine—well, that is ridiculous. But if you put in enough machines to actually increase productivity, a person will not be bored."

"The most desirable job in the plant is taking care of a robot," Engelberger says. "If you look at job enrichment, what does it mean? It means getting a job that's unstructured. If your job is structured—if you're obliged to do the same thing every minute or every two minutes or every hour—it certainly lacks enrichment."

The robot attendant's job is a varied one. "In most cases, a worker trained to take care of robots would have five in his charge," says Engelberger. "On a particular morning, he might come in, sit down, and read a paperback. Or he might be busy as a one-armed paperhanger because he's got to program one machine, service another, and change the dies on the die-cast machine."

In any event, the idea of working directly and continuously on a die-cast machine holds small appeal for today's young worker. Unimation's president speaks of one customer who has a die-cast operation: "As soon as the old artisans who immigrated from Europe retire, they put robots in their place. They've got 35 robots in the plant now, and they have *more* people working there because it's become an extremely competitive plant, and for every die-cast machine there are five other jobs doing secondary operations."

Over all, the American metalworking industry experiences an annual attrition rate of about 15 percent, Engelberger reports. "At any fair-sized company, they could introduce robots almost indefinitely on an attrition basis," he says. And those workers whose jobs are, indeed, eliminated by robots can be reassigned and upgraded. "There's not an instance that I know of where we have had a customer who has fired a person to put a robot to work," he adds.

The United Auto Workers, as might be surmised, have an interest in all this, and the 1977 UAW Skilled Trades Conference passed a resolution urging action to lessen the labor impact of robots and similar technology. Among other things, it called for the following:

- Training, retraining, and vocational education for workers affected by automation.
- A governmental requirement that new technology be installed in plants only when displaced workers can be absorbed into other jobs.
- Advance notice to the union when a work force is to be reduced because of technology, with the reduction to be done through attrition.
- Making sure that equipment maintenance and programming remain the responsibility of skilled workers.

Need for productivity

"I've found most of the union leaders I've talked with to be very logical," Engelberger says. "They say they believe not only in robots but in automation—that they know the country needs productivity. But they ask two things: that there be no massive displacement and that they share in the productivity benefits."

He continues, "I hasten to add that this is the way they talk when they're not in negotiations. If you know anything about negotiations, you'll know that irrationality is a great bargaining tool. These people didn't get to be heads of unions by being dummies, and they use irrationality when they're bargaining. When you're talking about something face-to-face betweentimes, they're not irrational."

Unimation's president is frequently asked if, in the long run, he doesn't actually eliminate jobs. "I say, 'In the long run, yes we do.' But I remember my grandfather telling my father that there was something immoral about a 40-hour workweek because he was used to a 70-hour week. And I don't know that there's anything written in stone that the week should be 40 hours—why not 20 hours?" Engelberger says.

The subject of robots, then, is really the subject of productivity. "I look at productivity and say that it's always good—*always* good," Engelberger declares. "There are all kinds of wonderful things you can do with the benefits; the question is, *what* do you do? Do we divide them up unevenly, give them only to stockholders—give them to everyone? Why not take the benefits and use them to clean up the country's water and air?"

It's a decision the politicians will have to make, he says, indicating that while it may be difficult, the United States is fortunate in having to make it. With no prospect of wealth to allocate, underdeveloped countries have no such decisions facing them, he observes.

"Although the name 'robot' has connotations that bring many things to mind, it's still just another way to increase productivity," says Cincinnati Milacron's John Fulmer. "This country has grown and prospered for the last 200 years because we came along with ways to increase our productivity. And we darn well better start paying attention to that because the Japanese and the Germans are improving significantly faster than we are."

Mention of the Japanese raises the question: What is a robot? The figure cited earlier as the global population of robots—5,000—is based on surveys that find some 2,000 robots in the United States, 1,000 in Europe, and 2,000 in Japan. But some Japanese sources place their number of robots as high as 13,000. One then enters into the usual international rivalries—with American experts poohpoohing the higher Japanese claims, saying they include mere transfer devices and hard automation.

"We've been working on it for two years, and we still don't have *one* definition of 'robot,' " says RIA Manager Donald Vincent. "There are a number of definitions floating around. Usually a robot's regarded as 'flexible automation, a machine that can be reprogrammed and do a variety of tasks.' "

But that really doesn't satisfy. Real-life robots may be neither the scheming malcontents of *R.U.R.*, nor the lovable factotums, R2D2 and C3PO, of *Star Wars*, yet terming them simply "flexible automation" doesn't convey the full splendor of their protean possibilities.

Maybe Joseph Engelberger has the right idea: "When people ask me what a robot is, I usually give a facetious answer, saying, 'I don't know how to define it, but I know one when I see one.' "

Perhaps others should develop Engelberger's intuitive sense. Whatever robots may be, it seems there are going to be a lot more of them around in time to come. ●

Presented at the Fifth International Symposium on Industrial Robots, September, 1975

Practical Applications Of A Limited Sequence Robot

By Ronald D. Potter
Auto-Place, INc.

INTRODUCTION

To many people, the word "robot" connotes a highly sophisticated or complex machine with an intelligence capable of approaching man's. We are fascinated by the complex actions that we can make these machines perform, but we tend to overlook the important role of robots in carrying out simpler industrial tasks. Therefore, we should examine the real purpose of an industrial robot.

THE PURPOSE OF AN INDUSTRIAL ROBOT

The real purpose of a robot in an industrial application is to perform a job in a safer or a more efficient manner than can be achieved with human labor. By removing danger, unfavorable conditions, fatigue, and monotony from human concern, the robot is fulfilling its humane purpose. By helping produce better quality parts at a greater rate of production for a longer period of time, the robot is fulfilling its technical objective of increased efficiency.

DEFINITION OF A LIMITED SEQUENCE ROBOT

There are jobs in all levels of industrial processing that warrant the use of robots. These jobs vary in the complexity of robot actions required. Many such processes require a robot capable of performing somewhat simple motions , such as picking up an oriented part and placing it into a machine. This robot requires only a limited number of sequential actions to perform the job. It does not require either the control or motion capabilities of a more sophisticated robot. For the purpose of this paper, I shall designate this simpler robot as a limited sequence robot , in order to distinguish it from a more complex robot.

A simple process that results in increased safety or efficiency is a very worthwhile application for a robot. The safety aspect is obvious. The efficiency aspect relates to productivity and profitability. From the human standpoint, the simple routine jobs tend to be the monotonous, boring jobs that don't provide much in the way of job enrichment for people. Psychologically, a more complex job may be more fulfilling for a person to perform, as he may develop a sense of satisfaction in the skills he has acquired. Man's emancipation from industrial "slavery" is largely dependent on his ability to enslave the robot; to make it do the jobs that are undesireable either mentally or physically for man. The utilization of a limited sequence robot for many of these jobs is an important phase which should not be overlooked for more "glamorous" applications.

AUTO-PLACE: A LIMITED SEQUENCE ROBOT

Auto-Place industrial robots provide up to 5 axes of motion that can be programmed into 24 sequential steps. Two sizes are available: Series 10 with a load capacity of 5 pounds, and Series 50 with a load capacity of 30 pounds. Over 500 of these robots have been employed in various industries in the past 6 years. Some of the processes in which Auto-Place robots have been used are metal stamping, powder metal compacting, plastic injection molding, transfer from indexing tables and conveyors, assembly, inspection, metal removal, hot glass handling, and many others. Several of these applications will now be discussed.

PRACTICAL APPLICATIONS OF AUTO-PLACE ROBOTS

Auto-Place industrial robots have been used in many applications where the temperature has been unsuitable for human labor. In one such application, the ambient temperature is 150°F, and the glass parts being transferred are at a temperature of 1100°F. The Auto-Place transfer process involves picking up 4 glass parts at a time from a furnace escapement and transferring them to fixtures on a rotary index table for tempering. Prior to the use of the robot, persons performing the transfer wore special heat resistant suits and worked alternately for 20 minutes per hour.

Another example of a high temperature application is shown in figure 3. This Auto-Place system, consisting of a Series 50 unit mounted on top of a machine base, transfers 4 hot glass parts at a time from an indexing pin conveyor to an annealing conveyor. The part temperature is approximately 300° F, and the ambient temperature is around 100° F.

Figure 4 shows an Auto-Place system utilizing a Series 10 unit to assembly telephone transmitter carbon cups on an oven conveyor. This system assembles 28 cups per minute in an ambient temperature of about 95° F. Besides removing personnel from this hostile environment, this system also produces parts at a much greater rate than had been achieved previously.

Figure 5 shows a Series 50 Auto-Place unit (partially hidden by the packaging station) that is used to remove 16 parts at a time from an injection molding machine. These parts are plastic proof coin cases, and the quality control requirements for this part are such that they cannot be handled by people.

The vacuum plate hand on the Auto-Place unit vacuums the parts on a thin lip, and the hand is equipped with sensors to insure that all parts are in place on the hand before the molding machine is recycled. Many Auto-Place applications utilize tactile sensors in the fingers to insure safety. After the Auto-Place unit has removed all 16 parts from the molding machine, it turns the parts 90°, rotates over to the packaging stations shown in the photograph, and loads the parts into the tubes. Four identical systems are now being used by this customer.

Figure 6 shows the use of Auto-Place robots in metal stamping presses. In this application, two Series 50 units per press load and unload a 10 pound part through a series of 4 presses. As in all metal stamping applications, one purpose of the robot is safety: an operator's hands do not enter the press area. A secondary purpose in this application is efficiency: with human loading factors, the part required 5 presses to process it, while with robot loading, only 4 presses were required. This is due to safety considerations in the design of the dies. An operator now picks up a part from a conveyor and places it on a fixture in front of the robot. He then depresses two palm buttons, which cycles the press. The unload robot, on the back side of the press, then removes the part from the press, while the load robot picks up a part from the fixture and loads it into the press. All safety interlocks between the robots and press have to be made before the operation can proceed.

In addition to these industrial applications, Auto-Place robots have been used in many munitions applications. Being completely pneumatic, these robots operate safely in an explosive atmosphere, and displace man from a dangerous environment.

The above example applications illustrate some of the ways in which a limited sequence robot can be put to use in industry to upgrade the human and technical aspects of a job. We have developed robots as tools to help upgrade industrial processes, and our task now is to exploit them to their fullest potential in this area.

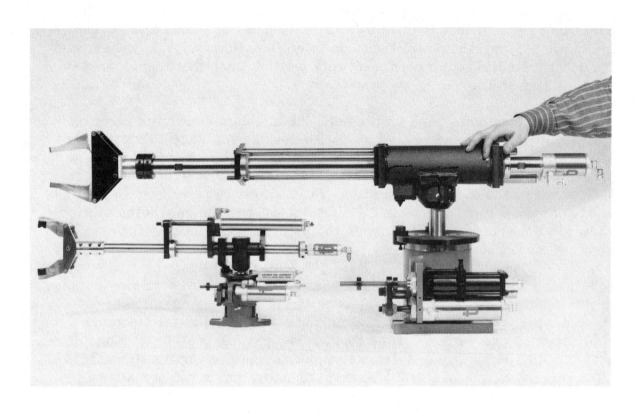

FIGURE 1   AUTO-PLACE SERIES 10 (5 LB CAPACITY) AND
SERIES 50 (30 LB CAPACITY) INDUSTRIAL ROBOTS

FIGURE 2   AUTO-PLACE PROGRAMMABLE PNEUMATIC
CONTROL SYSTEM (SHOWN OUTSIDE OF
ENCLOSURE)

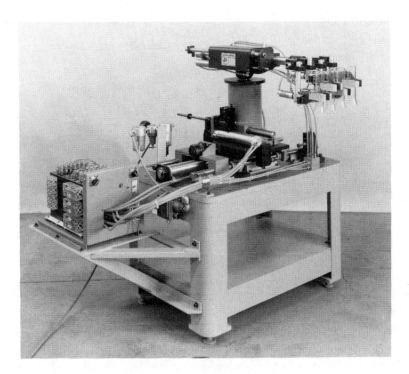

FIGURE 3   AUTO-PLACE SYSTEM USED TO TRANSFER HOT GLASS PARTS (SERIES 50)

FIGURE 4   AUTO-PLACE SYSTEM USED TO ASSEMBLE TELEPHONE PARTS ON AN OVEN CONVEYOR (SERIES 10)

FIGURE 5   AUTO-PLACE SYSTEM (WITH PACKAGING STATIONS)
USED TO UNLOAD PLASTIC INJECTION MOLDING
MACHINE (SERIES 50)

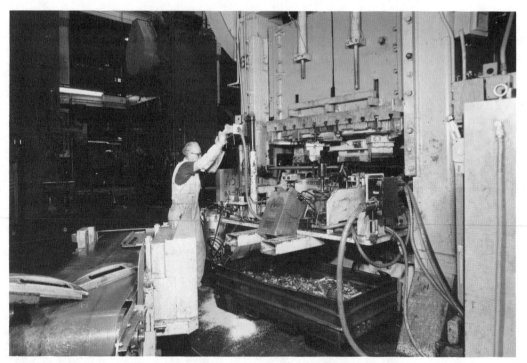

FIGURE 6   AUTO-PLACE SERIES 50 ROBOTS USED TO LOAD
AND UNLOAD METAL STAMPING OPERATION

FIGURE 7   AUTO-PLACE SERIES 10 ROBOT USED TO ASSEMBLE
SMALL AUTOMOTIVE PART ON A PALLETIZED
CONVEYOR

FIGURE 8   AUTO-PLACE SERIES 50 ROBOT WITH
PROTECTIVE COVERS (MOUNTED ON
ADJUSTABLE PEDESTAL)

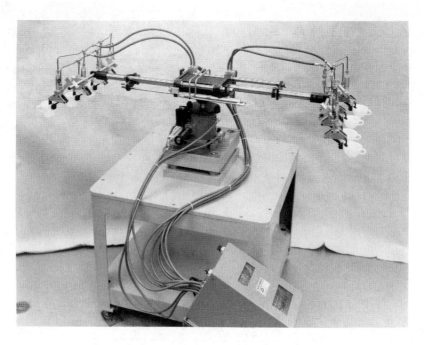

FIGURE 9   AUTO-PLACE SERIES 50 ROBOT WITH
                2 PARALLEL ARMS USED TO TRANSFER
                8 HOT GLASS PARTS

FIGURE 10   AUTO-PLACE SERIES 50 ROBOT WITH
                2 PERPENDICULAR ARMS USED TO
                TRANSFER GLASS BOTTLES INTO AND
                OUT OF A SANDBLASTING OPERATION

FIGURE 11   TWO AUTO-PLACE SERIES 50 ROBOTS
USED TO TRANSFER PARTS IN AND OUT
OF A GRINDING OPERATION

FIGURE 12   AUTO-PLACE SERIES 10 ROBOT USED TO
UNLOAD CARBIDE PARTS FROM A
COMPACTING PRESS

Presented at the 1978 International Engineering Conference, May, 1978

Industrial Applications For Electrically Driven Robots

By Brian Ford
ASEA, Inc.

A large number of installations have been made with the ASEA robot performing operations such as grinding, deburring, spot welding, arc welding, gluing, material handling, inspection, assembly, drilling, cutting, polishing and many others.

This paper will describe five applications shown in an accompanying 16 mm color film of installations in a factory.

These installations should be of particular interest to the small and medium production industries because they demonstrate methods to increase productivity, rely less on labor, achieve faster throughput, maintain higher machine utilization, improve quality and promote a better working environment.

INTRODUCTION

The five industrial robot applications that are designed are spot welding, measuring robots, shaft manufacturing, machinery castings and arc welding.

These applications put a high reliance on the accuracy, programmability, electrical drive system, special software programs and the microprocessor control of the robot.

1. SPOT WELDING

At SAAB-SCANIA an ASEA IRb-60 with 5 axes is used for spot welding of rear floor to SAAB 99.

The installation consists of an ASEA robot with welding equipment from ESAB mounted together on the same pedestal and a turntable provided with two rotary jigs. The turntable has two fixed positions, one load and leave station and the other a welding station. After manual loading of the jig, the turntable is indexed forward upon an order from the operator. The robot automatically receives a go-ahead signal when the jig is in position, whereupon the welding cycle is started.

In view of the size of the floor to be welded, the welding cycle is divided into two stages, the jig being turned between the two stages.

While the robot is welding, the operator removes the already welded floor and loads the jig with parts for the next floor.

The weight of the welding equipment applied directly on the

robot body is as follows:

Spot welding gun	59 kg	(load tilt)
Transformer	60 kg	(loads upper arm)
Transformer support	12 kg	(" " ")
Straight conductors	11 kg	(" " ")
Flexible conductors	3 kg	(load tilt)
Collectors	2 x 16.5 kg	(load upper arm and tilt equally)
Valves	7 kg	(loads upper arm)
Hoses	8 kg	(" " ")
Total	193 kg	

In addition to the load from the welding equipment, a load is also exerted on the robot if the welding electrode becomes welded to the workpiece.

To compensate for wear on the welding electrodes (max. 6-7 mm) built-in detecting equipment is used in the robot. This senses current alteration in the φ-and-tilt motors when the spot welding gun makes contact with the sheet metal. This is done in the first welding spot in each welding cycle. On the basis of the detection, the program is transformed in order to compensate the electrode wear. PTPL instructions are required to transform the program. The robot and the welding equipment cooperate so that after positioning the robot emits a go-ahead signal to the welding equipment. After welding, the welding equipment emits an acknowledgment signal to the robot which can then start the next positioning.

The turntable jigs are indexed and controlled by a number of input and output signals.

471 instructions are included in the program, 169 of these being positioning instructions.

It is important that the spot welding gun comes into contact with the sheet metal at the correct angle. It may be difficult to manage this always, since this robot only has five axes. This must be taken into account when the installation is set up initially.

Improvements have since been made to equip the IRb-60 with 6 axes of motion as original equipment.

Detecting equipment has been built in to compensate electrode wear.

The servo system must be re-trimmed in order to maintain good dynamic properties due to the considerable manipulating weight.

The application requires great rapidity in position. In order to achieve this, PTPL + WAIT must be used. This results in a positioning error which may cause problems in certain positions. This is particularly so when operating in positions in which the robot transmission is very stiff. In these positions the robot would move the floor sheet and jig with it in its efforts to get into the correct position. This problem is solved by the velocity reference being cleared when it is considered that the robot is in position.

In this case, the robot, the workpiece and the jigs will be exposed to very high stress. In order to limit this stress, the robot can be supplied with an overload protection means.

2. MEASURING ROBOTS

A new application is measuring car bodies. It has been engineer ed by Volvo at Gothenburg and is in use at Torslanda, Sweden, checking 120 points on the body at the end of a spot welding line to a repeatability within 0.5 mm.

Two 6 kg robots are mounted on tracks up to 5 m on either side of the body line, and they can traverse 5-3 m and reach the whole length of the car when it is in the measuring station on the line. (Specif. YB 121-102E.) Cars come down the line at a rate of 40 per hour, and to keep pace with the line, the robots check one quarter of the points on each car. Therefore, after every four cars all of the 120 points have been checked. Each robot checks about 15 points in the front section of one car, then another 15 points in the next car, and so on.

The system, calibrated against a master body, measures X, Y and Z departures from the nominal size at each point.

Deflections are measured by a three-dimensional probe on the robot arm, the robot itself moving to preset positions. This solution was preferred to putting a sensor on the robot arm to make contact with each point and to measure the displacement directly from the robot's resolvers.

A special computer is programmed to receive and interpret the data from the probe. Dimensional data can be analyzed statistically to identify trends and correct errors promptly.

The alternative to robot measurement would be a large measuring machine, which would be much slower in operation. Bodies would have to be taken off the line and it would be possible to inspect approximately one in 50, making the response time to an unfavorable trend or error much longer.

Points important for Volvo to check include shock absorber and engine mountings, window and door openings, and hood and trunk frames.

3. UABF SHAFTS

An ASEA IRb-60 is attending a machine group for production of shafts to ASEA gear boxes, series UABF.

Included in the machine group are:

> Shaft cutting
> BOEHRINGER lathe
> HURTH milling machine
> Nyberg & Westerberg grinding machine
> Measuring fixtures
> Removal conveyance

Cutting of the shaft

The machine is provided with a bar magazine. The bar is automatically fed forward and the material is cut to length, centered, and the first shoulder is turned.

The finished component is fed out onto a conveyor belt with the shoulder toward the feeding direction.

Lathe

The robot changes components in the lathe and starts rough machining. The robot turns to the control equipment and puts the finished shaft into a measuring fixture. Two diameters and one linear dimension are measured. The measurement results are stored by measuring equipment and transferred to the lathe when the rough machining is finished and fine machining is started. Should the shaft be outside the tolerance zone, no start signal is given for fine cutting. Neither does the robot receive a signal to retrieve the shaft and to proceed to milling and grinding.

The component is clamped with the turned end in the chuck (soft jaws) and with a pivot at the other end. Tools are provided to produce the following:

> 1 rough cut
> 1 fine cut
> 1 parting-off chock

A new (reserve) cut is changed after a preset number of shafts are turned. (Experience indicates approximately 150-200 shafts.) The lathe is provided with a protective gate.

Measuring fixtures

Two diameters are to be measured, preferably the shoulders where the fine cut is finally made.

The fixture is axially and radially self-orienting by means of

V-blocks and axial limit stops. It is easily adjustable for
different shaft sizes by means of modules (not adjusting screws).
The adjustment for the respective shaft is to remain repeatedly
the same at re-rigging (program-bound). The measuring equipment
is to be manually operable.

Grinding machine

The robot positions the shaft in the fixed pivot and provides a
signal for the tail pivot to move forward. When the tail pivot
is in position, a signal is given to the robot to release the
shaft, withdraw, and start the grinding cycle. After the grinding cycle is completed, the robot removes the shaft in the reverse sequential order as the insertion cycle.

Pinn mill

The component is clamped so that the pivot from the right
presses the shaft into a cone as shown in the figure below:

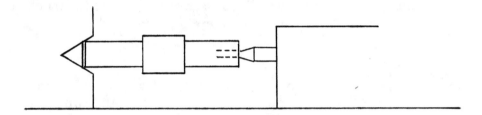

The machine is provided with a protective cover, automatically
opening and closing with "machine cycle start."

4. MACHINING OF CASTINGS

An ASEA IRb (5 axes) is used for the handling of castings. The
castings are to be machined (milled, drilled and tapped).

A special machine from Burkard & Weber is attended by the robot.

The gripper is holding the box internally. With one set of
fingers mounted on the gripper, two different internal diameters
can be gripped (which is necessary for these parts).

A buffer on the input is obtained with the use of plastic chains
for transport and storage of the parts.

In the robot program, steps are taken to pour coolants out of
the piece. A simple turning device is used to allow the robot
to get a new grip on the piece and put it upside down on the
output transport chain. A small deburring operation is also
performed.

5. ARC WELDING

Welding of front seats for cars in 0.9 mm thick steel sheets.
A total of 125,000 pieces per year are produced.

22 bead welds are located all over the workpiece.

In former production, there were problems with environment,
monotonous work, difficulties in obtaining qualified workers,
and uniform production quality.

The fixture used is of the same type as for manual welding, but
has automatic rotate and tilt motions. The operator initiates
the change of the piece.

The cycle time is 90 seconds per seat.

Wire with a diameter of 0.8 mm is used, and the wire feed rate
is 5 m per minute, which is equal to 110 ampere current. The
welding equipment is of type ESAB A 30 A system.

Production has been improved by 89% on a two-shift basis.

The welded object and principal layout before and after installation of the welding robot can be seen in the following illustrations.

CONCLUSION

The more sophisticated the industrial robot becomes, the more
likely it will be integrated into a manufacturing system - not
serving just one machine. Now that robots can handle tools
similarly to what a man can, a whole new area of application
requiring this technology is opened up.

In the not-too-distant future, there will be systems available
with vision, adaptive control, and sensory feedback as standard
equipment for utilization of the industrial robot.

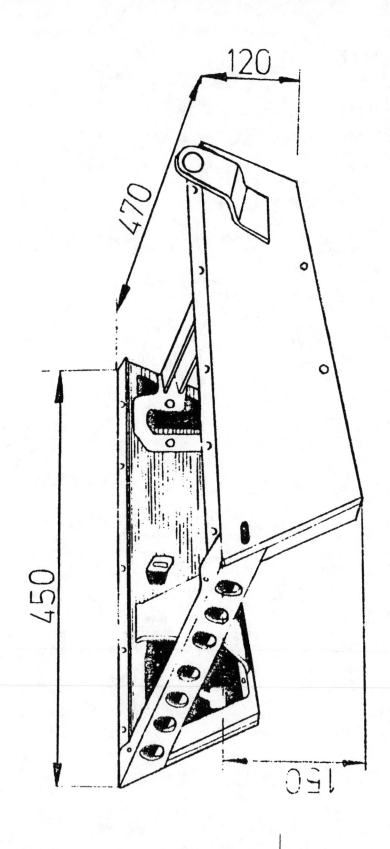

120

470

450

150

WELDED OBJECT

292

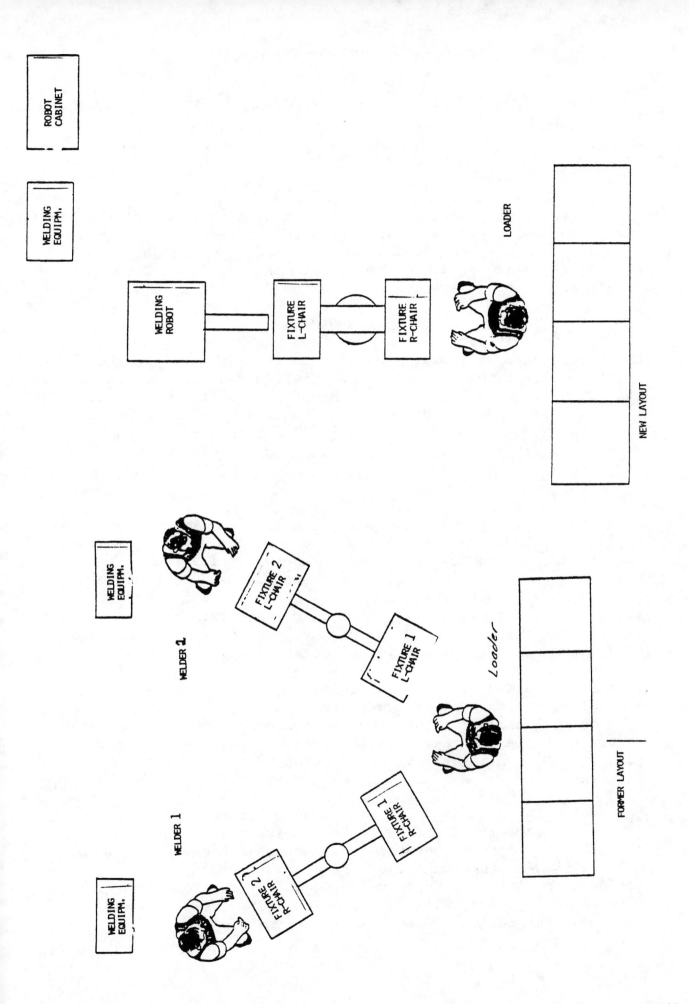

ROBOT CABINET

WELDING EQUIPM.

WELDING ROBOT

FIXTURE L-CHAIR

FIXTURE R-CHAIR

LOADER

NEW LAYOUT

WELDING EQUIPM.

WELDER 2

FIXTURE 2 L-CHAIR

FIXTURE 1 L-CHAIR

Loader

WELDER 1

FIXTURE 2 R-CHAIR

FIXTURE 1 R-CHAIR

WELDING EQUIPM.

FORMER LAYOUT

293

Presented at the Robots II Conference, October, 1977

Moving Line Applications With A Computer Controlled Robot

By Bryan L. Dawson
Cincinnati Milacron

A Computer Controlled Industrial Robot has tremendous flexibility. Many features may be incorporated, via software, into such a system. In particular, the feature of Stationary-Base Line Tracking is easily accomplished. This feature allows a fixed-base robot to execute programs on objects moving through its station.

In applying this feature, the robot is taught each program with the object in a stationary position in front of it. In the automatic mode, the robot will execute the taught programs on successive objects as they pass through its station, irrespective of their speed or direction of motion. The implementation of Stationary-Base Line Tracking is illustrated by various moving-line applications.

INTRODUCTION

Many manufacturing operations, especially those in the automobile industry, are carried out on continuously moving lines. Therefore a desirable capability of an Industrial Robot is that of working in conjunction with such lines. This capability has many advantages such as the economical gains through elimination of the initial cost and constant maintenance of shuttle systems, and the productivity gains due to elimination of station-to-station transfer times of shuttle systems.

The use of an Industrial Robot on moving lines does require, however, that its control system know the position of the robot at all times and that, due to forseeable but random occurrences, it can take decisive action. The use of a computer in the control allows such action to be easily accomplished.

This paper presents a discussion of a specific Computer Controlled Industrial Robot and its inherent application versatility. After a general description of the robot and its control system, the ability of the overall system to work in conjunction with continuously-moving conveyor lines, irrespective of their speed, will be discussed. This feature of the system is termed "Stationary-Base Line Tracking."

A COMPUTER CONTROLLED INDUSTRIAL ROBOT

An Industrial Robot, which represents the concept of Flexible
Automation, must satisfy the requirement of versatility, such
that it may lend itself to being adapted easily from one use
to another. Two factors that determine the degree of robot
versatility are:

> 1. The maneuverability of the robot arm.
> 2. The flexibility of the control system.

The flexibility of the control system is greatly increased
through the use of a computer.

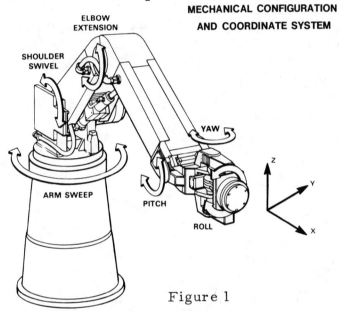

Figure 1

The Maneuverability of the Robot Arm

Figure 1 illustrates the mechanical configuration of the robot
arm. It is a manipulator with six rotary axes of motion.
Each axis is directly driven by its own hydraulic actuator,
without any intermediate power transmission devices such as
gears. Each actuator is driven by its own electrohydraulic
servo-valve, and signals representing the change in angular
position and velocity of the axis are fed back to the control
system by means of a resolver and tachometer. Figure 2
indicates the working range of the end of the arm of the robot.
The jointed arm configuration allows a large volumetric working range compared to the size of the structure and the floor
space required. The addition of an end effector such as a
spot weld gun or a gripper mechanism increases the effective
working range of the robot arm.

The six axis, jointed arm configuration with its associated maneuverability greatly enhances its use in stationary-base line tracking operations.

The Flexibility of the Control System

Realistically, the user of a robot is concerned mostly with the position of a given point on an end effector, for example the center of a pair of gripper jaws or the tip of an arc welding gun. This point of interest is defined as the Tool Center Point, and will be referred to as the TCP. Basically, the robot arm is a means of moving the TCP from one programmed point to another in space.

The control philosophy of the robot is built up around the Tool Center Point concept. The control directs the movement of the TCP in terms of direction, speed and acceleration along a defined path between consecutive points.

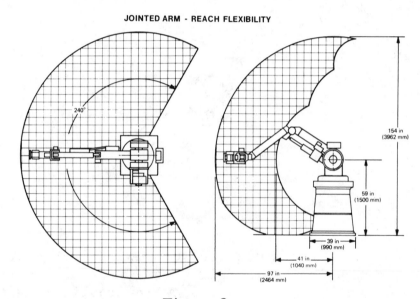

JOINTED ARM - REACH FLEXIBILITY

Figure 2

This method of control, termed a Controlled Path System, utilizes the mathematical ability of a computer to give the operator coordinated control of the predefined Tool Center Point in a familiar coordinate system during the teach operation. It also controls the TCP in terms of position, velocity and acceleration between programmed points in the replay or automatic modes of operation. The TCP is predefined by the operator who enters into the control a Tool Length. This represents the distance between the end of the robot arm and the TCP. In the Controlled Path System the coordinates of the TCP are stored as X, Y, and Z coordinates in space (Figure 1) and not as robot axis coordinates. It should be noted that during the automatic mode of operation the position of the TCP and the orientation of the end effector relative to the TCP are known at all times by the control system, thus making the implementation of Stationary-Base Line Tracking possible.

The mathematical ability of the computer also allows many other features to be written into the software of the system without the necessity of changing the control hardware. Various software routines can be assembled to provide the user with a choice of features. Some of these features of the Computer Controlled Industrial Robot are described in more detail in References 1, 2, 3, and 4.

TRACKING A MOVING LINE

Production operations in Industry are very often executed on moving line conveyor systems. The moving line concept has been shown to be very efficient in manufacturing processes. The automotive industry, in particular, has for many years used the moving line approach for automobile assembly. There-fore, an obviously desirable capability of an Industrial Robot is the ability to carry out operations on parts mounted on a continuously-moving conveyor. Such a capability is called Line Tracking.

There are many advantages to be gained by the use of a Line Tracking Industrial Robot, especially in plants where the moving line system is already installed. All these advantages lead to either increased productivity for the same cost or to the same productivity for less cost. These advantages are, of course, the aims of any manufacturing system. Some of these advantages are outlined below.

1. If the robot system to be used does not have line tracking capability, then expensive high-speed, heavy-duty shuttle systems must be installed to stop each part in a distinct location in the robot station.

2. In addition to the extra expense of installing such a shuttle system, maintenance becomes a critical factor. Many shuttle systems move parts between multiple stations. If the shuttle system should malfunction, many pieces of expensive equipment are made idle. If the equipment has the capability of working with a continuously moving

line a shuttle system is not required and there is less chance of idle equipment.

3. During initial installation of a shuttle system there is a period during which no useful work can be done. However, in a situation where line tracking equipment is to be used, each piece of equipment may be installed at any convenient time, minimizing the need to stop the line.

4. By using equipment with line tracking capability, the non-productive station-to-station time involved for shuttling parts from one station to another is eliminated. This time may now be used to do more operations on a given part with a particular piece of equipment, or to do the same operations with less equipment. Figure 3 illustrates the extra time available for a line tracking robot to operate on a part compared to the time available for a robot working on a shuttle type system. Naturally, the times used in the preparation of this type of graph will vary from installation to installation, but the trend is very clearly illustrated. For constant shuttle station-to-station times, the greater the number of parts per minute moving on the line, the greater the increase in productive time for a line tracking system as opposed to a shuttle system.

AVAILABLE PRODUCTION TIMES

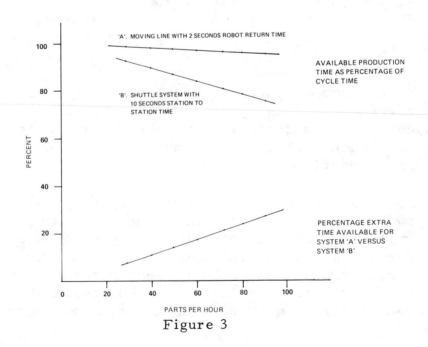

Figure 3

Two basic ways in which line tracking can be accomplished with an Industrial Robot are:

1. Moving-Base Line Tracking

   With this method, the robot is mounted on some form of transport system, e.g. a rail and carriage system, which moves parallel to the line and at line speed. This method requires the installation of the transport system which may not be possible or economical. If multiple robot systems are set up adjacent to one another alongside a moving line, there may be interference problems between adjacent stations. A powerful drive system is required for each transport device in order to return the robot back to its starting point from the other end of its tracking range in the fastest possible time.

2. Stationary-Base Line Tracking

   In this method of Line Tracking the robot is mounted in a fixed position relative to the line. Hence the name "Stationary-Base". This naturally constitutes an economical installation which requires less maintenance than is necessary with moving-base systems.

   Stationary-Base Line Tracking has another important advantage. It may not be possible for a robot that is working on a stationary part or a moving-base line tracking robot to reach as many points on the part as when the base of the robot is stationary with the part moving past it. For instance, there are cases with stationary-base line tracking where a single robot can work on the front, the middle, and the rear of a part as it progresses through the station.

FULL TRACKING CAPABILITY OF THE STATIONARY-BASE COMPUTER CONTROLLED INDUSTRIAL ROBOT

Full tracking capability of the robot allows it to perform its taught program on a part moving through its station, irrespective of the speed or position of that part. The positions of taught points, the orientation angles of end effectors around taught points and the velocities of motions between taught points will have the same values, <u>in relation to the part</u>, as they had when the program was taught. The Controlled Path system of the Computer Controlled Robot allows the full tracking capability to be easily implemented. Positions of taught points are stored in memory as coordinates in space and not as robot axis coordinates. During the running of the robot, the control knows exactly the position of the TCP and the orientation angles of the end effector around the TCP at all times. This latter feature is essential in order that the robot may take decisive action due to external influences on the system, because the TCP will be following paths in space between points which are different from those which it followed during the teach operation.

The system utilizes a position sensing device which is attached to the conveyor or part and which feeds a continuous signal to the control representing the position of the part relative to a fixed, but arbitrary, reference point. This signal is then used, in effect, to provide a "dynamic" zero shift of the taught program in the direction of motion of the moving part.

Figure 4, a block diagram of the computer control system, indicates the flow of information through the system during both the teach and automatic modes of operation.

Figure 4 also shows how the position sensor is interfaced to the computer in the control system. Usually, the robot is placed such that one of its major axes (X,Y,Z) is parallel to the moving line. The greatest utilization of the robot in terms of tracking range is made when its Y axis is parallel to the line.

TRACKING AND CONTROL SYSTEM DIAGRAM

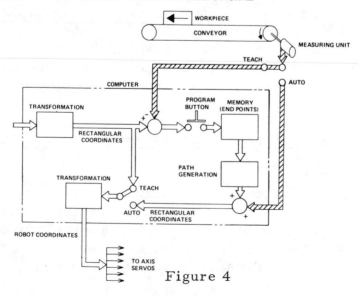

Figure 4

During the teaching operation the part is moved to a convenient position in front of the robot and stopped. Points are taught as normal but each coordinate in the direction of the line is modified by an amount equal to the current position sensor reading, prior to being stored in memory. Thus, the stored data are referenced to the start point of tracking. If it is desirable, for more convenient access, the part may be re-positioned at any time during the teaching operation.

In the automatic mode of operation, the stored points are used to generate the desired paths, which are then modified by the current position sensor reading. In this way, the control, in effect, changes the coordinates of taught points in the tracking direction by an amount equivalent to the distance between the position of the part at which the point was taught and the position of the part at which the point is replayed.

It should be noted that because point coordinates are stored as rectangular and orientation coordinates in space, and not as machine axis coordinates, only the translational coordinate in the direction of the moving line needs to be modified. All other coordinates remain the same and, thus, during the automatic mode of operation the three coordinates representing the orientation of the end effector relative to the part do not change. Hence, the end effector always replays a program-med point with the same orientation, relative to the part, as was taught. This means, for example, that if the robot was programmed to avoid an obstacle in a certain way during the teaching operation, then it will avoid that same obstacle in the same manner during automatic operation.

IMPLEMENTATION OF STATIONARY-BASE LINE TRACKING

The requirements for a typical robot installation to be used for a stationary-base line tracking application are:

1. A position sensor connected to the part or convey-or to indicate the position of the part. This sensor is electrically interfaced with the control.

2. A limit switch or other form of sensor which is actuated when the part is in a predefined position. This sensor signal, called "Target In Range", indicates to the control to start to use the in-formation provided by the position sensor to update the position of the TCP.

3. A series of limit switches or sensors which indicate to the control the style of part on which the robot is to operate. This permits the control to select the correct branch program for that part from its memory. These switch or sensor signals use a simple binary code to allow the control to select one of 15 different branch programs.

CONSIDERATIONS FOR STATIONARY-BASE TRACKING APPLICATIONS

If a sequence of operations to be performed on a stationary part is taught to a robot, the robot will replay the programmed points at the same positions, in space, at which they were taught. The points will always be within the range of the robot arm during replay because it is impossible to teach a point that is outside that range. However, when a robot is working on a continuously-moving part, taught points on the part that were within the range of the robot during the teaching operation may, due to a variety of circumstances, be outside that range during replay. Points that were taught with the part at one end of the range could be replayed with the part at the other end of the range. Hence, because the robot will now be replaying programs with modified paths between modified programmed points, there are certain considerations to be taken into account in the planning and programming of moving line tracking applications with a stationary-base robot. These are discussed below.

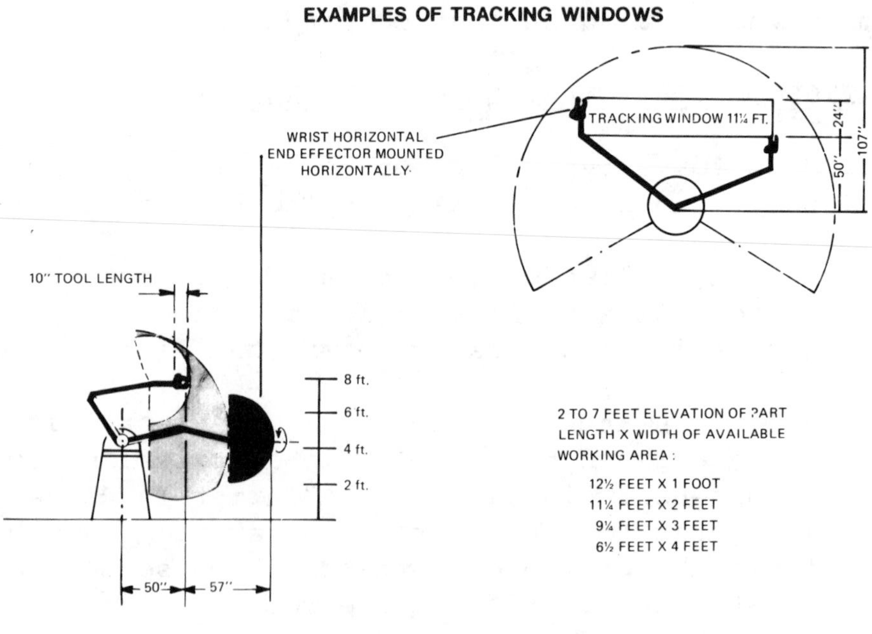

EXAMPLES OF TRACKING WINDOWS

WRIST HORIZONTAL
END EFFECTOR MOUNTED
HORIZONTALLY

TRACKING WINDOW 11¼ FT.

24"

50" 107"

10" TOOL LENGTH

8 ft.
6 ft.
4 ft.
2 ft.

50" 57"

2 TO 7 FEET ELEVATION OF PART
LENGTH X WIDTH OF AVAILABLE
WORKING AREA :

12½ FEET X 1 FOOT
11¼ FEET X 2 FEET
9¼ FEET X 3 FEET
6½ FEET X 4 FEET

Figure 5

1. Tracking Window

    The diagram in Figure 5 illustrates the robot's
    large tracking range, when used in tracking appli-
    cations in which the Y axis of the robot is set
    parallel to the moving line. As the diagram indi-
    cates, there are many parameters that influence
    the length of working range of the robot in the
    direction parallel to the moving line. This work-
    ing range of the robot parallel to the line is
    termed the "Tracking Window." The height of the
    part on the conveyor, the distance of the robot
    from the conveyor and the length and configuration
    of the end effector all play a part in determining
    the tracking window. Therefore, every tracking
    application must be considered separately in
    order that the robot is positioned correctly,
    relative to the conveyor, to ensure the optimum
    tracking window.

    Once the tracking window for a given sequence of
    operations has been established, it is entered
    into the memory of the control. The tracking
    window basically defines in memory the two limits
    in the tracking direction beyond which the robot
    will not attempt to reach. More than one tracking
    window may be defined for different segments of a
    tracking operation. These tracking window limits
    cause two separate types of action by the control.

    a. If the robot has just replayed a point, its
        control will check the position of the next
        point in the sequence. If the next point is
        inside the defined tracking window, the robot's
        TCP will move to that point. If the next
        point is outside the tracking window but moving
        towards it, the TCP will simply move with the
        line and, in effect, remain stationary relative
        to the part, until such time that the point
        moves inside the tracking window. Only then,
        will the TCP move to the next point. This
        ensures that the robot will never attempt to
        reach a point outside its range.

    b. If the next point in a sequence is outside
        the tracking window, but moving away from it,
        the robot will move with the line until such

time that the position of the TCP coincides
with the tracking window limit. At this time
the control will cause the robot to enter an
"Abort Branch" which will move the robot away
from the part along a predefined path. These
abort branches are discussed below.

2.   Abort Branches

A special software routine called "Abort Branch"
is included in the controls for stationary-base
line tracking robots. This function constantly
monitors the position of the robot's TCP and compares that position with the positions of the
limits of the tracking window. If, at any time,
during automatic operation, the robot's TCP
coordinate in the tracking direction coincides with
the coordinate of either limit of the tracking
window, the robot control immediately initiates an
abort branch. The abort branch will direct the
robot to exit from the part along a pre-taught
safe path relative to the part. The abort branch
is taught by the operator during the teaching
operation, and, depending upon the configuration
of the part on which the robot is working, can
contain a number of smaller branch programs. This
will ensure that the TCP of the robot will always
follow a safe and quick exit path from its operating position on the part, regardless of its
position in the taught program.

3.   Utility Branches

Occasionally, during the execution of a tracking
application, it will be necessary for the robot
to take some form of corrective action in response
to a signal that indicates the occurrence of some
malfunction of the peripheral equipment. Such
action may be taken with the computer controlled
robot by using a function called "Utility Branch".
A utility branch is taught in a similar manner to
an abort branch and if necessary, it can also
contain a number of smaller branches. The utility
branch is, however, initiated by an external signal
from the peripheral equipment rather than by an
internal signal from the control. As an example,
if, during a spot welding sequence, the gun tips

stick to a flange, immediate corrective action is
required. Upon receipt of the signal that indicates the "stuck tip" condition, the control will
immediately initiate the utility branch to take
corrective action regardless of where the robot
is in its sequence of operations. In this case
the action taken may be, for example, a twisting
motion to break the tips of the gun away from the
flange.

The Abort Branches and Utility Branches described above are
made possible by the fact that the control "knows" the position
of the TCP at all times. They ensure that, when the robot is
working with a moving line, logical decisions and actions are
made by the control to take corrective action in response to
the occurrence of random but foreseeable events. As with non-
tracking applications, other interface signals between the
robot and the peripheral equipment are easily implemented to
ensure that corrective action is taken by the robot in response
to other occurrences.

APPLICATIONS USING THE STATIONARY-BASE LINE TRACKING CONCEPT

The Computer Controlled Industrial Robot is being applied in
a number of different applications involving the concept of
Stationary-Base Line Tracking. Some of these applications
are described in this section.

   1. Spot Welding of Automobile Bodies

      Spot welding with the robot is being carried out
in automobile body assembly operations. Figure 6
shows the robot at work in such an operation.
The robot was taught programs to weld flanges
parallel to the direction of motion of the line,
and also perpendicular to the line direction in
both the horizontal and vertical planes. Abort
and utility branches were fully utilized in these
spot welding applications on moving car bodies.
Different abort branches were taught for different
weld areas of the body and, in a similar manner,
different utility branches were also taught. Both
the abort and utility branch concepts were fully
proven, the abort branches being initiated by
internal signals from the control and the utility
branches by signals indicating stuck spot weld gun
tips.

SPOT WELDING ON A MOVING LINE

Figure 6

2. <u>Transfer of Automobile Bumpers</u>

In this application two automobile bumpers are
transferred from a stop-go conveyor on which
they are built, to an overhead monorail conveyor
system. The overhead conveyor stops approximately
in sequence with the build conveyor but its
stationary position can vary widely from one cycle
to the next. It is therefore necessary to utilize
the tracking ability of the robot system to monitor
the position of the monorail conveyor so that the
bumpers may be placed correctly on its cradle.

3. <u>Automobile Body-Side Transfer</u>

In this application automobile body-sides are
picked up from a continuously-moving floor-
mounted conveyor, on which they are built, to an
overhead continuously-moving monorail conveyor.
This application requires that the robot system
has the ability to track two continuously-moving
lines, independently of one another. This
application not only illustrates the concept of
tracking but also the use of a heavy-duty robot

with a large working range. The body-side and the robot grippers together weigh approximately 250 lb. and the robot must reach to the floor-level conveyor to pick the body-side up and raise it to a height of approximately 10 feet to deposit it on the overhead conveyor.

4. Machine Tool Loading, Parts Retrieval and Storage

This application is not, perhaps, an example of the full tracking capability of the Computer Controlled Robot but does illustrate a novel use of the tracking ability. In this application, the robot is situated in the center of a machining cell comprising four different machine tools, an input station, an output station and a storage rack with 21 storage positions. These storage positions are arranged in three vertical columns, each containing seven positions.

Parts which enter the machining cell are picked by the robot which moves over to a position in front of the storage rack. A signal from a position sensing device is used by the robot control, operating in the tracking mode, to raise the robot TCP to the correct storage bin level whereupon the part is placed in storage by the robot. In this application the associated machine tools are being constantly monitored to keep the robot unloading parts, placing them in storage, extracting other parts from storage and loading them into machines. In its "spare time" the robot loads parts into the storage rack from the input station and takes parts from storage and loads them into the output station.

CONCLUSION

The six-axis computer-controlled robot system has been shown in the last few years to have excellent versatility. The mathematical ability of the computer to provide a Controlled-Path control system which allows total position, velocity and acceleration control of the robot's end effector has provided the means to implement the concept of Stationary-Base Line Tracking.

The "Name of the Game" in manufacturing operations is

Productivity: to produce the greatest number of parts, for the least cost, in the least time, with a minimum inventory. The use of a Stationary-Base Line Tracking Industrial Robot System in manufacturing operations will contribute greatly towards these objectives. Installation and maintenance costs will be reduced with this type of tracking system and, because less idle time is required than with a stop-go shuttle system, more production can be obtained.

The ability of the computer control to direct the robot to take decisive action due to external influences represents a great stride forward towards the completely automatic "Factories of the Future".

REFERENCES

1. "The Benefits of a Computer Controlled Robot", Merton D. Corwin, Society of Manufacturing Engineers, MS75-237 Fifth International Symposium on Industrial Robots, May, 1975.

2. "A Computerized Robot Joins the Ranks of Advanced Manufacturing Technology", Bryan L. Dawson, Proceedings of the Thirteenth Annual Meeting and Technical Conference of the Numerical Control Society, March 1976.

3. "Application Flexibility of a Computer-Controlled Industrial Robot," Richard E. Hohn, Society of Manufacturing Engineers, September 1976, MR76-603.

4. "Robot Decision Making", H. Randolph Holt, Society of Manufacturing Engineers, MS77-751, Robots II Conference November 1977.

Presented at the Fifth International Symposium on Industrial Robots, September, 1975

Further Development And New Applications
for the TRALLFA Robot

By Sjur Laerdal
TRALLFA Nils Underhaug A/S (Norway)

The TRALLFA Robot, originally characterized as a multi-purpose programmeable memory controlled Robot with continuous path type control, has (though well established within industrial coating applications with more than 300 installations world wide), during the last year, proven its suitability in a complete new application - Automatic Arc Welding.

Certain modifications had to be made to the servo system of the standard TRALLFA Robot to improve accuracy and further other special solutions had to be found. The result today is several installations performing complicated arc-welding jobs.
The programming procedure is the same as by spray-painting; simply leading the robot arm through the wanted (3-dimensional) movements and recording information on magnetic tape or disc memory.

Recent development of the TRALLFA Robot has led to the possible optional use of a computer-based control system with a magnetic tape or disc memory, which gives a more flexible solution with respect to e.g. program capacity, program search time and robot/conveyor synchronization.

A central computer control system capable of controlling up to 8 robots, has been developed. This development gives interesting perspectives for future applications of the TRALLFA Robot.

INTRODUCTION

Through years of experience in the robot field, the Trallfa company has been able to learn the different special requirements of robot users and based on this experience, which on second hand creates a better ability to foresee the future needs, the TRALLFA Robot has been further developed to meet this future.

The success of a robot system within industry will always depend on two main factors:

First: Quality - not only to be understood as basic product quality, but also the qualities or usefulness in actual applications.

Second: Basic system price and operational costs.

The rapid development on the hardware side of the computer technology - with the resulting price decrease on processing power - seems for once to match the demands for lower price with a higher quality. The Trallfa company has on this background decided to take advantage of the obvious possibilities presented by computer technology and has now introduced a "new" generation of computer-controlled robots, in addition to the well tried direct-memory controlled robots.

The TRALLFA Robot can thus be presented with a whole range of different possibilities of control systems, where price and particular user-requirements will decide the control option to be chosen. The new "generation" of control systems has been based on a micro-processor unit with a magnetic disc, "Floppy Disc", as standard memory element. In this concept one may also have, for use in multi-robot systems, a central computer with larger central memory capacity, on-line connected to the local robot computers which in turn receive data and undertake the local processing.

The future robot-user will, as he grows more familiar with industrial robots (IR), be more capable of specifying what he really needs and he will not seriously be looking for a "once-and-for-all" solution to all his automation problems. He should, however, be given the possibility to adapt his robot to future demands. To give him this possibility has been one of the main goals in the development of the new concept of the TRALLFA Robot control system.

The TRALLFA Robot has a continuous-path type of control, and as such it is particularly useful in jobs where the human skill is performing 3-dimensional curve motions as a necessity. Traditionally, TRALLFA Robot was concentrated in the spray-painting field. It is obvious, however, that in several other industrial applications, the continuous path control is a must.

Besides all kinds of spray gun applications of coatings Trallfa
has for some time been working on a complete new application
for the TRALLFA Robot - Industrial Arc Welding.
This application has proven to be very suitable for the TRALLFA
Robot, and considering that it is only half a year ago since
the Welding Version of the TRALLFA Robot, TR-3000W, was introduced, a remarkable number of installations have taken place
within industry already.

The TR-3000W is basically the same as the TR-3000 model, and
this fact helps to create a very flexible robot system along
the same lines of development, as mapped up by the new control
system briefly described above.

A spray-painting robot may with minor modifications be operated
as a welding robot, and users of painting and welding robots
can have robots from the "same family" which considerably
easens training and maintenance.

THE APPROACH TOWARDS A COMPUTER ADAPTED ROBOT SYSTEM

In this chapter I will explain the fundamentals of the TRALLFA
Robot system and discuss the new computer based system along
with the reasons why Trallfa has adapted this solution.

Basic System Fundamentals

The original TRALLFA Robot might be characterized as a "multi-
purpose programmeable memory-controlled robot". It is a tool-
carrying robot originally designed to replace the human spray
gun operator in industrial application processes, and its 5,
alternatively 6, degrees of freedom allow it to emulate the
movements of the human hip, should, elbow and wrist.
The TR-3000 (Fig. 1) consists of:

- a Manipulator

- a hydraulic Power Supply

- a Control Unit, which may contain different
  alternatives with respect to memory and computer
  control systems.

Details of specifications are shown in TABLE 1.

The Manipulator

The manipulator arm has the ability to position a tool in space
by means of its five or six continuously moveable servo links.
(Fig. 2) Each of the five links is controlled by a hydraulic
actuator containing its own position measuring system which
will supply analog signals corresponding to every position an
actuator may have. The measuring system, consisting of a
resolver and a potentiometer, indicates the position of the
hydraulic actuator. The transducers are driven by a steel-tape

fixed to the piston rod over a gear system in the interior of the actuator.

The Hydraulic Power Supply

Hydraulic fluid under pressure is applied to the manipulator from the hydraulic power pack via two hoses. The power pack contains fluid reservoir, a motor/pump unit, the pump being of the inner gear type, and a valve manifold. The fluid level and the fluid temperature are watched by sensors which will trip at too high temperature or a too low fluid level. The pump motor cannot be started as long as one of the sensors is tripped, thus forming a "fail to safe" system.

The Control Unit

The control unit contains the main junction box, the electronics rack, the control panel and the "RPS" or "SCT" tape reader or, as we will see later on, the computer/disc system. From the control panel all functions of the robot, including switched output functions can be controlled. To ease operation, a remote control panel which is an exact copy of the lower right half of the main control panel, is delivered with the robot as a separate unit.

System Fundamentals

The movements of the TRALLFA Robot are controlled by an analog/digital control system. For teaching purposes, two auxiliary handles are attached to the front of the robot arm. One is fitted with a trigger operated micro-switch and the complete handle is fastened to the bracket which holds the automatic spray or welding gun. To start the magnetic tape at the beginning of a programming cycle, the other handle is utilized. It is fitted with a trigger micro-switch as well, which, when depressed, will start the magnetic tape. The trigger is held depressed during the entire teaching sequence, and any movements made are automatically recorded. Releasing this trigger will end the program and cause the tape to reset to the starting position.

The TRALLFA Robot is programmed for work by leading the robot arm with the attached tool manually through the desired operation. The movements produced during the teaching sequence will be recorded in digital form on magnetic tape or disc. Several output functions can be taught within a program for any period of time, and in any sequence desired.

Teaching procedures are relatively simple and can be performed by virtually anyone, but should be performed by a person with skill in the actual application, and not necessarily in the robot technique.

Servo System

During the teaching sequence the signals from the hydraulic actuators will, in the electronics of the control unit, be converted to groups of digital information. Every 1/80 of a second the digital information of each actuator will simultaneously be sampled and recorded on a continuously moving magnetic tape. This process repeats automatically as long as the magnetic tape is moving. When the robot is placed in the REPEAT mode and the hydraulic system is energized, the tape is normally started by an external signal, after which it will transmit the information to the control unit for comparison to the actual positions of the actuators. Any positional error between the tape and the actuators will cause the corresponding servo valves to drive the actuators and align them with the desired position. As the positional data changes, the actuators will follow, reproducing the movements recorded.

Computer-based System

Based on special requirements from different robot-users and some interesting prospects with potential users, and further in order to meet future demands and user flexibility in general, we have for some time been looking for a more advanced control system for the TRALLFA Robot.
Due to the rapid development in computer techniques and the resulting price decrease on computer components, this technique had become a realistic alternative for robots in practical applications. But in the abundance of different possibilities - which solution is to be chosen?

Three general goals were set for the new system to be chosen:

1. The new control system had to be compatible with the existing robot system without having to make any drastic changes in the fundamentals of the existing well tried design, due to demand for exchangability with the "old" system.

2. The price to the user of the new system had to be within certain limits.

3. The new control system had to be easy to operate and had to offer acceptable solutions with respect to some specific areas, especially programming capacity, synchronization and cooperation in multi-robot installations.

The idea was first developed due to a specific requirement of having a central control of several robots working on the same job. Such a system is actual within the automobile industry, e.g. by painting or underbody-coating. One single robot will in such cases not be sufficient. In one actual project 8 robots will be working in two groups of 4 units each. (Fig.3) Some characteristics of such an application are:

- large programming capacity is a necessity
- changes in car models must be taken care of
- different car models will follow randomly
- the conveyor speed will have variations
- total conveyor stop will occur and these coincidences should not cause any rejects or permit a collision with the robot manipulator
- up to 10-15 different car models must have their programs available on-line
- possibilities of storing large number of programs for other models off-line.

This basic idea resulted in a concept where a central mini-computer with a disc memory was interfaced to the single robots and supplied these with data. This system had certain advantages:

- a centralized control system offered an easier operation of the robots
- by using a disc with one fixed and one exchangeable package a large quantity of data may be stored off-line
- by the use of real-time clock a full synchronization with conveyor speed can be reached. This follows through integration, i.e. measuring of the physical distance between the starting sensor and the starting point, instead of time. In this case the movements of the robot will automatically be synchronized with variations in conveyor speed.
- through different techniques the demands for storage of data may be reduced considerable by processing the data during the programming procedure, e.g. only a part of the registered data need to be stored. During playback a simple interpolation may follow.

The above describes briefly the original system concept. This concept, however, was re-evaluated in order to have a system offering the same advantages, but which at the same time gave more flexible solutions with respect to the total robot program. Would this be possible within acceptable economic limits?

The occurring thought was to adopt the known DNC-concept. In such a concept, a central computer with a disc is connected to several local computers, which receive sets of data and undertake the local processing. Such a multi-prosessor system offers obvious advantages:

- it is modular and with an easily expandable structure.

- greater safety: The single robots may be operated on a stand-alone basis with its own memory.

- the communication between a central computer and local computers simplifies the interfacing.

- extended use of standard components

- a greater deal of the processing may follow locally - total processing capacity increases

- greater capacity for future developments

- communication robot to robot is possible via the central computer.

The optimal solution in this case was found after research among several possibilities and is based on a micro-processor unit as the core of the new control system. This computer can operate on a stand-alone basis, with e.g. a floppy disc memory. Other memories may be interfaced to the computer, as shown in the diagram (Fig. 4). In case of a multi-robot system the robots also may be connected to a central computer. With this system we have reached a solution which matches up with the initial goals.

1. The new system is within the scope of the already existing robot system and creates a valuable addition to the existing line of control possibilities of the TRALLFA Robot, whereas the customer do not have to buy a more sophisticated and expensive solution than he really needs, but may, if the application or situation and the requirements in any way should change, have computer control to his existing unit without having to buy a complete new robot. An optimum of flexibility is thus obtained.

2. The price of one TR-3000, e.g. with CPU (Micro-processor) and Floppy-Disc drive on a stand-alone basis, is only slightly higher than the price for TR-3000 with the RPS-tapereader, and thus also the goal with respect to price is achieved. (This advantageous price can be reached through a close cooperation with the supplier of the micro-computer unit.)

3. The TRALLFA Robot is, with the new control system, just as easy to operate as ever, and offers the following advantages:

- programming capacity:

    Because of the processing possibilities the relatively great quantity of data necessary to operate the CP-system may be reduced in several ways, e.g. by reducing the sampling speed. I will illustrate this as follows:

one TR-3000 with a double disc memory (Floppy Disc)
has a total program capacity at a sampling speed of
80 per second, of about 4 minutes on each disc (totally
8 minutes).

With a reduction in sampling frequency to 5 samples per
second the robot will have a total capacity on both discs
of 128 minutes. See diagram Fig. 5 showing different
steps in sampling frequency, and the resulting increase
in program capacity by a double disc memory.

- Continuous programming (actual by welding).

- Type of programming/control: Teach-in programming.

  The control is originally CP, the robots may, however,
  with the new system also be operated PTP (point-to-point).
  (Important by straight line arc welding or handling)

- Robot/conveyor synchronization:

  Full-synchronization possible.

- Program search time: Max. 1 second

- Program copying:

  Available for both TR-3000 SCT, RPS and TR-3000 with
  Computer control/Floppy Disc.

- Central Computer Control:

  Available for control of max. 8 robots.

- Detection Systems:

  The system offers a great number of possibilities for
  different detection devices (image scanning).

With this robot program, it is our opinion that the foundation
has been laid for tomorrow's practical industrial robot applications, which without too complex technology may comply with
the user's requirements of today and which may be extended
according to his changing demands by adding available options.

APPLICATIONS FOR THE TRALLFA ROBOT SYSTEM

Coating Applications

TRALLFA Robot has for years (since 1969) been covering this
field, and with more than 300 robots installed within industry
and more than 1 million hours of working experience, we have
reached a firm position as the leading robot manufacturer in
this field. The areas of application covers:

- All kinds of Spray-Painting (also electrostatic and powder spraying)
- Enamelling (also part dipping and sprinkling of enamel)
- Spraying of reinforced glass and polyester
- Underbody coating
- Flame-spraying (thermo-spraying)
- Spraying of porcelain
- Other spray gun applications of coatings.

Industrial Arc Welding Applications

Trallfa has developed a new TR-3000W model specially designed for Arc Welding operations (Fig. 6). The control system is basically the same as for the spray painting version. Programming the W-version is performed by leading the robot arm with the attached welding equipment manually through the desired operation, welding the first object in the manner the operator is used to.

The control of welding current, voltage, cleaning of welding gun and smoke absorption has been taken care of in a semi-automatic unit interfaced to the robot control system. This enables the operator to pre-select welding parameters, and to change these during operation, to obtain optimal welding conditions. The available output functions in the control system can be used during programming to operate positioning welding fixtures.

Use of TRALLFA Robot for industrial arc welding applications has in practice proven to give excellent return compared to manual operation, as robot manipulation of welding equipment increases utilization by 2-3 times. This has necessitated the design of special robot adapted welding equipment.

The first robot for welding was installed in the end of November 1974 and has since (installation was made in 1-2 days) operated continuously on a 2-shift basis. Welding equipment may be a CO_2 semi-automatic welding machine with a special welding gun designed to stand continuous welding (compared to standard welding guns which only stands an intermittence of up to 60%). The robot has, in actual installations, proven to have an intermittence of about 90%.

The new control system described above may besides common CP-holders also be controlled PTP. This is an advantage by possible straight-line welding. By a reduced sampling frequency as described, or/and combination with other memory devices like e.g. a tape station, one may have welding programs which goes on for hours. The return on investment by such installations is obvious.

Some problems have arisen by adopting the TRALLFA Robot for welding applications, such as the improvement of accuracy and the temperature compensation (due to changing oil temperatures). Special servo cards have been designed, and the compensation for changes in oil temperature has been taken care of by modifying the measuring system within the linear actuators. The result is an accuracy on a standard TR-3000W which fully complies with the requirements of most welding applications. By other means, e.g. shortening of the horizontal arm, the accuracy may be further improved..

Different modes of waving may be pre-selected and is adjustable from 0 up to 60 mm. Practical experience has shown that it is a great advantage that programming is performed in a teach-in manner and that the welding is done under real welding conditions (control of the warping of the welding object!).
From a sociological point of view, it is in our opinion also important that programming is performed by the man actually working with an application.

CONCLUSION

The computer based system gives a lot of room for further development, only to be mentioned; different means of detection devices, the servo loop placed in the software, etc.

As far as application is concerned, TRALLFA Robot has besides coating and welding applications been put sucessfully to work in other applications like deburring of castings and polishing. Other tasks will follow in the future - a new development may be actual tomorrow - something happens every day in the field of robot technology and in the field of different applications. I guess that is what makes industrial robots so facinating.

TABLE 1

S P E C I F I C A T I O N S

MOVEMENT	ARM	Vertical	2040 mm
		Horizontal	3150 mm
		Radial	975 mm
	WRIST	Bend	210°
		Swing	210°
		Rotation	210° (optional)
	HAND	Grip	On/Off (optional)
CONTROL	MODE		C.P. or P.T.P.
	PROGRAMMING		Teach-In
	MEMORY CHARACTERISTICS		- Direct Memory, SCT or RPS Tapereader - Micro Processor/Disc
	PROGRAMMING CAPACITY		- SCT 85 sec.(105 optional) - RPS 12 min.(20 min.optional) Max.60 diff.programs - Micro Processor/Disc (Floppy) See Fig. 5
FUNCTIONS	POWER		220/400 V 3-phase 60Hz ∂ 4.5kVA Other voltages available
	COOLING		- Air-Cooling standard - Water-Cooling available
	OPERATING TEMP. (recommended)	Manipulator	150°F
		Hydraulic	150°F
		Control Console	90°F
	WEIGHT CARRYING CAPACITY	Normal	5-10 kg
		Maximum (low speed)	30 kg
	SPEED ARM	Maximum	1.7 m/sec.

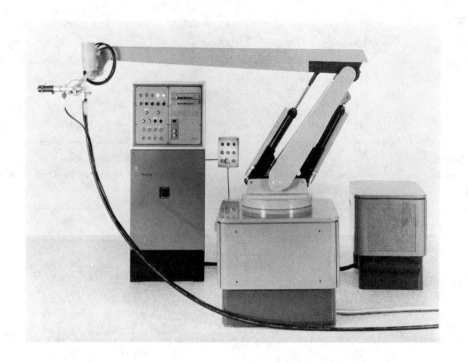

Fig. 1 TRALLFA ROBOT TR-3000

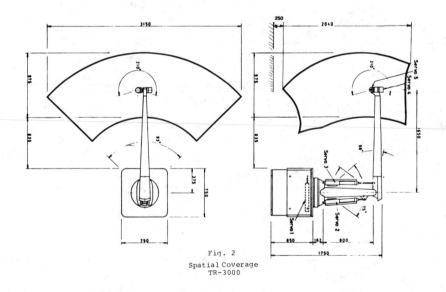

Fig. 2
Spatial Coverage
TR-3000

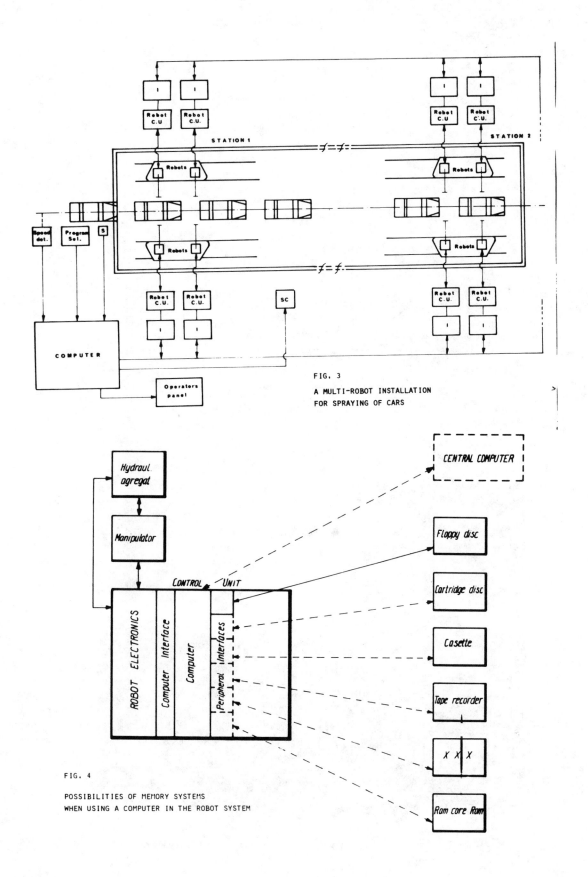

FIG. 3

A MULTI-ROBOT INSTALLATION
FOR SPRAYING OF CARS

FIG. 4

POSSIBILITIES OF MEMORY SYSTEMS
WHEN USING A COMPUTER IN THE ROBOT SYSTEM

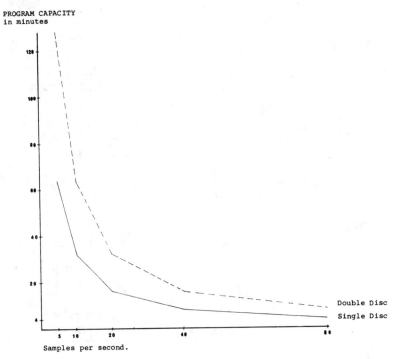

PROGRAM CAPACITY
in minutes

Samples per second.

Double Disc

Single Disc

Fig. 5   PROGRAM CAPACITY WITH DIFFERENT SAMPLING FREQUENCIES

Fig. 6 TR-3000 W INDUSTRIAL ARC WELDING
         INSTALLATION.

CHAPTER 7
ADVANCED TECHNOLOGY

Commentary

Volume I closes with this chapter on advanced technology for robots. The current status, future directions, research efforts and potential of sensory feedback and adaptive control are explored.

Included are several robot-vision systems and applications: a fixed-camera, simple vision; a mobile-camera, real-time robot control system; a bin-picking camera/robot system; and a robot-mounted camera system for part identification and acquisition.

Other technology includes touch and force sensors, adaptive control and passive-compliance insertion devices. The use of sensory information for robot programming and error correction is also discussed.

Use of Sensors in Programmable Automation*

Charles A. Rosen and David Nitzan
Stanford Research Institute

Introduction

If previous studies[1-4] are any indication, there is a national need to develop advanced automation techniques to increase industrial productivity and to enhance the well-being of workers by eliminating hard, dangerous, and dull jobs. Although an ultimate goal might be a completely automated factory in which design, planning, scheduling, and manufacturing are all under computer control, automation will probably evolve slowly with the development of the appropriate technology.

One day, however, repetitive jobs in labor-intensive operations will be performed by computer-controlled machines supervised by a small, highly trained group of operators who will set up and program each job, modify procedures to fit the particular circumstances, change over for new batches or models, maintain the equipment, and cope with breakdowns and stoppages. Thus, the system will "time-share" the operators, augmenting their capacity to do useful work by relieving them of the need to perform relatively low-level jobs that can best be done by machines.

In this paper we are concerned with the application of sensor-mediated programmable automation to material-handling, inspection, and assembly operations in batch-produced, discrete-part manufacturing. These operations, which are still highly labor-intensive, have been estimated to represent over half the total cost of product manufacturing costs in the United States.[1]

*This work was funded in part by Grant GI-38100X1, the National Science Foundation.

Programmable automation consists of a system of multidegree-of-freedom manipulators (commonly known as industrial robots) and sensors under computer control, which can be programmed, primarily by software, to perform specified jobs in the manufacturing process and can be applied to new (but similar) jobs by reprogramming. The capabilities of such programmable systems differ sharply from those of conventional "fixed" or "hard" automation systems, in which special-purpose machines are designed to perform only specific repetitive tasks. This is particularly important where production runs are small and where different models may have to be produced frequently.

Most industrial robots in use today are point-to-point manipulators, which perform a variety of "pick and place" material-handling jobs and spot welding. The remaining industrial robots are continuous-path manipulators, which are used in arc welding, paint spraying, and so on. A major limitation of these manipulators is their primitive sensory feedback.[5] In particular, commercially available industrial robots have neither contact sensors (force, torque, and touch sensors) as aids to manipulation, nor noncontact sensors (visual and range sensors) as aids to recogniton, inspection, or manipulation of workpieces. Extensive research in machine intelligence, a branch of computer science, has provided tools, techniques, and concepts that are directly relevant to this class of problems. This work originated in a number of "hand-eye" and robot programs[6-13] that developed and demonstrated fairly complex systems of integrated effectors and sensors to guide a manipulator by using computer-processed contact and noncontact sensory feedback.

However, machine intelligence research has been primarily directed at finding general methods that demonstrate principles, and relatively little attention has been paid to questions of computational cost and program complexity—questions that are of major importance to industrial applications. For example, the goal of most visual scene-analysis programs has been the identification of all the objects in a scene, regardless of orientation, with parts of some objects occluded by other objects, and under varying and rather difficult illumination conditions. Under these conditions computer programs that can barely cope with fairly simple scenes are huge, difficult to code and modify, and time consuming; hence, they require large computers. On the other hand, recognition of a relative small number of parts, one at a time against a controlled background and under controlled illumination, can be accomplished by algorithms that will run quickly on a small computer.

Experienced machine-intelligence scientists have just begun to apply their knowledge to the constrained and therefore simplified problems of programmable automation in the factory. A number of research and development programs[14-18] are now in being with the specific goals of exploring, applying, and expanding machine-intelligence techniques and concepts to programmable automation. Concurrently, inexpensive yet powerful minicomputers and microcomputers are now becoming available, making it possible to control integrated sensor-manipulator systems that can be economically justified.

Needs for sensors

Extending the present capabilities of "pick-and-place" industrial robots will require a considerable improvement in their capacity to perceive and interact with the surrounding environment. In particular, it is desirable to develop sensor-mediated, computer-controlled interpretive systems that can emulate human capabilities. To be acceptable by industry, these hardware/software systems must perform as well or better than human workers. Specifically, they must be inexpensive (provide an acceptable return on investment), fast (comparable to average speed of human workers), reliable (error and failure rates considerably lower than those of humans), and suitable for the factory environment.

Sensor needs can be broadly divided into three areas of application: visual inspection, finding parts, and controlling manipulation.

Visual inspection. Here we are concerned only with an important subset of visual inspection: the qualitative and semiquantitative type of inspection performed by human vision rather than by measuring instruments. Such inspection of parts or assemblies includes identifying parts; detecting of burrs, cracks, and voids; examining cosmetic qualities and surface finish; counting the number of holes and determining their approximate locations and sizes; assessing completeness of assembly; and so on. Sensory methods being developed for augmenting industrial robot systems can also be applied effectively to inspection that requires accurate mensuration, but these applications will not be considered here.

Visual inspection occupies many workers in factories. *Explicit* inspection, performed by workers whose sole job is to inspect parts, subassemblies, and assemblies, has been estimated to represent approximately 10 percent of the total labor cost for all durable goods.[19] This cost is second only to that of assembly operations, which is approximately 22 percent of the total cost. It is fair to assume that the majority of these inspection tasks are done visually.

Implicit visual inspection is performed by assemblers to ascertain that the assembled workpieces are the correct ones, are complete, and have not been damaged. This task represents a small but essential part of the assembly function. Adding defective or wrong workpieces to an assembly that may have acquired a considerable value will produce costly scrap or require expensive correction later.

The wide variety of significant characteristics that are routinely examined visually by humans indicates the complexity of the processing that must be performable by automated systems. It is evident that a large library of computer programs will have to be developed to cope with the numerous classes of inspection. To avoid the lengthly and costly programming for every new inspection job, this library must be made generally available to all manufacturing industries.

Finding parts. Where fixed or hard automation is justified for high-volume mass production, workpieces must be positioned and oriented with considerable precision, usually at high cost for special jigging. For material-handling and assembly operations in the unstructured environment of the great majority of factories, it will probably be necessary to "find" workpieces—that is, to determine their positions and orientations and sometimes also to identify them. It is possible to preserve workpiece orientation throughout the manufacturing process by suitable jigging or special palletizing. However, the cost entailed with this approach may not be justified, especially if batches are small or product modifications are frequent.

For example, present industrial robots cannot cope with the problem of picking up parts, one at a time, from a bin containing many randomly oriented parts. Such bins are used for temporary storage and for transportation of parts from station to station in many factories. It is highly unlikely that a very expensive replacement for this function will be implemented in most factories, especially for small batches or for parts that emerge from processes that inherently produce disorder, such as tumble polishing and plating.[20]

The inability of existing industrial robots to adapt to random positions and orientations of parts greatly limits their usefulness in material-handling and assembly operations. Further, if robot systems are to replace people currently doing these jobs, then these systems must also be able to perform the kind of qualitative and semiquantitative inspection tasks described previously. Thus, it is necessary to augment existing robots with visual sensors to be able to determine the identity, position, and orientation of parts and to perform visual inspection as needed. Up to the present, such sensors have been optical or electro-optical and have consisted of television cameras (vidicon and solid state) and several types of scanning systems that measure intensity and range data.

Controlling manipulation. Manipulation of workpieces and tools for material handling and assembly jobs requires many basic operations, such as grasping, holding, orienting, inserting, aligning, fitting, screwing, turning, and so on. In a completely structured environment it may be possible to perform all these operations in a feed-forward manner with no sensory control or correction needed.

It is instructive to note that human manipulation, being imprecise, depends almost entirely on sensory feedback to control both simple and complex manipulative operations. In general, the human worker makes use of both noncontact (visual) sensing and contact (force, torque, or touch) sensing. There is little doubt that a blind worker can perform many manipulative tasks using contact sensing alone. However, he cannot easily perform most inspection functions, cannot readily cope with unexpected instrusions into his work space, and would generally require a far more structured situation to equal the performance of a sighted worker.

On the basis of these observations, it appears useful to consider the use of both contact and noncontact sensors in manipulator control and to try to assess where each sensor is most appropriate. One approach is to divide the sensory domain into coarse and fine sensing, using noncontact sensors for coarse resolution and contact sensors for fine resolution. For example, in acquiring a workpiece that may be randomly positioned and oriented, a visual sensor may be used to determine the relative position and orientation of the workpiece rather coarsely, say, to one tenth of an inch. From this information the manipulator can be positioned automatically with a precision matching that of the visual sensor. The somewhat compliant fingers of the manipulator hand, bracketting the workpiece, will now be close enough to effect closure, relying on touch sensors to stop the motion of each finger when a specified contact pressure is detected. After contacting the workpiece without moving it, the compliant fingers have flexed no more than a few thousandths of an inch before stopping. The touch sensors have thus performed fine resolution sensing and have compensated for the lack of precision of both the visual sensor and the manipulator. This task, which is quite common, illustrates the relative merits of each sensory modality and the advantages of using both. It appears likely that the combined use of these sensors and the associated computer hardware and software will be cost effective within a short time.

Other common applications for contact sensors, which implicitly entail fine resolution or precision sensing, include:

- Collision avoidance, using force sensors on the links and hand of a manipulator. Motion is quickly stopped when any one of preset force thresholds is exceeded.

- Packaging operations in which parts are packed in orderly fashion in tote boxes. Force sensors can be used to stop the manipulator when its compliantly mounted hand touches the bottom of the box, its sides, or neighboring parts. This mode of force feedback compensates for the variability of the positions of the box and the parts and for the small but important variability of the manipulator positioning.

- Insertions of pegs, shafts, screws, and bolts into holes. Force and torque sensors can provide feedback information to correct the error of a computer-controlled manipulator. Again, one may first use visual sensors with relatively coarse resolution to find the hole, bring the peg to an edge of the hole (perhaps partially inserted), and then align the peg with the hole by moving in a direction that minimizes the measured binding force and torque.

By reducing the field of view, it is possible to increase the resolution of noncontact sensors to approach that of contact sensors. This method may be too slow because of the large number of fields required to cover a given area of interest and the excessive amount of computation. However, this limitation may be overcome by using a fixed, wide field of view to find the target and a movable, narrow field of view to obtain high resolution. Alternatively, mounting a small optical sensor on the hand of a manipulator will provide reasonably high resolution over the small

but very important field of view close to where it is most needed. It would then be possible for the manipulator to follow especially identified lines, or to control its motion based on the location of holes or fiduciary marks. It is likely that this "eye-in-hand" mode of operation can be applied successfully to situations that require positioning of intermediate precision.

Contact sensors

As discussed earlier, contact sensing of force, torque, and touch can be usefully combined with visual sensing for many material-handling and assembly tasks. For certain classes of manipulation, however, visual sensors will not be used.[21,22] Here are some examples:

- The workpieces cannot be seen because of occlusion or lack of sufficient light.
- The visual sensors are busy doing other tasks.
- Instances in which processing of contact sensory data is simpler or faster than that of visual sensory data.

The functions of contact sensors in controlling manipulation may be classified into the following basic material-handling and assembly operations:[21,23]

- Searching—detecting a part by sensitive touch sensors on the hand exterior without moving the part.
- Recognition—determining the identity, position, and orientation of a part, again without moving it, by sensitive touch sensors with high spatial resolution.
- Grasping—acquiring the part by deformable, roundish fingers, with sensors mounted on their surfaces.
- Moving—placing, joining, or inserting a part with the aid of force sensors.

Force and torque sensors. There are basically three methods for sensing forces (and torques) for controlling a manipulator, depending on their location:

- Measuring the forces acting on the joints of the manipulator without adding special sensors.
- Measuring the forces acting between the last link of the manipulator and its hand by means of a wrist force sensor.
- Measuring the force exerted by the manipulator hand on a workpiece by means of a separate pedestal sensor.

Measurement of joint forces and torques. The force and torque acting on each joint of a manipulator can be sensed directly. If the joint is driven by an electric dc motor, then sensing is done by measuring the armature current; if the joint is driven by a hydraulic motor, then sensing is done by measuring the back pressure. Two examples in which joint forces were measured follow.

Inoue[24] programmed a manipulator to insert a peg into a hole, using force sensing at the manipulator joints. Paul[25] programmed the Stanford arm to assemble a water pump consisting of a base, a gasket, a top, and six screws. Joint forces were computed from measurements of motor currents. His program included compensation for gravity and inertial forces. Force feedback was used to locate holes for inserting two pins that were later used to align the gasket.

Sensing joint forces directly has the advantage of not requiring a separate force sensor. However, the force (or torque) between the hand and its environment is not measured directly. Thus, the accuracy and resolution of this measurement are adversely affected by the variability in the inertia of the arm and its load and by the nonuniform friction of the individual joints.[26]

Wrist force sensors. A wrist force sensor measures the three components of force and three components of torque between the hand and the terminal link of the manipulator. Basically, a wrist force sensor consists of a structure with some compliant sections and transducers that measure the deflection of the compliant sections along three orthogonal axes as a result of the applied force and torque. There are different types of force transducers, such as strain gage, piezoelectric, magnetostrictive, magnetic, and others.[21-27] The most common of these is the strain-gage transducer, which is inexpensive, reliable, and rugged.

A wrist sensor employing four beams with ball joint mounts and strain gages was built at Draper Laboratory and found by Groome[28] to have a resolution of only 4 binary bits because of hysteresis in the ball joints.

Goto[29] built a compliant wrist force sensor with cross springs and strain gages and used the sensor to control insertion of a 1/2-inch-diameter polished cylinder into a vertical hole with 7- to 20-micron clearance in less than 3 seconds. The insertion operation was based on two types of compliance: active compliance, whereby sensed forces are used to command the manipulator to correct positional or orientational errors, and passive compliance, whereby such errors are corrected by the reaction forces between the workpieces.

Figure 1 illustrates a typical strain-gage wrist force and torque sensor. The sensor, with seven-bit resolution, was built at Stanford Research Institute[14] for the Stanford arm. The sensor is made of a milled 3-inch-diameter aluminum tube, having 8 narrow elastic beams with no hysteresis. The neck at the end of each beam transmits a small bending torque and thus increases the strain at the other end of the beam where two foil strain gages (shown as black rectangles) are cemented. The two strain gages are connected to a potentiometer circuit whose output is proportional to a force component normal to the strain-gage planes, and is automatically compensated for variation in temperature.

The wrist sensor measures the three components of force and three components of torque in a Cartesian coordinate system (*x, y, z*) attached to the manipulator hand. As shown in Figure 2, the hand coordinates *x*, *y*, and *z* are also called, respectively, "lift," "sweep," and "reach," when referring to forces or translations, and "turn," "tilt," and "twist" when referring to torques or rotations.

A wrist force sensor made of milled aluminum cylinder and foil strain gages cemented to beams was built by Watson and Drake.[30] Using extensional strain gages and strain-gage shear bridges, they were able to use only three beams.

A hand with sensors for a Unimate manipulator performing material handling was built by Hill and Sword.[14,31] The sensors included a wrist force sensor, using compliant elements and potentiometers to sense the relative displacement of the hand, as well as touch and proximity sensors. Figure 3 shows a sequence of actions illustrating the use of this hand in orderly packing of water pumps into a tote box: The Unimate moves rapidly to a previously taught starting position (a) and then moves slowly down until a threshold *z* force is sensed (b), up 1/2 inch, along the −*x* direction until a threshold *x* force is sensed (c), along the +*x* direction 1 inch, along the +*y* direction until a *y* threshold is sensed, along the −*y* direction 1/2 inch, and down until a threshold *z* force is sensed (d); it then opens the fingers to release the pump and moves quickly to acquire the next pump (e).

Pedestal force sensor. A pedestal force sensor, forming a base for assembly operations, was also built by Watson and Drake.[22,30] The sensor measures the three components of force and three components of torque that are applied to a workpiece mounted on the pedestal.

The pedestal force sensor is shown in Figure 4. Its frame consists of three aluminum plates, about 40 cm square and 2.5 cm thick. The middle plate is connected to the top plate by four strain-gage force transducers that are set vertically and to the bottom plate by four similar transducers that are set horizontally. The sensor has a 4000:1 dynamic range (12-bit resolution). It was used at Draper Laboratory to perform peg-in-hole experiments.

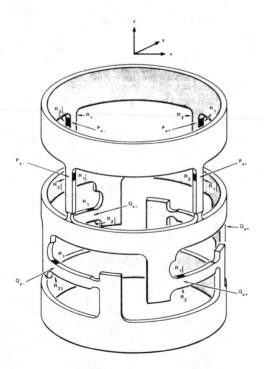

Figure 1. Strain gage wrist sensor.

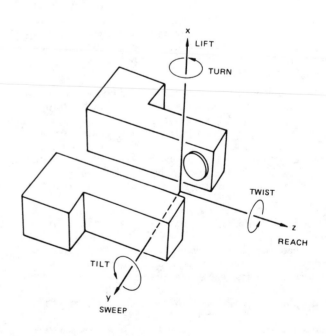

Figure 2. Hand coordinates.

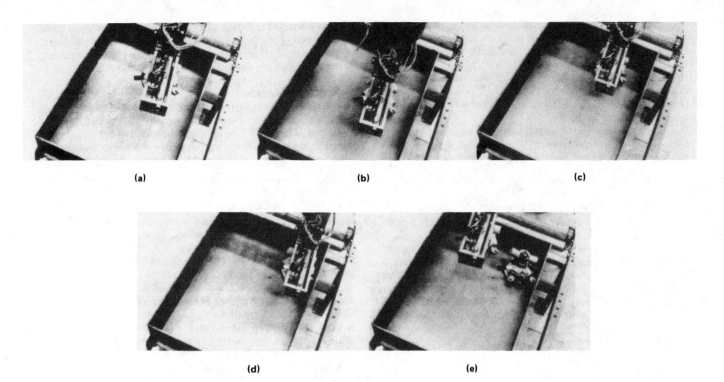

(a)   (b)   (c)

(d)   (e)

Figure 3. Force sensor control of manipulator hand in pump packing.

Touch sensors. Touch sensors are used to obtain information associated with the contact between the finger(s) of a manipulator hand and objects in the workspace. They are normally much lighter than the hand and are sensitive to forces much smaller than those sensed by the aforementioned force sensors.

Touch sensors may be mounted on the outer and inner surfaces of each finger. The outer sensors may be used to search for an object and possibly determine its identity, position, and orientation. Outer sensors may also be used for sensing unexpected obstacles and stopping the manipulator before any damage can occur. The inner mounted sensors may be used to obtain information about an object before it is acquired and about grasping forces and workpiece slippage during acquisition.

Touch sensors may be classified into two types, binary and analog.

Figure 4. Pedestal force sensor (courtesy of Draper Laboratory).

Binary touch sensors. A binary touch sensor is a contact device, such as a switch. Being binary, its output is easily incorporated into a computer controlling the manipulator.

A simple binary touch sensor consists of two microswitches, one on the inner side of each finger. Paul[25] used such a sensor to determine whether a part was present or absent and to center the hand over it during automated assembly of a water pump. With the addition of a position potentiometer that measures the distance between the two fingers, this primitive sensor can also be used to classify a small set of parts by determining some of their dimensions.

Ernst,[6] who pioneered sensor-mediated manipulation under computer control, built a two-fingered hand with both binary and analog touch sensors. However, he later found that binary touch sensors were far more useful and he used the analog touch sensors primarily as binary sensors.

Goto[23] built a hand with two fingers, each having 14 outer contact sensors and 4 inner, pressure-sensitive, conductive-rubber sensors. He used the touch information to acquire blocks randomly located on a table and pack them tightly on a pallet.

Garrison and Wang[32] built a gripper with 100 pneumatic snap-action touch sensors located on a grid with 0.1- by 0.1-inch centers. The sensors consisted of contact terminals, a thin metal sheet with elastic shallow spherical domes, and a flexible insulating rubber sheet on the outside. Physical contact is sensed when external pressure exceeds a preset threshold, causing the activation of a snap-action switch consisting of a dome and a terminal.

Analog touch sensors. An analog touch sensor is a compliant device whose output is proportional to a local force. Analog touch sensors are usually mounted on the inner surface of the fingers to measure gripping forces and to extract information about the object between the fingers.

329

Hill and Sword[26] built a manipulator hand with a wrist force sensor and analog touch sensors; the hand is shown in Figure 5. Seven outer sensing plates and a matrix of 3-by-6 inner sensing buttons are mounted on each finger (or jaw). The force on each sensor acts against a compliant washer and displaces a vane that controls the amount of light received by a phototransistor from a light-emitting diode.

Takeda[33] built a touch sensing device for object recognition. The device consists of two parallel fingers, each with an array of 8-by-10 needles that are free to move in a direction normal to each finger and a potentiometer that measures the distance between the fingers. As the fingers close, the needles contact the object's contour in a sequence that depends on the shape of the object. Software was developed to use the sensed touch points to recognize simple objects, such as a cone.

Contact sensing is still in a highly experimental stage. As yet there is no commercial line of contact sensors that have been proved in industrial application. At the same time, it has become quite evident to manipulator users and developers alike that contact sensing will become a valuable addition to programmed manipulation in the near future.

Noncontact sensors

As previously discussed, noncontact sensors are potentially useful in identifying and finding parts in sensor-controlled manipulation and in visual inspection.

The major categories of noncontact sensors that have been used with robot systems are proximity sensors, electro-optical imaging sensors, and range-imaging sensors. These are described separately below.

Proximity sensors. A proximity sensor is a device that senses and indicates the presence or absence of an object without requiring physical contact. As described by *Machine Design*,[34] five of six major types of proximity sensors now available commercially are radio frequency, magnetic bridge, ultrasonic, permanent-magnet hybrid, and photoelectric. A relatively new solid-state sensor (made by Micro Switch Incorporated) is based on the Hall Effect. These noncontact sensors have widespread use, such as for high-speed counting, protection of workers, indication of motion, sensing presence of ferrous materials, level control, reading of coding marks, and noncontact limit switches.

The modern photoelectric proximity sensor, a relatively new version of the old photoelectric tube and light source, appears to be well adapted for controlling the motion of a manipulator. This sensor consists of a solid-state light-emitting diode (LED), which acts as a transmitter of infrared light, and a solid-state photodiode, which acts as receiver; both are mounted in a small package. As shown in Figure 6, the sensitive volume is approximately the intersection of two cones in front of the sensor. This sensor is not a rangefinder because the received light is not only inversely proportional to the distance squared but is also proportional to the target reflectance and the cosine of the incidence angle, both

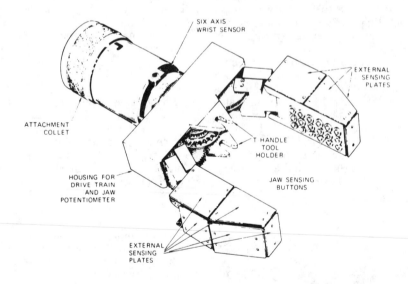

Figure 5. Hand with wrist force sensor and analog touch sensors.

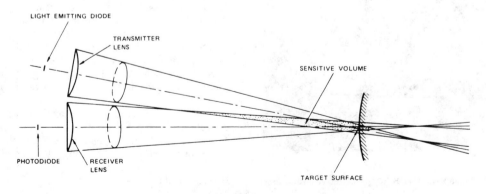

Figure 6. Proximity sensor.

of which may vary spatially. However, if the reflectance and incidence angle are fixed, then the distance may be inferred with suitable calibration. Usually, a binary signal is generated when the received light exceeds a threshold value that corresponds to a predetermined distance. Furthermore, the sensor will detect the appearance of a moving object in a scene by sensing the change in the received light. Such devices are sensitive to objects located from a fraction of an inch to several feet in front of the sensor.

A proximity sensor, interfaced to a control computer, was used to stop the motion of an industrial robot in its approach to within a predetermined distance of a given solid surface.[14] If a "stop" signal to the moving hand is initiated before contact with the surface is made, there is time to stop the industrial robot without damage. This is far superior to the use of a mechanical limit switch for manipulator protection.

A more interesting application of multiple photoelectric proximity sensors used to control the positioning of a manipulator is described by Johnston.[35] Lateral positioning of a hand was controlled by signals from two sensors to center the hand over the highest point of an object. Bejczy[36] described the potential use of proximity sensors for three-dimensional control of the hand and suggested several control algorithms.

Electro-optical imaging sensors. Until recently, electro-optical imaging sensors have provided the most commonly used "eyes" for industrial robots and visual inspection. Standard television cameras, using vidicons, plumbicons, and silicon target vidicons, have been interfaced with a computer and have provided the least expensive and most easily available imaging sensors. These cameras scan a scene, measure the reflected light intensities at a raster of, say, 320-x-240 pixels (picture elements), convert these intensity values to analog electrical signals, and feed this stream of information serially into a computer—all within 1/60 of a second. These signals may be either stored in the computer core memory for subsequent processing or processed in real time "on the fly," with consequent reduction of memory requirements.

In the past few years several solid-state area-array cameras, competitive with the above vacuum tube type of television cameras, have become commercially available. These small, rugged, and potentially reliable cameras are fabricated using modern large-scale-integration silicon technology and will probably become the dominant electro-optical sensors for industrial applications. The photo-active chip of an area-array camera consists of photodiodes, usually charged-coupled devices,[37] whose number at present varies from 32x32 to 320x512 for different requirements of resolution. These cameras operate in a raster-scan mode, similar to that of the vidicon television cameras, and produce two-dimensional images of scenes.

A one-dimensional solid-state camera, using a linear diode array that varies from 16 to 1872 elements, is also available commercially. This device can perform a single linear scan and is very useful for sensing objects that are in relative motion to the camera, such as workpieces moving on a conveyor belt. An example is shown in Figure 7, where a connecting rod moves past the viewing station and top-view and side-view linear scans are performed by two linear diode arrays, each scan initiated by a repetitive signal from a position sensor (incremental encoder) that is coupled to the moving conveyor belt. For each scan, values of light intensity at a fixed number of discrete points are measured, converted into electrical signals, and sent to a computer. These signals are either processed in real time or stored in memory until the image of the entire workpiece is obtained for subsequent processing.

Another large class of electro-optical sensors, which differ in several important characteristics from the above cameras, has been used primarily in advanced "hand-eye" artificial intelligence research projects.[7-9] These sensors include the image dissector camera, the cathode-ray flying spot scanner, and the laser scanner. They have been described and compared with the more common vidicon type of television cameras in reviews by Earnest[38] and Chien and Snyder.[39] These electro-optical sensors can be programmed to image selected areas of the field of view in a random access manner, as contrasted with the prescribed "raster scan" acquisition of the ordinary television camera. In many instances this method of operation permits the acquisition, storage, and processing of only the relevant data in a field of view. The image dissector,

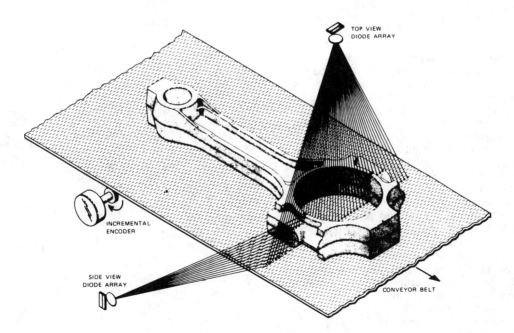

Figure 7. Obtaining orthogonal views of a moving part by two linear diode arrays.

however, has low sensitivity, requiring high levels of illumination, and is relatively expensive; for these reasons it has not attained widespread acceptance. Nevertheless, one commercial company[40] has adapted the image dissector for visual inspection and recognition and alignment of parts, and claims to be able to make noncontact measurements to a precision of less than 1 micron and to recognize and determine position of parts at speeds of 1000 frames per second.

As noted previously, recognition of parts and determination of their positions and orientations are often requirements for manipulating parts in material-handling and assembly operations. Researchers working in "hand-eye" programs have made use of electro-optical imaging sensors to identify, locate, and manipulate simple objects. A few examples follow:

• At Stanford University,[41] a television camera fitted with color filters was used to identify four colored blocks. Using a computer program that extracted edges and vertices, the position and orientation of each block was determined, thus providing the information needed to stack the blocks according to certain rules.

• At the University of Nottingham,[42] the identity, position, and orientation of flat workpieces were determined, one at a time, from a top-view image obtained by a television camera. The camera and a manipulator were mounted on a turret in the same fashion as lens objectives are mounted on a common turret of a microscope. After the identity, position, and orientation of each workpiece had been determined, the manipulator rotated into a position coaxial with the original optical axis of the camera lens and acquired the workpiece.

• At Hitachi Central Research Laboratory,[43] prismatic blocks moving on a conveyor belt were viewed, one at a time, using a vidicon television camera. A low-resolution image (64x64 pixels) was processed to obtain the outline of each block. A number of radius vectors from the center of area of the image to the outline were measured and processed by a minicomputer to determine the identity, position, and orientation of each block. The block was then picked up, transported, and stacked in an orderly fashion by means of a simple suction-cup hand whose motion was controlled by the minicomputer.

• At General Motors Research Laboratories,[44] an experimental system was devised to mount wheels on an automobile. The location of the studs on the hubs and the stud holes on the wheels were determined using a television camera coupled to a computer, and then a special manipulator mounted the wheel on the hub and engaged the studs in the appropriate holes. Although this experiment demonstrated the feasibility of a useful task, further development is needed to make this system cost-effective.

• At Osaka University,[45] a machine-vision system, including a television camera coupled to a minicomputer, has been developed to recognize a variety of industrial parts, such as gasoline-engine parts, when viewed one at a time on a conveyor. Resolution of 128x128 elements digitized into 64 levels of gray scale were used. In lieu of the usual sequence of picture processing, extraction of relevant features, and recognition, the system makes use of predetermined part models that guide the comparison of the unknown part with stored models, suggesting the features to be examined in sequence and where each feature is located. This procedure reduced the amount of computation required. Further, by showing sample parts and indicating important features via an interactive display, an operator can quickly train the system for new objects, the system generating the new models automatically

from the cues given by the operator. The system is said to recognize 20 to 30 complex parts of a gasoline engine. Recognition time and training time were 30 seconds and 7 minutes, respectively.

• At Stanford Research Institute[11,46] a hardware/software system under minicomputer control has been developed that determines the identity, position, and orientation of each workpiece placed randomly on a table or on a moving conveyor belt and, using a Unimate industrial robot, acquires that workpiece and moves it to its destination. The electro-optical sensors employed include a solid-state 100x100 area-array camera and a solid-state 128x1 linear array camera. A workpiece is recognized by using either a method based on measuring the entire library of features ("nearest-reference" classification) or a method based on sequential measurement of the minimum number of features that can distinguish one workpiece from the others ("decision-tree" classification). Selection of the distinguishing features for the second method is done automatically by simply showing a prototype to the viewing station of the system. The decision-tree classification method was applied to recognition of different workpieces, such as foundry castings, water-pump parts, and covers of electrical boxes. For example, showing the water-pump parts in Figure 8 resulted in automatic generation of the decision tree in Figure 9, where x_1, x_3, x_4, and x_7 are a subset of the set of features x_1 through x_7 (perimeter; square root of area; total hole area; minimum, maximum, and average radius from center of area to outline; and the ratio x_1/x_2), which are invariant to the position and orientation of a part. The feature selected at each tree node is the most distinguishing feature for dividing the group of part candidates into two subgroups. The subgroup to be followed during recognition time will depend on the measured value of that feature. This process is repeated recursively until a terminal node is reached and the unknown part is identified. In all cases, either training (tree building) or recognition was achieved in much less than 1 second.

• At Stanford Research Institute, Agin[14,46] developed an interactive programming system to aid the program-

Figure 8. Water pump parts.

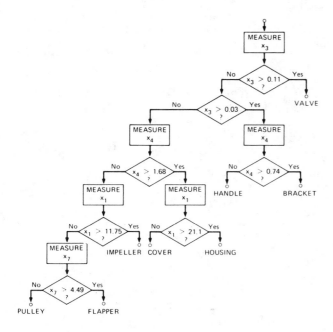

Figure 9. Decision tree for water pump parts.

Figure 11. Display of lamp base inspection.

mer in constructing an inspection strategy for each inspection job. The operator has the use of a graphics terminal, a light pen, and a library of image-processing and feature-extraction routines. A typical inspection job, which at present is done by human vision only, is to ascertain that the number and placement of electrical contacts on lamp bases, such as shown in Figure 10, is correct. An acceptable lamp base has two contacts that are separated and positioned with specific tolerance with respect to an insertion guide pin on the side. The result of the automated inspection is shown in Figure 11, where crosses indicate rejection and a plus sign indicates acceptance. Note that the inspection was performed while two lamp bases were touching. It is estimated[46] that lamp bases can be inspected at a rate of about 6 per second, using a PDP-11/40 minicomputer with a linear diode array camera.

Figure 10. Lamp bases.

Range-imaging sensors. A range-imaging sensor measures the distances from itself to a raster of points in the scene. Although range sensors are used for navigation by some animals (e.g., the bat), hardly any work has been done so far to apply range image to control the path of a manipulator. This situation may change in the future however, as the technological and economical difficulties currently entailed with range-imaging sensors are overcome.

Different range-imaging sensors have been applied to scene analysis in various research laboratories.[47-59] These sensors may be classified into two types, one based on the trigonometry of triangulation and the other based on the time of flight of light (or sound).

Triangulation range sensors are further classified into two schemes, one based on a stereo pair of television cameras (or one camera in two locations),[47-50] and the other based on projecting a sheet of light by a scanning transmitter and recording the image of the reflected light by a television camera.[51-54] Alternatively, the second scheme may transmit a light beam and record the direction of the reflected light by a rocking receiver. The first scheme suffers from the difficult problem of finding corresponding points in the two images of the scene.[47-49] Both schemes have two main drawbacks: missing data for points seen by the transmitter but not by the receiver and vice versa,[55] and poor accuracy for points that are far.[50]

The above drawbacks are eliminated by the second type of range-imaging sensor using a laser scanner, which is also classified into two schemes: one based on transmitting a laser pulse and measuring the arrival time of the reflected signal,[56] and the other based on transmitting amplitude modulated laser beam and measuring the phase shift of the reflected signal.[59] A simplified block diagram of the latter sensor is shown in Figure 12. The transmitted

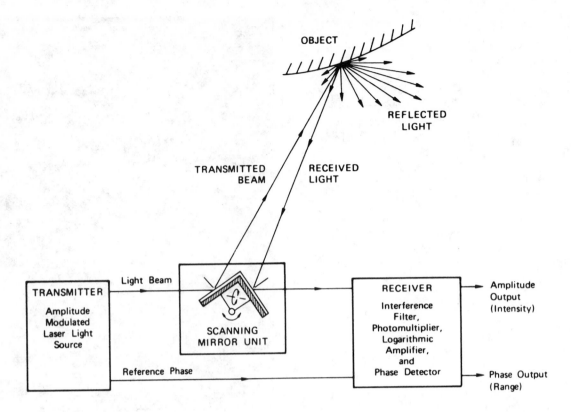

Figure 12. Simplified block diagram of a range imaging sensor.

beam and the received light are essentially coaxial.

Range-imaging sensors have been applied so far primarily to object recognition. However, they are also very suitable for other tasks, such as finding a factory floor or a road, detecting obstacles and pits, and inspecting the completeness of subassemblies.[58,59]

References

1. R. H. Anderson, "Programmable Automation: The Future of Computers in Manufacturing," Report ISI/RR-73-2, Information Science Institute, University of Southern California, Marina del Rey, California, March 1973.

2. N. M. Kamrany, "A Preliminary Analysis of the Economic Impacts of Programmable Automation upon Discrete Manufacturing Industries," Report ISI/RR-73-4, Information Science Institute, University of Southern California, Marina del Rey, California, October 1973.

3. C. A. Rosen, "Robots, Productivity, and Quality," 1972 ACM National Conference, Boston, Massachusetts, August 1972.

4. Automation Research Council, "Proceedings of the Discrete Manufacturing Industries Workshop," Report No. 3, February 1974; and "Proceedings of the Technology Workshops," Report No. 5, February 1974.

5. F. J. Engelberger, "Economic and Sociological Impact of Industrial Robots," Proc., 1st Int. Symp. on Industrial Robots, April 1970, pp. 7-12.

6. Ernst, H. A., "MH-1, A Computer-Operated Mechanical Hand," AFIPS Conf. Proceedings, SJCC, San Francisco, May 1-3, 1962, pp. 39-51.

7. M. L. Minsky, "An Autonomous Manipulator System," Project MAC Progress Report III, M.I.T., July 1966; and subsequent Project MAC, Artificial Intelligence Memo Series, M.I.T., Cambridge, Massachusetts.

8. J. McCarthy et al., "A Computer with Hands, Eyes, and Ears," AFIPS Conf. Proc., FJCC, 1968, pp. 329-338.

9. J. A. Feldman et al., "The Stanford Hand-Eye Project," Proc. 1st. Int. Joint Conf. on Artificial Intelligence, Washington, D.C., May 1969, pp. 521-526.

10. N. J. Nilsson, "A Mobile Automation: An Application of Artificial Intelligence Techniques," Proc., 1st Int. Joint Conf. on Artificial Intelligence, May 1969, pp. 509-520.

11. M. Ejiri, T. Uno, H. Yoda, T. Goto, and K. Takeyasu, "A Prototype Intelligence Robot That Assembles Objects from Plan Drawings," IEEE Trans. on Computers, Vol. C-21, No. 2, 1972.

12. H. G. Barrow and S. H. Salter, "Design of Low-Cost Equipment for Cognitive Robot Research," in Machine Intelligence, Vol. 5, B. Meltzer and D. Michie, eds., American Elsevier Publishing Co., 1970.

13. A. K. Bejczy, "Machine Intelligence for Autonomous Manipulation" in Remotely Manned Systems, Exploration, and Operation in Space, E. Heer, ed., California Institute of Technology Pub., Pasadena, California, 1973.

14. C. A. Rosen et al., "Exploratory Research in Advance Automation," Reports 1 through 5, prepared by Stanford Research Institute under National Science Foundation Grant GI38100X, December 1973 to January 1976.

15. J. Nevins et al., "Exploratory Research in Industrial Modular Assembly," Reports 1 through 3, prepared by Draper Laboratory under National Science Foundation Grant No. GI-39432X and ATA74-18173 A01, June 1, 1973, to August 31, 1975.

16. T. O. Binford et al., "Exploratory Study of Computer-Integrated Assembly Systems," National Science Foundation Progress Report September 15, 1974, to November 30, 1975.

17. W. B. Heigenbotham et al., "Visual Feedback Applied to Programmable Assembly Machines," *Proc. 2nd Int. Symp. on Industrial Robots*, Chicago, May 1972, pp. 63-76.

18. P. M. Will and D. D. Grossman, "An Experimental System for Computer-Controlled Mechanical Assembly," *IEEE Trans. on Computers*, September 1975, pp. 879-888.

19. J. L. Nevins, N. Kamrany, and D. E. Whitney, "Productivity Technology and Product System Productivity Research," Report R-928, Vol. 2, prepared by Draper Laboratory under National Science Foundation Grant No. APR 75-15334, 1976.

20. D. D. Grossman and M. W. Blasgen, "Orienting Mechanical Parts by Computer-Controlled Manipulator," Correspondence, *IEEE Trans. on Systems, Man, and Cybernetics*, September 1975, pp. 562-565.

21. T. D. Binford, "Sensor System for Manipulation," *Proc. 1st Conf. on Remotely Manned Systems (RMS), Exploration, and Operations in Space*, E. Heer, ed., 1973, pp. 283-291.

22. J. L. Nevins et al., "A Scientific Approach to the Design of Computer Controlled Manipulators," Report No. R-837, Draper Laboratory, Cambridge, Massachusetts, August 1974.

23. T. Goto, "Compact Packaging by Robot with Tactile Sensors," *Proc. 2nd Int. Symp. on Industrial Robots*, IIT Research Institute, Chicago, Illinois, 1972.

24. H. Inoue, "Computer Controlled Bilateral Manipulator," *Bull. Japanese Soc. Mech. Eng.*, Vol. 14, 1971, pp. 199-207.

25. R. C. Bolles and R. Paul, "The Use of Sensory Feedback in a Programmable Assembly System," Stanford Artificial Intelligence Project Memo No. 220, Stanford University, Stanford, California, October 1973.

26. J. W. Hill and A. J. Sword, "Manipulation Based on Sensor-Directed Control: An Integrated End Effector and Touch Sensing System," *Proc., 17th Annual Human Factor Society Convention*, Washington, D.C., October 1973.

27. H. N. Norton, "*Handbook of Transducers for Electronic Measuring Systems*," Prentice Hall, Inc., Englewood Cliffs, New Jersey, 1969.

28. R. C. Groome, "Force Feedback Steering of a Teleoperator System," Report No. T-575, Draper Laboratory, Cambridge, Massachusetts, August 1972.

29. T. Goto, T. Inoyama, and K. Takeyasu, "Precise Insert Operation by Tactile Controlled Robot HI-T-HAND Expert-2," *Proc. 4th Intl. Symp. on Industrial Robots*, Tokyo, Japan, November 1974, pp. 209-218.

30. P. C. Watson and S. H. Drake, "Pedestal and Wrist Force Sensors for Automatic Assembly," *Proc. 5th Int. Symp. on Industrial Robots*, IIT Research Institute, Chicago, Illinois, September 1975, pp. 501-511.

31. C. A. Rosen and D. Nitzan, "Development in Programmable Automation," *Manufacturing Engineering*, September 1975, pp. 26-30.

32. R. L. Garrison and S. S. M. Wang, "Pneumatic Touch Sensor," *IBM Technical Disclosure Bulletin*, Vol. 16, No. 6, November 1973.

33. S. Takeda, "Study of Artificial Tactile Sensors for Shape Recognition—Algorithm for Tactile Data Input," *Proc. 4th Int. Symp. on Industrial Robots*, Tokyo, Japan, November 1974, pp. 199-208.

34. "Proximity Switches," *Machine Design*, April 26, 1973, pp. 33-37.

35. A. R. Johnston, "Proximity Sensor Technology for Manipulator End Effectors," *Proc., 2nd Conf. on Remotely Manned Systems*, California Institute of Technology, Pasadena, California, 1975.

36. A. K. Bejczy, "Algorithmic Formulation of Control Problems on Manipulation," *Proc., 1975 Intl. Conf. on Cybernetics and Society*, IEEE Systems Man and Cybernetics Society, San Francisco, 1975.

37. G. F. Amelio, "Charge-Coupled Devices," *Scientific American*, February 1974.

38. L. D. Earnest, "Choosing an Eye for a Computer," Memo No. 51, Stanford Artificial Intelligence Project, Stanford University, Stanford, California, April 1, 1967.

39. R. T. Chien and W. E. Snyder, "Hardware for Visual Image Processing," *IEEE Trans. on Circuits and Systems*, Vol. CAS-22, pp. 541-551, June 1975.

40. M. G. Dreyfus, "Visual Robots," *Industrial Robot J.*, December 1974, pp. 260-264.

41. J. Feldman et al., "The Use of Vision and Manipulation to Solve the Instant Insanity Puzzle," *Proc., 2nd Int. Joint Conf. on Artificial Intelligence*, The British Computer Society, September 1971.

42. W. B. Heginbotham, P. W. Kitchin, and A. Pugh, "Visual Feedback Applied to Programmable Assembly Machines," *Proc., 2nd Int. Symp. Industrial Robots*, IIT Research Institute, Chicago, Illinois, May 1972, pp. 77-88.

43. "Hitachi Hand-Eye System," *Hitachi Review*, Vol. 22, No. 9, pp. 362-365.

44. Olsztyn, J. T., et al., "An Application of Computer Vision to a Simulated Assembly Task," *Proc., 1st Int. Joint Conf. on Pattern Recognition*, 1973, pp. 505-513.

45. M. Yachida and S. Tsuji, "A Machine Vision for Complex Industrial Parts with Learning Capacity," *Proc., 4th Int. Joint Conf. on Artificial Intelligence*, 1975, pp. 819-826.

46. G. J. Agin and R. O. Duda, "SRI Vision Research for Advanced Industrial Automation," *Proc., 2nd U.S.A.-Japan Computer Conference*, 1975, pp. 113-117.

47. M. D. Levine, D. A. O'Handley, and G. M. Yagi, "Computer Determination of Depth Maps," *Computer Graphics and Image Processing*, Vol. 2, October 1973, pp. 131-150.

48. D. A. O'Handley, "Scene Analysis in Support of a Mars Rover," *Computer Graphics and Image Processing*, Vol. 2, December 1973, pp. 281-297.

49. M. J. Hannah, "Computer Matching of Areas in Stereo Images," Memo AIM-239, Stanford Artificial Intelligence Laboratory, Stanford University, Stanford, California, July 1974.

50. D. Nitzan, "Stereopsis Error Analysis," Technical Note 71, Artificial Intelligence Center, Stanford Research Institute, Menlo Park, California, September 1972.

51. Y. Shirai and M. Suwa, "Recognition of Polyhedra with a Range Finder," *Proc. 2nd, Int. Joint Conf. on Artificial Intelligence*, September 1971, pp. 80-87.

335

52. G. J. Agin and T. O. Binford, "Computer Description of Curved Objects," *Proc., 3rd Int. Joint Conf. on Artificial Intelligence,* August 1973, pp. 641-647.

54. Y. Shirai, "A Step Toward Context-Sensitive Recognition of Irregular Objects," *Computer Graphics and Image Processing,* Vol. 2, December 1973, pp. 298-307.

55. D. Nitzan, "Scene Analysis Using Range Data," Technical Note 69, Artificial Intelligence Center, Stanford Research Institute, Menlo Park, California, August 1972.

56. A. R. Johnston, "Infrared Laser Rangefinder," NASA New Technology Report No. NPO-13460, Jet Propulsion Laboratory, Pasadena, California, August 1973.

57. C. A. Rosen, "Combined Ranging and Color Sensor," U.S. Patent No. 3,945,729, March 1976.

58. R. O. Duda and D. Nitzan, "Low-Level Processing of Registered Intensity and Range Data," *Proc. 3rd Int. Joint Conf. on Pattern Recognition,* November 1976, pp. 598-601.

59. D. Nitzan, A. E. Brain, and R. O. Duda, "The Measurement and Use of Registered Reflectance and Range Data in Scene Analysis," *Proc. IEEE,* Vol. 65, February 1977, pp. 206-220.

Charles A. Rosen is a senior scientific adviser in the Artificial Intelligence Center at SRI International. His present major interest is programmable automation. Since 1957 he has been with SRI, where he initiated and managed major research programs in microelectronics, pattern recognitions and artificial intelligence. From 1950 to 1957 he conducted and managed research in solid state devices, specializing in the use of semiconductor, magnetic, and piezoelectric effects at General Electric. Prior to that, he was involved with aircraft instruments, communication equipment, spot-welding systems, and university teaching.

Rosen received the BEE from Cooper Union, New York, the ME from McGill University, and the PhD from Syracuse University. He is a Fellow of the IEEE, and a member of APS, AAAS, AND RESA. He serves on the Advisory Committee on Technical Innovations at the National Academy of Sciences.

David Nitzan is a senior research engineer in the Artificial Intelligence Center at SRI International. Since 1971 he has been working on programmable industrial automation and machine perception using a laser range/reflectance sensor. During 1970-71 he taught electrical engineering courses at the Technion in Haifa. From 1959 to 1970 he developed models for magnetic flux switching in toroidal and multipath cores, and computer programs for analyzing electronic circuits including magnetic cores.

Before joining SRI in 1959 Nitzan was an associate in electrical engineering at UC Berkeley. Earlier experience included field engineering for the Bechtel Corporation and the designing of electrical installations for the Sollel Boneh Company of Haifa. He has a BS from the Technion and an MS and PhD from UC Berkeley, all in electrical engineering. He is a senior member of the IEEE and a member of the Research Society of America.

Presented at the Ninth International Symposium on Industrial Robots Conference, March 1979

CONSIGHT: A Practical Vision-Based Robot Guidance System

By Mitchel R. Ward, Lothar Rossol, Steven W. Holland
General Motors Research Laboratories
and

Robert Dewar
General Motors Manufacturing Development

A vision-based robot system capable of picking up parts randomly placed on a moving conveyor belt is described. The vision subsystem, operating in a visually noisy environment typical of manufacturing plants, determines the position and orientation of parts on the belt. The robot tracks the parts and transfers them to a predetermined location. This system can be easily retrained for a wide class of complex curved parts and demonstrates that future systems have a high potential for production plant use.

INTRODUCTION

In many manufacturing activities, parts arrive at work stations by means of systems that do not control part position. Since present robots require parts to be in precisely fixed positions, their use is precluded at these work stations. To automate these part handling operations, intricate feeding devices that precisely position the parts are required. Such devices, however, are often uneconomical and unreliable.

Robot systems equipped with vision represent an alternative solution. This paper describes CONSIGHT, a vision-based robot system for transferring unoriented parts from a belt conveyor to a predetermined location. CONSIGHT:

- determines the position and orientation of a wide class of manufactured parts including complex curved objects,

- provides easy reprogrammability by insertion of new part data, and

- works on visually noisy picture data typical of many plant environments.

As a result of these characteristics — and because the vision subsystem does not require light tables, colored parts or other impractical means for enhancing contrast — CONSIGHT systems have a high potential for production plant use

From the outset CONSIGHT was targeted for plant installation. The experimental version, described here in detail, has been succeeded by a production prototype. And, although the prototype was implemented on different hardware and had to be reprogrammed, the overall structure of the system has remained intact. At each phase careful consideration was given to succeeding phases and close cooperation was maintained among the groups responsible for the different versions of CONSIGHT.

Much of the motivation for this work was provided by early vision and robot tracking research done at SRI International.

A FUNCTIONAL OVERVIEW

CONSIGHT functions in two modes: a setup mode and an operational mode. During setup, various hardware components are calibrated and the system is programmed to handle new parts. Once calibrated and programmed for a specific part, the system can be switched to operational mode to perform part transfer functions.

Part transfer operates as follows: The conveyor (Fig. 1) carries the randomly positioned parts past a vision station which determines each part's position and orientation on the moving belt. This information is sent to a robot system which tracks the moving parts on the belt, picks them up, and transfers them to a predetermined location. The above sequence operates continuously with no manual intervention except for placing parts on the conveyor.

Fig. 1. CONSIGHT conveyor, camera, and robot arrangement

CONSIGHT is capable of handling a continuous stream of parts on the belt, so long as these parts are not touching. The maximum speed limitation is imposed by neither the vision system nor the computer control, but by the cycle time of the robot arm.

Calibration for CONSIGHT is a simple procedure taking about 15 minutes and may be performed by an operator who need not understand or be aware of the mathematical procedures involved. Calibration need be done only during initial setup or after physically moving the robot, conveyor or camera.

Programming CONSIGHT for new parts is also a simple procedure which retains much of the "teaching by showing" concept of current robot systems.

A SYSTEM OVERVIEW

Organization

CONSIGHT is logically partitioned into independent vision, robot, and monitor subsystems. Intersystem communication was designed to allow easy substitution of new subsystems with minimal impact. For example, the vision subsystem reports only a unique point (i.e., x and y coordinate) for each part, and an orientation. Neither the location of the point on the part, nor the reference from which to measure orientation, is specified. Other vision modules [2,3] can be substituted with little effect on the control program, the robot system, the part programming methods, or the calibration methods. Equally important, another robot, a different monitor subsystem, or different control computers could be substituted without a great effect on other CONSIGHT subsystems. This modular organization has been very beneficial during the implementation of the production prototype (described later), however, this is not to say that replacing one robot with another is by any means a trivial task.

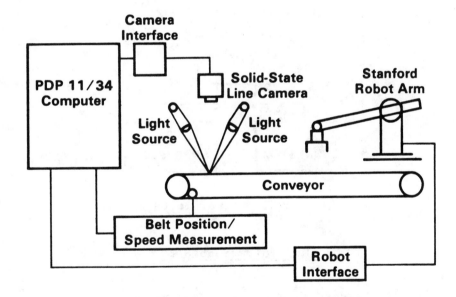

Fig. 2. CONSIGHT hardware schematic

Hardware

The hardware of the experimental system consists of a a Digital Equipment Corporation PDP 11/34 computer operating under the RSX-11S real-time executive, a Reticon RL256C 256x1 line camera, a Stanford Arm made by Vicarm [4], and a belt conveyor which is instrumented with an encoder for position and speed measurement (Fig. 2).

Software

The software organization for CONSIGHT reflects the three major modules of the system.

The monitor coordinates and controls the operation of CONSIGHT and also assists in calibration and reprogramming for new parts. The monitor queues part data and the system is thus capable of handling a continuous stream of parts on the belt.

The vision subsystem uses structured light [5,6,7] in which two projected light lines, focused as one line on the belt, are displaced by objects on the belt. The line camera, focused on the line, detects the silhouette of passing objects. When it has seen the entire object, the vision subsystem sends to the monitor the object's position and a belt position reference value.

The robot subsystem executes a previously "taught" robot program to transfer the part from the conveyor to a fixed position. It accepts information concerning the part's location on the moving belt and uses this data to update the "taught" program. It then monitors belt position and speed to track the part along the moving belt, pick up the part and transport it to a predetermined location.

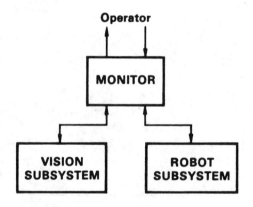

Fig. 3. CONSIGHT software organization

THE VISION SUBSYSTEM

The vision subsystem detects objects passing through its field of view and reports their position and orientation to the monitor program. Parts may follow in an unending stream. It is also permissible for several parts to be within the field of view simultaneously. Parts which are overlapping or touching each other are ignored and allowed to pass by the robot for subsequent recycling.

The vision subsystem employs a linear array camera. The linear array images a narrow strip across the belt perpendicular to the belt's direction of motion. Since the belt is moving, it is possible to build a conventional two-dimensional image of passing parts by collecting a sequence of these image strips. The linear array consists of 256 discrete diodes, of which 128 are used in the system described here. Uniform spacing is achieved between sample points (both across and down the belt) by use of the belt position detector which signals the computer at the appropriate intervals to record the camera scans of the belt.

The two main functions of the vision subsystem are object detection and position determination.

Object Detection

A fundamental problem which must be addressed by computer vision systems is the isolation of objects from their background. If the image exhibits high contrast, such as would be the case for black objects on a white background, the problem is handled by simple thresholding. Unfortunately, natural and industrial environments seldom exhibit these characteristics. For example, foundry castings blend extremely well with their background when placed on a conveyor belt. Previous approaches for introducing the needed contrast, such as the use of flourescent painted belts or light tables [8] would severely restrict the number of potentially useful applications of vision-based robot systems. We developed a unique lighting arrangement which accomplishes the same results without imposing unreasonable constraints on the working environment.

A slender tungsten bulb and cylindrical lens are used to project a narrow and intense line of light across the belt surface. The line camera is positioned so as to image the target line across the belt. When an object passes into the beam, the light is intercepted before it reaches the belt surface (Fig. 4). When viewed from above, the line appears deflected from its target wherever a part is passing on the belt. Therefore, wherever the camera sees brightness, it is viewing the unobstructed belt surface; wherever the camera sees darkness, it is viewing the passing part (Fig. 5).

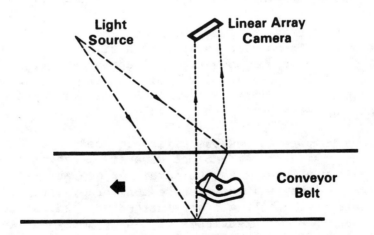

Fig. 4. Basic lighting principle

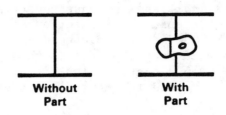

Without Part　　　**With Part**

Fig. 5. Computer's view of parts

Unfortunately, a shadowing effect causes the object to block the light
before it actually reaches the imaged line, thus distorting the part
image. The solution is to use two (or more) light sources all directed at
the same strip across the belt (Fig. 6). When the first light source is
prematurely interrupted, the second will normally not be. By using
multiple light sources and by adjusting the angle of incidence appropriately, the problem is essentially eliminated.

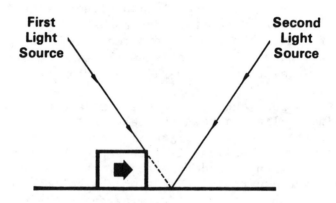

First Light Source　　　**Second Light Source**

Fig. 6. Improved lighting arrangement

The described lighting arrangement produces a height detection system.
Any part with significant thickness will be "seen" as a dark object on a
bright background. The computer's view is a silhouette of the object.
Note that while the external boundary will appear sharp, some internal
features, such as holes, are still subject to distortion or occlusion due
to the shadowing effect. These internal features may optionally be
ignored or recorded under software control.

It is the vision computer's responsibility to keep track of objects
passing thru the vision system. Since several pieces of an object or even
different objects may be beneath the camera at any one time, the conti-
nuity from line to line of input must be monitored. The conventional
binary image segmentation schemes are the 4-connected and the 8-connected
algorithms. Both of these algorithms result in ambiguous situations [9]
when deciding on the inside/outside relationship between some binary
regions. A clever solution to that problem is the use of 6-connected
regions [10], that is, connectivity is permitted along the four sides of a

picture element and along one of the diagonals. At the expense of a minor directional bias in connectivity determination, the inside/outside ambiguity is resolved. In addition, the algorithms which implement the segmentation for 6-connected regions remain simple and symmetric with respect to black and white, since the 6-connected algorithm artificially introduces the pleasing properties gained through hexagonal tessalation.

The 6-connected binary segmentation algorithm is readily adapted for run-length coded input, that is, where only the transition points between black and white segments are recorded. This is a significant advantage. The straight forward binary segmentation algorithm requires that the intensity of the neighbors for each pixel be examined. The execution time is therefore "order n squared" where "n" is the linear camera resolution. Since the number of black/white transitions across a line is essentially independent of the resolution for these types of images, the execution time is reduced to "order n" for the algorithm using the run-length coding scheme.

Once the passing objects have been isolated from the background, they may be analyzed to determine their position and orientation relative to a reference coordinate system.

Position Determination

For each object detected, a small number of numerical descriptors is extracted. Some of these descriptors are used for part classification -- that is, deciding which part it is. Others are used for position determination.

For position specification, we describe the part's position by the triple (x, y, theta). The x and y values are always selected as the center of area of the part silhouette. For most parts, this represents a well-defined point on the part. There is no convenient method for uniquely assigning a theta value to all parts. However, one useful descriptor for many parts is the axis of the least moment of inertia (of the part silhouette). For long thin parts, this can be calculated accurately. The axis must still be given a sense (i.e., a direction) to make it unique. This is accomplished in a variety of ways and is part specific. The internal computer model for the part specifies the manner in which the theta value should be computed. For example, one method available for giving a sense to the axis value is to select the moment axis direction which points nearest to the maximum radius point measured from the centroid to the boundary. Another technique uses the center of the largest internal feature (e.g., a hole) to give direction to the axis. Several other techniques are also available.

Parts which have multiple stable positions require multiple models. Parts whose silhouettes do not uniquely determine their position cannot be handled.

Reprogramming the vision system for a new part requires entering a description of a new model. Each model description includes information to determine if a detected object belongs to the class defined by the model and also prescribes how the orientation is to be determined.

The vision system sees the world through a narrow slit. As objects pass by the slit, statistics concerning that object are continuously updated. Once these statistics have been updated, the image line is no longer required. Consequently, storage need only be allocated for a single line of the two-dimensional image, offering a major reduction in memory requirements.

The block of statistics describing an object in the field of view is referred to as a component descriptor. The component descriptor records information for every picture element which belongs to that component. This includes the following.

1. external position reference
2. color (black or white)
3. count of pixels
4. sum of x-coordinates
5. sum of y-coordinates
6. sum of product of x- and y-coordinates
7. sum of x-coordinates squared
8. sum of y-coordinates squared
9. min x-coordinate and associated y-coordinate
10. max x-coordinate and associated y-coordinate
11. min y-coordinate and associated x-coordinate
12. max y-coordinate and associated x-coordinate
13. area of largest hole
14. x-coordinate for centroid of largest hole
15. y-coordinate for centroid of largest hole
16. an error flag

Considerable bookkeeping is required to gather the appropriate statistics for passing objects and to keep multiple objects segregated. The primary data structure used for this purpose is the "active line." The active line records the location of each black and white segment beneath the camera and also the objects to which they are connected. For every new segment which extends a previously detected object, the statistics are simply updated. If the segment is the start of a new object which is not in the active line, it will be added to the active line and have a new component descriptor initialized for it. When an object in the active line is not continued by at least one segment in the new input line, it must have passed completely through the field of view. The block of statistics is then complete and may be used to identify the object and compute its position and orientation.

Fig. 7 illustrates the coordinate reference frames used by the programs. It is convenient to consider the x-origin of the vision system to be permanently attached to the belt surface moving to the left. Normally, this would cause the current x-position beneath the camera to climb toward infinity. To avoid this, all x distances are measured relative to the first point of object detection. This in turn induces a complication. The complication occurs when two appendages of one object begin as two separate objects in the developing image. It then becomes necessary to combine the component descriptors for the two appendages into a single component descriptor which reflects the combination of the two. Moments, however, have been referenced to two different coordinate systems (i.e., the x-origin for each was taken as the point of first detection). The required shifting of moments is accomplished by applying the Parallel Axis Theorem.

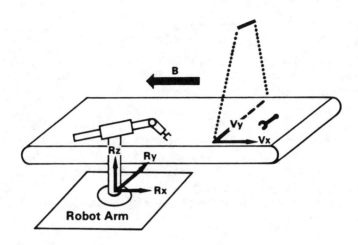

Fig. 7. CONSIGHT coordinate systems

Once parts have passed completely through the field of view, final position determination is made. The required computations proceed asynchronously with respect to the processing of the new lines of picture data. The results must be provided to the monitor subsystem before the part travels past the robot's pickup window. To coordinate the scheduling of these final computations, a queue of completed component descriptors is maintained. Component descriptors are removed from the head of the queue and processed as time permits.

Since the belt on which the parts rest is moving, the vision system records the current belt position whenever a new object appears in the developing image. This belt position reference value is obtained from the belt position/speed decoder. Since the leading edge of each part defines the origin of the coordinate system to be used for that part, the position of that part at some future time can then be readily determined by checking the current belt position and adjusting for any belt travel since the initial reference was recorded.

THE ROBOT SUBSYSTEM

The robot programming subsystem is implemented as two independent tasks. One task is required for robot program development and is necessary only during the programming and teaching phase. The second task, the run-time control system, is required both during the teaching phase and during robot program execution. It interprets and executes a robot program and controls the robot hardware. The execution of this task is controlled by special requests sent from other tasks.

A robot program consists of statements specifying: a position to which the robot should move (setpoint), an operation the robot should perform, or the environment for subsequent execution. Positions to which the robot moves are either taught by moving the robot manually and recording the position or are programmed by entering the specific Cartesian coordinates of a point in space from the keyboard.

In addition to this basic programming support, tracking and real-time program modification were developed for CONSIGHT. Tracking provides the ability to execute a robot program relative to some moving frame of reference. Program modification provides the ability to modify, in real-time, the robot program from another program and thereby dynamically modify the robot's path.

Tracking is implemented by defining new reference coordinate systems called FRAMEs [11]. Normally the robot operates in a Cartesian coordinate system [R] with its origin at the base of the robot (Frame 0). The robot's Cartesian position is described by a matrix [P] which defines the position and orientation of the hand in [R]. The arm solution program then determines a joint vector [J] from [P].

$$[P] \quad \rightarrow \quad [J]$$

If however, we want to define [P] relative to a different coordinate system (frame) whose position in [R] is defined by a transformation [F], then the solution program must perform the following:

$$[F] \, [P] \quad \rightarrow \quad [J]$$

Frames provide a means of redefining the frame of reference in which the robot operates. The robot may be programmed relative to one frame of reference and the resulting program executed relative to a different frame of reference. For example, a robot may be programmed to load and unload a testing machine. If the testing machine is moved, the entire program can be updated by simply redefining the frame specifying the position of the testing machine without re-programming each individual position point in the program. The overall effect is to translate/rotate every position to which the robot moves.

In addition to having a position, a frame is defined with a velocity and a time reference. This position, velocity, and time reference are used to predict the frame's position. Each time the run-time system performs an arm solution, (i.e., transforming the position matrix into the corresponding joint angles), it first computes a predicted position for the current frame.

Program modifications are special asynchronous requests sent to the run-time system from other tasks. Via these requests, an external program can modify a robot's programmed path, read the robot's position, start/stop robot program execution, and interrogate status -- all while the robot is operating. These requests greatly expand the capability of the robot system without the use of a powerful robot programming language. Much of the logical, computational, and input/output capabilities of an algorithmic language (Fortran) are retained for programming and controlling the robot external to the normal robot programming and control system.

In CONSIGHT, the part position determined by the vision subsystem defines the position and orientation of a frame and the belt direction and speed define the frame velocity. The approach, pickup, and departure points are all programmed relative to this frame. The robot subsystem does not directly interface to the belt encoder for belt position and velocity data, but receives the data via a request in the same way that the vision data is furnished. Thus, the rate at which the belt position and velocity data are updated is controlled by the monitor program and is a function of the expected variability of the belt speed. The approach, pickup, and departure points are dependent on the type of part being picked up as well as its position and must be modified for each cycle of the robot.

In the production prototype system the Stanford robot was replaced by a production robot with a tracking capability but which does not include the Frame concept. Thus, some of the monitor programs had to be modified to include some of the functions provided throught the use of the frames.

THE MONITOR

The monitor coordinates the operation of the vision and robot subsystems during calibration and part programming as well as during the operation phase. Calibration is required during the initial setup or whenever the camera, the robot or the conveyor have been moved. Part programming is required only when modifying the system to handle a new part.

Calibration

Calibration is the process whereby the relationship between the vision coordinate system [V] and the robot coordinate system [R] is determined (Fig. 7). In particular we want to compute the position of a part [r] in [R] given the part position [v] in [V]. Taking into account belt travel, this computation is represented by the following equation:

$$[r] = [T] \ [v] + s \ b \ [B]$$

In the equation above, [T] is the transformation between [V] and [R], s is a scale factor relating belt distance to robot distance, b is the distance of belt travel, and [B] is the belt direction vector relative to [R]. Thus, [v] and b are the independent variables, and [T], s, and [B] are the unknowns to be determined by the calibration procedure.

To determine [B] and s, a calibration object is placed on the belt within reach of the robot hand. The hand is manually centered over the calibration part, the hand position is read, and a belt encoder reading is taken. The conveyor belt is started and the part is allowed to move down the belt. The robot hand is again centered over the calibration part. A second hand position and belt encoder reading are taken. The monitor system can now compute the belt direction vector [B] and a scale factor s converting belt encoder units to centimeters.

Determining the coordinate transformation between [V] and [R] completes the calibration. The procedure assumes that the plane [Vx, Vy] is parallel to the plane [Rx, Ry] and that scaling is the same in both the x and the y directions.

A calibration object is again placed on the belt and allowed to pass by the vision station. The object position [v1] is determined by vision and a belt reading is taken. The part moves within reach of the robot and the conveyor is stopped. The robot hand is centered over the calibration object, the robot position [r1] is read and the distance of belt travel b1 is computed. This procedure is repeated with the calibration object placed at a different position on the belt resulting in a second set of data, [v2], b2, and [r2]. Combining these two sets of data points into the form above yields:

$$[r1,r2] = [T] [v1,v2] + s [B] [b1,b2]$$

This gives us 4 equations for determining the 4 unknowns in the transformation matrix [T].

Part Programming

Part programming is the process whereby the system is taught to recognize and pick up a new part. To do this, the vision subsystem must have been separately programmed to recognize the new part as described earlier in the vision section. To program the robot for a new part is to define the gripper position for part pickup. Normally the pickup position is offset from the part reference position determined by vision. Thus, once the vision subsystem locates a part and computes its position, the actual robot pickup position must still be computed.

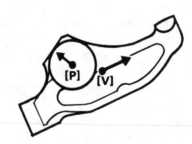

Fig. 8. The pickup problem

Fig. 8 illustrates this problem in a general way. The part position and rotation [v] have been computed by the vision subsystem. The robot needs to know the gripper position and orientation [p] as a function of [v]. The problem is further complicated since [v] is computed with respect to the vision coordinate system [V] and [p] must be computed with respect to the robot's coordinate system [R].

The procedure consists of passing a part by the vision station where [v] is computed and a belt reading is taken. The part moves down the belt to within reach of the robot and the belt is stopped. The monitor subsystem computes the current position and orientation of the part in the robot's coordinate system [R] and sends this position and the part direction to the robot subsystem as the definition of a frame. The robot hand is then placed on the part at the desired position for the grasp and pick up. This hand position is computed by the robot subsystem relative to the newly defined frame and is sent to the monitor subsystem. Thus, the pickup offset is determined relative to the part position and orientation as determined by the vision subsystem.

Later, during the operation phase the procedure is reversed. The part position [v] is again computed relative to [R] and used to define a frame in which the robot operates during the pickup phase of the robot program. The pickup position is then simply the offset determined by the above procedure. The robot subsystem automatically determines the desired robot position at any instant from the frame position, the pickup offset, and the belt position and velocity.

The Operation Phase

The final function of the monitor subsystem is to control the operation of the CONSIGHT system. Fig. 9 illustrates the overall logic of this portion of the monitor. Although the logic is straightforward, the monitor system deals with two rather subtle problems.

The first involves gripper orientation. Since the robot has a limited range of motion in all of its joints, it has a limited workspace. In particular the outermost joint, which controls hand orientation when the hand is pointing down, has only 330 degrees of rotation. However, in order to pick up randomly oriented parts, the gripper must be capable of handling all orientations. Since the gripper is symmetric about 180 degrees, we can effectively achieve all rotations by rotating the outer joint by $\pm$180 degrees for some orientations. The monitor system determines when the orientation needs to be changed for a particular pickup position. If the pickup is modified, the subsequent setpoints may have to be similarly modified so that all parts arrive at the same orientation.

The orientation problem described above arises whenever the robot path is dynamically defined. The general problem, for grippers that are not symmetric or when the gripper is not pointing straight down, is not solvable with commercially available robots. Robots that provide wrists with greater flexibility (greater joint range) are required.

Error recovery is the second problem. The possibilities for parts out of reach, impossible robot positions, collisions, etc. increase significantly when the robot's path is being changed for each cycle. Errors detected by either the vision or the robot subsystems are reported to the monitor. The monitor then takes action, usually restarting the robot. Like the orientation problem above, however, the dynamic nature of the system makes a good general solution to error handling difficult. Further research and experience with installed systems is needed to determine how to best approach the error handling problem.

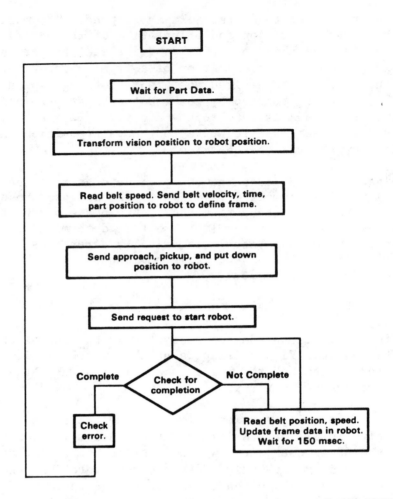

Fig. 9. Overview of the operational phase of CONSIGHT

PROTOTYPE PRODUCTION SYSTEM

The previous sections described the overall structure and concepts of CONSIGHT and presented details of the implementation of the experimental system. The next phase was to develop a prototype production system. The objectives of this phase were to:

1. Production harden the system. Replace laboratory components with industrial components.

2. Reduce ultimate system cost.

3. Improve reliabilty of system.

4. Reduce the robot cycle time to 5 seconds for a simple pickup, transfer, putdown operation.

To achieve the above objectives required major hardware changes and recoding of software to reflect the hardware changes and to improve the speed. Because CONSIGHT was developed for plant installation from the outset, few functional changes in software were required. Fig. 10 is a schematic of the system. The overall hardware structure resembles the software structure of the experimental system.

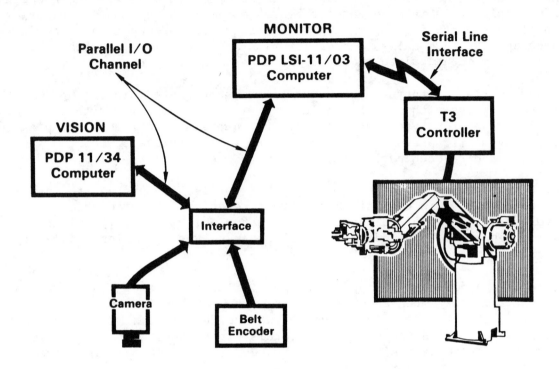

Fig. 10. Production prototype hardware

The T3 robot, manufactured by Cincinnati Milacron, is computer-controlled and has a line tracking capability similar to that developed for the experimental system. However, in order to meet the functional and operational specifications for CONSIGHT, two enhancements were made to the standard robot. The normal maximum operating speed of 1270 mm/sec was increased to 2500 mm/sec. Secondly, an adaptive branch function was provided for external modification of the robot program [12]. Both of these features were added to the system by Cincinnati Milacron at GM's request and are now commercially available.

The system interface is a hardware device which accepts data from the camera, automatically thresholds it and sends only the transition points between black and white segments to the vision program. Additionally the interface passes position data from the vision computer to the monitor computer and interfaces to the belt encoding device.

The single control computer of the experimental system has been replaced by a PDP 11/34 vision computer, an LSI-11 monitor computer, and a robot controller which is an integral part of the T3. These changes have greatly improved the cycle time and the reliability of the system.

The monitor subsystem is implemented on an LSI-11 microcomputer. The software is more complex for the production system than for the experimental system. The part programming (described earlier) is implemented in the monitor whereas in the experimental system the frame concept in the robot system automatically handled much of the mathematics. The queueing of multiple parts is also handled by the monitor whereas in the experimental system standard operating system requests were used for data queueing. Operationally, the calibration and operational portions of the monitor are identical to those of the experimental system.

The production prototype system was first demonstrated picking up symmetric parts with only a single part on the conveyor. Further developments of the prototype have included part queueing, part sorting, pick up of non-symmetric parts, refinement of the vision hardware, robot checkout and reduced cycle time.

The gripper orientation problem described earlier is more severe on the T3 since the roll axis has only 240 degrees rotation. Manufacturing Development has added an additional actuator which will flip the gripper by 180 degrees. This additional roll range solves the gripper orientation problem when the gripper is pointing down.

The error handling of the prototype is similar to that of the experimental system. Normally when an error occurs or an unrecognized part is detected, the part is skipped and allowed to fall into a bin at the end of the conveyor. Although not elegant, this error handling technique is not unusual in production situations where humans also, on occasion, are unable to perform their assigned tasks.

Work is continuing on further refining and packaging of the vision hardware and software, developing special robot hands, and extensive system checkout. The current prototype system is operating at a cycle time of 5 seconds and at belt speeds of up to 20 cm/sec.

SUMMARY

CONSIGHT differs from previous vision-based robot systems in its potential for practical production use. The vision subsystem, based on structured light, does not require high scene contrast. Both the vision and the robot subsystems are easily reprogrammed for new parts, in fact, the simplicity of "teaching-by-doing" is retained in spite of the complexities of the vision-controlled robot motions.

The advent of low-cost computer-controlled robots, such as the PUMA arm [13], will initiate a new era of sensor-controlled robot systems such as CONSIGHT.

REFERENCES

[1] S. Holland, L. Rossol, M. Ward, "CONSIGHT-I: A Vision-Controlled Robot System for Transferring Parts from Belt Conveyors," General Motors Research Laboratories Symposium "Computer Vision and Sensor-based Robots", September, 1978.

[2] W. Perkins, "Model-based vision system for scenes containing multiple parts," 5th International Joint Conference on Artificial Intelligence, pp. 678-684, Cambridge, August, 1977.

[3] M. Baird, "Sequential image enhancement technique for locating automotive parts on conveyor belts," 5th International Joint Conference on Artificial Intelligence, pp. 694-695, Cambridge, August, 1977.

[4] M. Ward, "Specifications for a computer controlled manipulator," GM Research Publication GMR-2066, February, 1976.

[5] G. J. Agin, "Representation and discrimination of curved objects," Stanford University Artificial Intelligence Project Memo, AIM-173, October 1972.

[6] M. Oshima, Y. Shirai, "A scene description method using three-dimensional information, Progress Report of 3-D Object Recognition," Electrotechnical Laboratory, Japan, March, 1977.

[7] R. J. Popplestone, et al., "Formation of body models and their use in robotics," University of Edinburgh, Scotland.

[8] G. J. Agin, "An experimental vision system for industrial application," Proceedings, 5th International Symposium on Industrial Robots, p. 135, Chicago, September, 1975.

[9] R. Duda, P. Hart, Pattern classification and scene analysis, Wiley-Interscience Publication, p. 284, 1973.

[10] G. J. Agin, "Image processing algorithms for industrial vision," Draft Report, SRI International, March, 1976.

[11] R. Paul, "Manipulator path control," 1975 International Conference on Cybernetics and Society, pp. 147-152, September 1975.

[12] H. R. Holt, "Robot decision making", 2nd North American Industrial Robot Conference, November, 1977.

[13] R. Beecher, "PUMA: Programmable universal machine for assembly," General Motors Research Laboratories Symposium "Computer Vision and Sensor-based Robots," September, 1978.

Reprinted from Robotics Today, Fall, 1979

Real-Time Robot Control with a Mobile Camera

Here is a system

that uses visual feedback

to control a manipulator

in real time.

A solid-state TV camera

attached to the

end-effector obtains

the needed information

GERALD J. AGIN
Senior Computer Scientist
SRI International

Existing robot technology is clearly in need of sensory feedback to extend its limited capabilities. Visual feedback can minimize the need for jigs and fixtures and ease workpiece tolerance requirements. Visual feedback controlling a manipulator in real time can allow it to work on a moving line without requiring precise control of the line.

For some time, SRI has been interested in the use of visual feedback to control a manipulator in real time. The approach has been to place a small solid-state TV camera in the manipulator end-effector and use its visual feedback to guide the hand to a given target, *Figure* 1. This method, called "visual servoing," may be applied to a large variety of tasks in material handling, inspection, and assembly.

Since control is in real time, a key point in the approach is to make use of binary images to achieve fast and reliable image processing. Special consideration must then be given to lighting and contrast in the image, but the reward for this is fast operation. In some applications, projected light patterns are used to obtain information about range or depth. The real-time nature of the servo system requires consideration of the dynamics of mechanical components and leads to questions of stability and speed of response.

STATIC ANALYSIS

The function of computer vision in a visual servoing application is to determine the spatial relationships that exist between the camera, tool, and workpiece. Some of the possible configurations that may be useful for servoing, as well as the constraints that govern possible placement of the camera with respect to the tool and workpiece, are discussed here.

Perspective Depth Measurement. A camera maps a three-dimensional world into a two-dimensional image. Inevita-

bly, some information is lost in this imaging process. It is impossible to determine the coordinates of a single point in three dimensions from its image; two coordinates may be measured that constrain the point's position to a ray in space, but the position

1. Unimate robot uses visual servoing to track an object on a moving belt. When pictures taken by camera located in hand indicate a position error of 0.05" or less, the robot reaches down and grasps the object.

along that ray cannot be inferred from those coordinates.

Suppose that a Cartesian coordinate system were affixed to the lens of a TV camera, with the X-axis pointing toward the camera's right, the Y-axis pointing upward, and the Z-axis directed along the camera's principal ray. From the image of a single point its x and y coordinates may be established (given some assumption about depth), but the z coordinate is indeterminate.

When images are allowed to consist

of many points, some information lost in imaging may be recovered. Various components of the scene may give perspective cues that permit estimation of depth—the third dimension Z.

The size of any solid object's image is related to its distance from the camera, but size also depends on other factors such as focus and the setting of the binary threshold. To escape this dependency on unrelated factors, it is preferable to measure the distance between two points instead of the size of a single object. This distance should be measured between points whose image locations are not dependent on threshold or focus. To a first approximation the centers of stripes, spots, holes, or similar markings satisfy such a requirement.

It is useful to examine the resolution and accuracy of this method to measure depth. Sensitivity to a change in Z depends upon the sensitivity of the camera/image processing system to changes in X and Y. Consider the situation illustrated in *Figure 2*. A pair of marks is observed at a known depth Z_1, resulting in some perceived spacing between the marks in the image. Since the value of Z_1 is known, we can make use of calibration information to calculate the actual separation W_1. Now suppose the marks are moved to a new, unknown depth Z_2. The spacing of the marks in the image will change, so that a new apparent spacing W_2 may be calculated at the original depth Z_1. We are interested in the ratio

$$\frac{\Delta W}{\Delta Z} = \frac{W_2 - W_1}{Z_2 - Z_1}$$

which relates sensitivity in depth to sensitivity in the other two dimensions. $\Delta W/\Delta Z$ is related to the angle ϕ as shown in the figure. In the symmetrical case shown,

$$\frac{\Delta W}{\Delta Z} = -2 \tan (1/2\ \phi)$$

in the limit as ΔZ approaches 0. For small ϕ, we may substitute the approximate relationship

$$\frac{\Delta W}{\Delta Z} \approx -\tan \phi$$

$\Delta W/\Delta Z$ relates resolution or sensitivity in depth to resolution in the other two dimensions. If ϕ is small, it takes a large change in Z to produce a small change in W.

If a camera lens with longer focal length were used in an effort to improve depth resolution, the resolution in X would be improved, but the field of view would also be narrowed. It would be necessary to bring the two marks closer together and proportionately narrow the angle ϕ. The resulting depth resolution would remain the same.

Rotations of rigid bodies may be inferred from the image as well, with rotation about the Z-axis the most directly measurable of all. Rotations about the X-axis or the Y-axis also produce effects in the image, but these effects are more subtle. To actually measure these tilts involves measuring depth at several locations in the image.

Triangulation Depth Measurement. An alternate method of obtaining information about the third dimension involves triangulation—the use of two or more distinct points of view. The use of two separate cameras constitutes stereo. Comparing pictures taken sequentially from a single camera whose position has changed with respect to the object is referred to as motion parallax.

Another way of achieving triangulation is to place a camera at one point and a projector that beams a particular pattern of light upon the object to be sensed at another point. The geometric considerations are similar to those for stereo, but the image processing is different in that only the pattern of projected light is analyzed. The use of patterned light offers some unique benefits not obtainable with conventional imagery.

The simplest light pattern is a single point. This pattern can be generated by a laser beam or by the use of conventional optics to image a pinhole with a light source behind it. For explanatory purposes, assume that the laser or projector is to the left of the camera, on the minus X-axis.

The geometric relationships of the projector, camera, and workpiece are illustrated schematically in *Figure 3*.

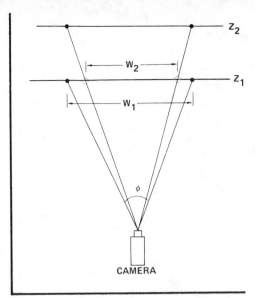

2. *Drawing shows the relationship of depth to perceived spacing. Rotations of rigid bodies may also be inferred.*

We derive the following equation:

$$\frac{\Delta X}{\Delta Z} = \frac{X_2 - X_1}{Z_2 - Z_1} = \tan \phi$$

in the limit as ΔZ approaches zero. The situation is similar to that of *Figure 2*, except that in this case the angle ϕ is independent of the camera's field of view. Resolution in depth may be made arbitrarily fine (if the effects of a finite spot size are disregarded) by using camera lenses that give high magnification.

If, instead of a spot, a vertical bar of light is projected, additional information about the scene can be obtained. The vertical bar of light may be generated by passing a laser beam through a cylindrical lens or by using conventional optics to image a slit with a light source behind it. Analysis of a vertical bar is similar to scanning a single point up and down and measuring Z at several different Y-values, except that all the points are read with a single picture.

Note that a horizontal bar of light would give no information at all. If a single beam were scanned from side to side and many images of the spot were input, depth could be measured at many points along a horizontal line. But, in reading of all the points simultaneously in a single image, the knowledge of which point in the image corresponded to which position of the beam is lost. The same situation arises whenever the lens center of the camera lies

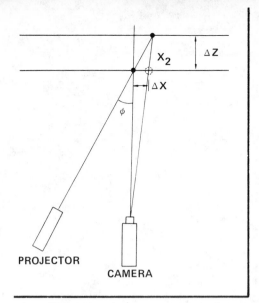

3. *Drawing shows measurement of depth by triangulation, with projector beaming light pattern at object.*

in the plane of light generating the bar.

If the target is a flat surface, the camera will perceive a straight line. The horizontal position of the line in the image is a function of the z-coordinate or depth of the surface. Any tilt of the line away from the vertical corresponds to a rotation of the plane about the X-axis. If the target, instead of being a featureless flat surface, has a recognizable shape, other information can be inferred. Specifically, if there is a horizontal edge or corner, the image of the bar of light will have a discontinuity, the position of which is a function of the object's y-coordinate.

Two parallel bars of light will give information about rotation of a plane about the Y-axis, as determined by the perceived spacing between the lines. As more parallel bars are added, coverage of the scene increases, but it becomes more difficult to determine which perceived line corresponds to which projected line. For determining the position and orientation of a plane in space, two parallel vertical bars cannot be improved upon.

The Tool. The object of visual servoing is usually to apply a tool to a workpiece. Frequently, however, the tool occludes part of the scene. Ideally, the camera should measure the position of the tool relative to the workpiece, but with real-time binary image processing systems this is not usually possible. If the target cannot be sensed because the tool intervenes, the tool must be attached rigidly to the camera in a known relative position.

There are two distinct modes of visual servoing. In "point mode," servoing is used to bring a tool to some specific location—for instance, to insert a bolt into a hole. If the target is in motion the servo system should track it, so that the relative velocity of the camera and tool, with respect to the workpiece, is zero. In "line mode," the objective is to follow a path at some specific nonzero velocity—in sealing, gluing, or seam following, for example.

When the camera moves with respect to the target in line mode, additional geometric information about the position and orientation of the object may be obtained from that motion. If the manner in which the camera moves is known, successive images of a groove can give the orientation of the groove where a single image cannot.

Once visual servoing achieves the desired relationship between the camera and the workpiece, the tool can be placed in the same relationship. In point-mode servoing this involves a separate motion. In line-mode servoing the tool can simply follow the camera.

DYNAMIC ANALYSIS

The camera and its controlling computer are only one element in a system that tries to maintain a specified relationship between the camera and workpiece. The other major component is the Unimate robot, considered in combination with its controlling LSI-11 computer. The vision and manipulation subsystems interact with each other in unusual and sometimes unpredictable ways. Stability and speed of response, as well as accuracy, are important. A simplified block diagram of the control system is shown in *Figure* 4.

Image processing and overall control take place in the PDP-11/40 minicomputer. Visual analysis of a scene is made in two phases: picture taking and image processing. Picture taking involves formation of the image on the camera diode array chip and reading of the signal into an image buffer in the PDP-11/40. This takes 10 to 66 milliseconds. Image processing is slower, requiring 100 to 500 milliseconds.

Analysis of the image data produces an "error signal" that drives the rest of the servo system. This error is a vector quantity, consisting of position in one, two, or three dimensions, and may also include orientation. The control algorithms (also residing in the PDP-11/40) convert the error signal(s) into commands to the manipulation subsystem. Commands are transmitted through a DR-11 parallel interface.

The programs to control the Unimate robot, which run in the LSI-11 microcomputer, and the communication protocols have been described previously. [1,2] This facility is used to specify both positions and velocities.

Position and Velocity Determination. If the visual target is known to be stationary, only positions need be commanded. The PDP-11/40 derives position information in the camera's own coordinate frame, and the LSI-11 microcomputer performs the necessary coordinate transformations to move in the appropriate direction.

In following a path at fixed velocity, the ability of the LSI-11 software to deal with moving coordinate systems is utilized. The Unimate control is commanded to maintain a fixed velocity (again relative to the camera's coordinate frame); thereafter position changes relative to the moving coordinate system can be commanded.

In maintaining position with respect to an object that moves at an unknown velocity, position and velocity are controlled simultaneously. This is possible because velocity applies to the moving coordinate system, whereas position applies to the manipulator with respect to that coordinate system.

Twenty times per second the LSI-11 computer transforms the instantaneous commanded position to a set of joint positions (encoder set point values) that are passed on to the Unimate hardware control. A complete and rather complicated servo system moves the robot hydraulically until the actual readings of the joint encoders agree with the commanded set points. The response is nonlinear, nonisotropic, and generally unpredictable. The robot will achieve its commanded position in a reasonable time, but some joints move faster than others, and the speed of response of some joints may depend on the positions of other joints.

As the Unimate moves the TV camera relative to the workpiece, it causes a change in the perceived scene, thereby closing the control loop. The simplest way to visually servo a manipulator is to take a single picture, estimate the error, wait a sufficient time for motion to terminate, and then repeat the process. When the target is stationary and speed of response is not critical, this approach can give adequate results. This is the control algorithm used for bolt-insertion experiments.[2,3]

Compensating for Delays. When a faster servo response is desired, or when the target may be in motion, it is desirable to take pictures as often as possible. If, for each picture, an incremental move were commanded to precisely cancel the error signal, the delays in the system would quickly cause a highly unstable response. A way to defeat instability is to command smaller moves. The correction applied to the Unimate will be the product of the error estimate and a constant called *beta*. *Beta* is always less than 1 and may be adjusted to give overdamped, underdamped, or critical response to a step function in the target's position.

If the target is in motion, the same error signal will control the velocity of the moving coordinate system. Each time an error is sensed along the $+X$ direction, for example, the product of that error and another constant, which we call *gamma*, will be added to the coordinate system velocity. *Gamma* is

generally independent of *beta* and should be adjusted for critically damped response.

It has been determined experimentally that the value of *beta* that achieves critical damping is linearly related to the time interval between successive pictures. A simple scene with minimal picture processing may be analyzed in as little as 150 milliseconds; however, with increasing sophistication of the processing used, with a more complex scene, and with extraneous noise in the image, processing time may increase to as much as 500 milliseconds. To achieve the best response, *beta* must be smaller for the shorter cycle times.

To take advantage of this relationship, a semiautomatic adjustment algorithm, in which the time between pictures is measured using the PDP-11/40's real-time clock, has been implemented for *beta*. Then, instead of using a constant *beta*, one multiplies the error estimate by the ratio of elapsed time to a hypothetical "Unimate positional response time." (If that ratio turns out to be greater than unity, we let *beta* = 1.0.) A value of 500 milliseconds works best for the response time. A similar relationship appears to hold for *gamma*, but further study is needed.

Compensating for Arm Motion. The proportional control algorithm is deficient in that no account is taken of arm motion during picture processing. There are basically two ways to accom-

plish this. The first is to use some mathematical model of the arm. The other way is to use the Unimate control program in the LSI-11 to measure the actual position of the arm.

A somewhat smoother response is achieved by using the following simple predictive arm model: keeping track of the most recent incremental motion command sent to the robot and assuming that during picture processing the arm has moved some proportion of that commanded distance. One assumes that the arm is at a different position from where it was when the picture was actually taken, and compensates for this in estimating the position error. If 20 to 30% of the previous motion command is "cancelled out" in this way, the response curves show fewer small random excursions.

It is sometimes better to actually measure position than to merely estimate it. The Unimate's LSI-11 computer may be interrogated to find out the precise, actual position of the Unimate's end-effector at any given time by reading the joint encoders and converting them to Cartesian values in the camera's coordinate system. Furthermore, this information can be obtained at almost the exact time the picture is taken. Thus, at some instant both the absolute position of the camera and its visually measured position error can be known. Regardless of processing delays, the Unimate is commanded to a new absolute position relative to the measured one.

A number of problems associated with servoing in position-measuring mode must be solved before it can be used generally. One problem is that small orientation errors tend to grow and cause drift. Since motion is relative to measured positions, any tilt in the hand's axis as it moves becomes incorporated into the new target position. If the error is systematic, it builds up. The details of moving in several degrees of freedom relative to an actual position, while constraining other degrees of freedom, have not yet been resolved.

Another problem stems from the delays in communication between the PDP-11/40 and the LSI-11. At present excessively long, they are caused by

4. *Block diagram of visual-servo control system shows its two major elements—the camera, with its controlling PDP-11/40 computer, and the robot, with its LSI-11 computer.*

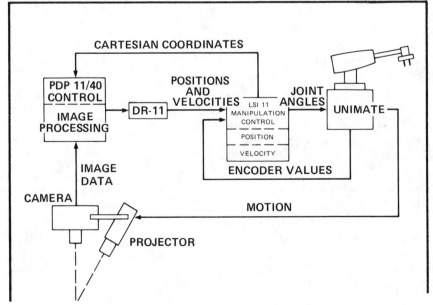

the scheduling program in the LSI and the large number of messages that need to be exchanged between the two computers. This program can be solved by improving the scheduling algorithm or by doing more of the calculations in the LSI-11.

EXPERIMENTAL RESULTS

The use of visual servoing for insertion of bolts has already been documented.[2] A related mode, wherein a movable table is positioned relative to a fixed camera, has also been demonstrated.[3] In all the bolting experiments tried so far, the target has been stationary and the system has achieved a placement accuracy of better than 0.05″ (1.2 mm) by taking one or two pictures.

An adaptation of the experiment shown in *Figure* 1 was employed for simulation of spot welding on a moving line. *Figure* 5 shows the Unimate set up to perform this experiment. The target was a piece of 1/4″ (6.3 mm) plywood, supported vertically on the moving belt. A slit projector cast a bar of light upon the edge of the plywood. The servo control used the image of this bar to control both the position and the orientation of the Unimate, relative to the edge. Servoing was constrained to a fixed height above the belt, with the camera's principal axis pointing horizontally. In addition to the camera, the Unimate's end-effector included a simulated spot welding gun.

After positioning itself relative to the edge, the arm moved by dead reckoning to place the gap in the tool over the plywood. The size of the gap relative to the plywood required that the maximum error be 6 mm. The Unimate could achieve this accuracy about 90% of the time. It is unclear how much of the overall inaccuracy is attributable to servoing inaccuracy, to inaccuracy of the dead reckoning move, to nonrigidity of the tool, to variation in belt speed, or to rocking of the plywood on the belt.

In another experiment visual servoing was applied to following a curved path in three dimensions at constant velocity. Such a mode of operation could apply to gluing, sealing, and seam following. The path to follow was created by taping together pieces

5. *Experiment using robot for simulated spot welding on a moving line achieves 90% accuracy.*

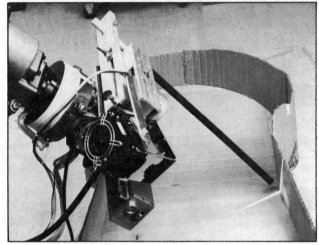

6. *Unimate with simulated tool, shown here in servoing position, follows a curved path in three dimensions.*

of cardboard to form a large "tub" with a groove curving in three dimensions. *Figure* 6 shows the tub with the Unimate in servoing position.

The slit projector was directed to generate a V-shaped pattern on the groove. This image was used to directly control camera position in two dimensions and to establish an orientation in which the axis of the camera bisected the "V". As the camera and simulated tool moved forward, successive corrections were made to reorient the direction of motion (in two degrees of freedom) to remain parallel to the groove. Thus, in this demonstration, the servo system controlled all of the Unimate's six degrees of freedom. A space intrusion problem caused by the size and bulk of the support hardware, and a generally jerky response prevented this demonstration from being completely successful. Further investi-

gation will be necessary to overcome these difficulties. ∎

REFERENCES

1. Rosen, C. *et al.*, "Machine Intelligence Research Applied to Industrial Automation," Sixth Report, SRI International, Menlo Park, CA, November 1976.

2. Rosen, C. *et al.*, "Machine Intelligence Research Applied to Industrial Automation," Seventh Report, SRI International, Menlo Park, CA, August 1977.

3. McGhie, D. and Hill, J. W., "Vision-Controlled Subassembly Station," SME Technical Paper MS78-685, 1978.

—Adapted from "Real-Time Control of a Robot with a Mobile Camera." Presented at the 9th International Symposium and Exposition on Industrial Robots, March 13-15, 1979, Washington, DC, sponsored by the Society of Manufacturing Engineers and the Robot Institute of America. This work was supported by the National Science Foundation under Grant APR75-13074, and by sixteen industrial affiliates.

Presented at the Robots II Conference, October, 1977

Workpiece Transportation By Robots Using Vision

By Robert K. Kelley
J. Birk
V. Badami
University of Rhode Island

Workpieces transported to fixtures by robots may not have precise alignment with the robot hand. Practical methods to avoid obstacles without requiring the complexity of geometric modeling are presented. An algorithm has been developed which computes the arm joint values needed to compensate for this misalignment. The algorithm relates changes in the features of images from two TV cameras to changes in the position and orientation of a workpiece. During an instruction phase, known position and orientation perturbations are automatically applied to a sample workpiece to establish the correspondence between these changes. Also, safe ways to grasp the workpiece and collision free trajectories are defined. During production as each workpiece is handled, position and orientation variations are noted to permit the actual relation between the workpiece and robot hand to be computed. The workpiece can then be transported along a safe trajectory and brought to the fixture without a collision.

INTRODUCTION

Industry is looking to put robots to work at new jobs, especially in batch manufacturing and assembly. Computer-based manipulators (robots) have the desirable attributes which are needed for such tasks. They are easily reprogrammed and of general purpose design. Their workpiece grasping systems include tongs, electromagnets and vacuum pickups. However, robot hands typically do not have the sensory capability of human hands. They cannot measure workpiece slip, for example, when acquiring a workpiece. For this and other reasons, workpieces which are transported by robots to fixtures or other goal sites may not be precisely aligned with the robot hand. The problem is to transport this kind of robot held workpiece to its goal site without a collision. That is, without regrasping the workpiece, compensate for the misalignment and avoid collisions enroute and with goal site structures as the robot places the workpiece.

WORKPIECE TRANSPORTATION ALGORITHM

An algorithm has been developed which computes the robot arm joint values needed to compensate for workpiece misalignment with the robot hand. In general, misalignment can be in any of three degrees of freedom in position and three degrees of freedom in orientation. Throughout this paper, the term "pose" will be used to refer to both position and orientation.

The workpiece transportation algorithm will be described by first considering the activities which take place during the production phase. Then the instruction phase activities which preface the production activities and make them possible are described.

Production Phase

During production the pose of each workpiece is checked as it is handled. Pose variations are calculated, corrections are computed and alignment is performed before the workpiece is transported along a safe trajectory to the fixture. A diagram of a typical production layout is shown in Fig. 1.

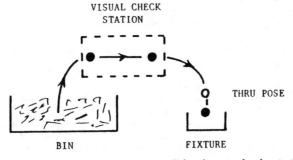

Fig. 1. Typical production layout with visual check station.

Workpiece pose check.--Each workpiece is brought to the visual check station. Initially, the robot assumes prespecified arm joint values to bring the workpiece to a predefined check pose.

Two television cameras are used to extract image feature values from the pair of workpiece images. For example, image features such as the center of gravity and direction of the minimum moment of inertia axis may be extracted from the binary image of a workpiece. These features are drawn on the binary image of a sample workpiece in Fig. 2. These feature values are compared with those expected for the workpiece having the check pose. If

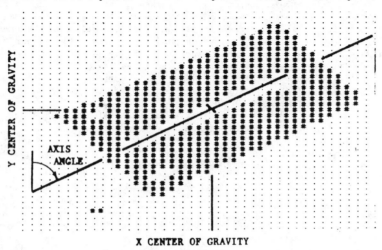

Fig. 2. Binary image of workpiece with features indicated.

the feature variations correspond to acceptably small misalignments in the workpiece pose, the workpiece can be transported to the fixture. If the feature variations indicate larger workpiece misalignments, the workpiece pose must be corrected before the workpiece can be brought to the fixture.

Workpiece pose misalignment is computed in terms of perturbations to the workpiece pose. The perturbations are calculated by weighting the image feature variations. In matrix form this calculation is expressed as

$$\overline{\Delta p} = [Q] \, \overline{\Delta f}$$

where:
$\overline{\Delta p}$ = perturbation vector for the workpiece pose
$[Q]$ = weighting matrix relating feature variations to pose perturbations
$\overline{\Delta f}$ = variation vector for image features.

Thus, each constituent of workpiece misalignment can be obtained. Tests can be applied to each degree of freedom individually or in combination to determine whether or not the misalignment is acceptably small.

Using the results of the workpiece pose check, one of the following actions is appropriate:

1. Proceed to transport workpiece along a safe trajectory to the fixture when the pose is satisfactory.

2. Correct the workpiece pose and then
   a. transport along a safe trajectory if the effect of the correction does not need to be observed, or
   b. repeat the workpiece pose check if the effect of the correction needs to be determined.

3. Discard that workpiece and acquire another when the workpiece pose misalignment is too large.

Workpiece pose correction.--Workpiece pose perturbations define a correction which can be applied to the current estimate of the robot hand to workpiece relation (Ref. 1). Thus $\overline{\Delta p}$ defines a correction transformation matrix, T_c. The robot hand to workpiece relation is denoted by a transformation matrix, ^hT_w. (The reader is directed to Ref. 2 where a detailed discussion of representing kinematic relations as transformation matrices is given.) The corrected estimate is computed by the matrix product (see diagram in Fig. 3)

$$({}^hT_w)_c = ({}^hT_w) \ (T_c) \ .$$

Since the robot base to workpiece relation for the check pose at the visual check station is known, the corrected pose for the robot hand can be obtained. Let $({}^oT_w)_{check}$ denote the workpiece check pose at the visual check station. Since

$$({}^oT_w)_{check} = ({}^oT_h)({}^hT_w), \ {}^oT_h = ({}^oT_w)_{check}({}^hT_w)^{-1}$$

defines the robot base to hand relation. This is illustrated in Fig. 4.

The robot arm joint values which correspond to this hand pose are found by solving equations which are specific to different robot kinematic configurations. (See Ref. 3, Appendix 7, for a typical solution.) When the arm joint values are adjusted to these new values, the workpiece pose should be closer to the nominal one.

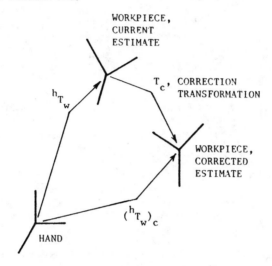

Fig. 3. Diagram for correction of ${}^hT_w$ estimate.

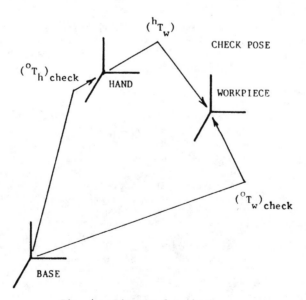

Fig. 4. Diagram for check pose.

361

Safe trajectory transport.--To transport a workpiece to a fixture, a selection is made from among the predefined safe trajectories. The selection is made on the basis of the deviation of $(^hT_w)_c$ from $(^hT_w)_j$ associated with a safe trajectory. If this deviation is small for the jth legal way to grasp the workpiece, then it is assumed that the robot arm structures will not collide with the workstation when the workpiece is moved along a safe, collision free trajectory.

The safe trajectory is specified as a sequence of poses (called thru poses) through which the workpiece progresses as it is transported from the visual check station to the fixture. The kth workpiece thru pose along the trajectory specifies a corrected robot hand pose. During instruction, $(^oT_w)_k$ is defined. During production it is required that the workpiece achieve the same pose. (See diagram in Fig. 5.)

INSTRUCTION

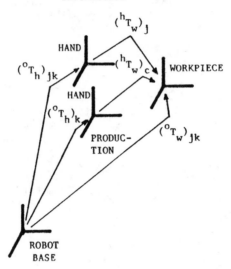

Fig. 5. Diagram for reestablishing k^{th} workpiece thru pose.

Thus

$$(^oT_w)_k = (^oT_h)_k \, (^hT_w)_c \, ,$$

where $(^hT_w)_c$ does not change as the workpiece moves along the trajectory. This is shown in the diagram in Fig. 6. Hence

$$(^oT_h)_k = (^oT_w)_k \, (^hT_w)_c^{-1}$$

specifies the corrected robot hand pose to achieve the desired workpiece thru pose on the trajectory. Notice $(^hT_w)_c^{-1}$ is computed only once for each trajectory and $(^oT_h)_k$ is solved for the robot arm joint values to achieve each thru pose on the trajectory.

Instruction Phase

During instruction a sample workpiece is employed to obtain the data which is needed to accomplish the production phase activities. In particular, legal ways to grasp the workpiece in the fixture are specified, safe, collision free trajectories between the fixture and the visual check station are defined by thru poses, workpiece check poses at the visual check station are defined, and the correspondence between workpiece pose perturbations and image feature changes is determined experimentally.

Legal ways to grasp workpiece.--To describe legal ways to hold a workpiece, a sample workpiece is placed in the fixture. Manually, the robot hand is brought to the workpiece and the workpiece is grasped without removing the workpiece from the fixture. The arm joint values are recorded.

The arm joint values are used to compute the robot base to hand relation ^oT_h. For the first legal way to grasp the workpiece, the robot hand to workpiece relation $(^hT_w)_1$ is prespecified. For example, the workpiece coordinate system might be aligned with the hand coordinate system and dis-

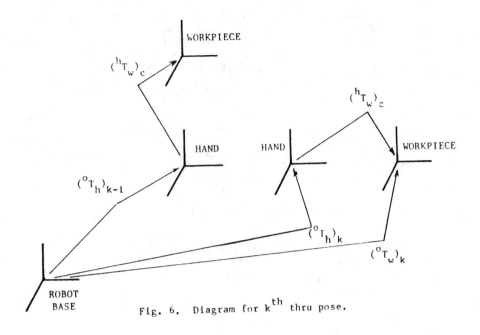

Fig. 6. Diagram for k^{th} thru pose.

placed to a point near the center of the workpiece. Thus

$$(^{O}T_{w})_{fixture} = (^{O}T_{h})_{1} \ (^{h}T_{w})_{1}$$

describes the workpiece in the fixture.

To specify a second way to legally grasp the workpiece in the fixture without a collision, the robot hand is brought to the workpiece until the workpiece can be grasped, and then arm joint values are recorded. These arm joint values specify $(^{O}T_{h})_{2}$. Since the workpiece is in the fixture (see diagram in Fig. 7), the robot hand to workpiece relation is obtained from

$$(^{h}T_{w})_{2} = (^{O}T_{h})_{2}^{-1} \ (^{O}T_{w})_{fixture} \ .$$

These steps are repeated for each $(^{h}T_{w})_{j}$ which corresponds to a distinct way to grasp the workpiece.

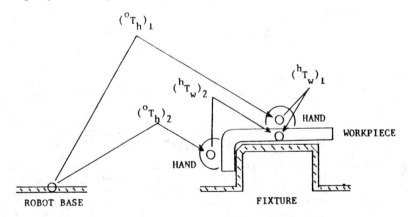

Fig. 7. Defining ways to grasp workpiece.

Specification of safe trajectories.--For each legal way to grasp a workpiece, a safe trajectory is specified. This trajectory is one through which the workpiece passes as it is transported between the visual check station and the fixture. (A common safe trajectory might be sufficient for all legal ways to grasp a workpiece.)

The safe trajectories are specified by grasping the workpiece while in the fixture in one of the legal ways, for example, the first way. The workpiece is then removed from the fixture to a thru pose by a programmer using manual controls to manipulate the robot hand. Typically the last thru pose is close to the fixture and differs only by a small displacement. Arm

363

joint values are recorded. This procedure is repeated to define sufficient thru poses to guarantee a collision free path between the visual check station and the fixture. The programmer should select trajectories which maximize clearance between the arm and the goal site. Clearance between the workpiece and the goal site should also be as large as possible as this will not be controlled during production since the workpiece trajectory, not arm joint trajectory, is made to resemble trajectories defined during instruction.

The set of arm joint values at each thru pose along a trajectory define a workpiece pose. For example, referring to Fig. 8, at the kth pose along a trajectory (which begins with the pose nearest the visual check station) the arm joint values define $(^OT_h)_{jk}$ where j is used to indicate the way the workpiece is grasped. Then the workpiece pose at the kth thru pose is

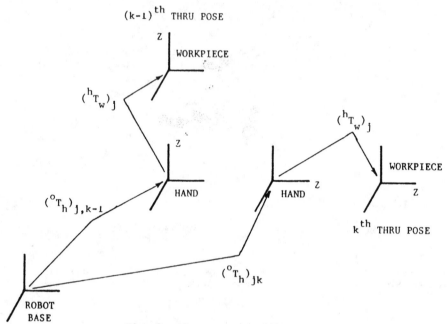

Fig. 8. Thru pose specification.

defined by the matrix product

$$(^OT_w)_{jk} = (^OT_h)_{jk} \, (^hT_w)_j \, .$$

Check pose.--A check pose for the workpiece at the visual check station is defined for each legal way the robot can grasp the workpiece. (Ideally, a common pose would suffice.) Two properties which are desirable for the check pose are:

1. That the check pose have as many arm joint values as possible in common with those for the workpiece fixture pose to minimize the introduction of pose errors as a result of mechanical motions. This is particularly true of rotary joints for which small angular errors give rise to significant displacement errors when multiplied by the length of the arm links. (This consideration should also influence the selection of thru poses when specifying trajectories.)

2. That the feature values extracted from the pair of television images of the workpiece at the check pose possess sufficient sensitivity and range to characterize the pose perturbations of interest.

The second property can be facilitated by the proper placement of the pair of television cameras.

Pose perturbations and feature changes.--For each distinct check pose, the correspondence between perturbations to the workpiece pose and the changes in the image feature values is determined. This is done experimentally by sequentially perturbing the workpiece coordinate system one degree of freedom at a time and computing the feature variations which result.

If the coordinate system is perturbed in position by ΔX,ΔY,ΔZ, and in orientation by Δα,Δβ,Δγ (see Fig. 9 for definitions), then the correspondence

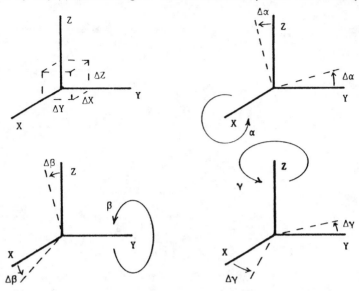

Fig. 9.   Coordinate system perturbations.

can be represented as a weighting matrix $[R]$ so that

$$\overline{\Delta f} = [R]\, \overline{\Delta p}$$

where $\overline{\Delta f}$ and $\overline{\Delta p}$ are feature variation and pose perturbation vectors as defined previously, and $\overline{\Delta p} = (\Delta X, \Delta Y, \Delta Z, \Delta\alpha, \Delta\beta, \Delta\gamma)^t$.   The columns of $[R]$ are obtained by sequentially applying

$$\overline{\Delta p_X} = (\Delta X, 0, \ldots, 0)^t,\ \overline{\Delta p_Y},\ \ldots,\ \overline{\Delta p_\gamma}\ .$$

To be able to determine the pose perturbations which correspond to feature variations, a weighting matrix $[Q]$ as mentioned previously is needed to satisfy

$$\overline{\Delta p} = [Q]\, \overline{\Delta f}$$

where: $[Q] = [R^t R]^{-1}\, R^t$, pseudo matrix inverse (more than six features)

or $[Q] = [R^{-1}]$, matrix inverse (six features).

EXPERIMENTS

The configuration shown in Fig. 10 is being used to establish an experimental basis for applying the transportation algorithm described in the preceding section.   During our experiments, image analysis and workpiece pose correction are performed alternatively as the $^h T_w$ estimate is refined.

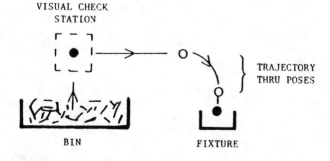

Fig. 10.   Experiment configuration.

The investigation is being conducted at a single visual check station which features a compact visual zone.   The visual zone is defined by the fields of view of the two television cameras.   These cameras have a right angle relationship between the two principal rays as shown in Fig. 11.

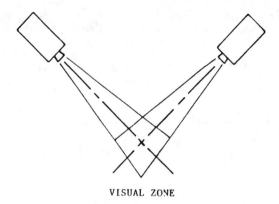

VISUAL ZONE

Fig. 11. Right angle camera geometry.

The URI Mark III robot which is used to perform the actual workpiece manipulations was designed and fabricated at the university. The robot has six degrees of freedom, three orthogonal linear joints, and three rotary joints. The robot and its design is discussed in Refs. 3, 4, and 5.

Our image analysis is presently based on closed circuit television camera images which are converted to binary format by applying a computer controlled intensity threshold to each analog picture element sample. Modules to automatically determine the perturbation weighting matrices $[R]$ and $[Q]$ have been written and are operational. This software includes feature extraction, robot kinematics, arm joint value solution, and mathematical computation modules. Almost all of the instruction phase software has been validated but has not yet been integrated. Specifically, the module for the specification of the legal ways to grasp a workpiece is working. The production phase software shares a large amount of the instruction phase software. The module for pose correction is being assembled with the largest portion of the production phase software awaiting integration and validation.

Example: Pose Correction

To show the nature of the pose correction algorithm, an artificial example will be presented which uses some data collected on the sample workpiece whose binary image was shown in Fig. 2. A single television camera was used. The camera had its principal optical axis collinear with the workpiece coordinate system Z axis. Thus the angles α and β are rotations about the axes in a plane parallel to the image plane and γ is an angle of rotation in the image plane.

The workpiece was rotated independently about each of the three workpiece coordinate axes. Data points for the change in a feature value (direction of the minimum moment of inertia axis) are plotted in Fig. 12.

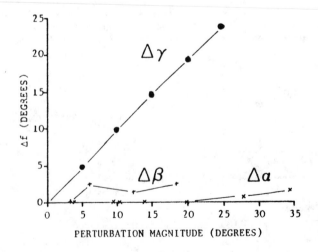

Fig. 12. Experimental data.

Notice that for angular perturbations within 20°, there is no correspondence between Δf and $\Delta\alpha$, only a very noisy correspondence between Δf and $\Delta\beta$, and a nearly one to one correspondence between Δf and $\Delta\gamma$. The $[R]$ matrix in this case is a row vector where:

$$R_1 = avg\ (\Delta f_1)/avg\ (\Delta\alpha)$$

$$R_2 = avg\ (\Delta f_2)/avg\ (\Delta\beta)$$

$$R_3 = avg\ (\Delta f_3)/avg\ (\Delta\gamma)$$

where avg is the arithmetic average of the quantities. For this data,

$$[R] = (0,\ 0.144,\ 0.973)$$

Because this is an underdetermined system, the pseudo inverse is given by

$$[Q] = R^t\ [R\ R^t]^{-1}\ .$$

The computation yields the column vector

$$Q = \begin{bmatrix} 0. \\ 0.149 \\ 1.006 \end{bmatrix}$$

Thus, the pose correction is based on

$$\begin{bmatrix} \Delta\alpha \\ \Delta\beta \\ \Delta\gamma \end{bmatrix} = \begin{bmatrix} 0. \\ 0.149 \\ 1.006 \end{bmatrix} (\Delta f)$$

Assume that the true workpiece pose can be represented by $[0,\ 0,\ 20^\circ]_{true}$ and it is desired to bring the pose to $[0,\ 0,\ 0]$. (Actually the angles must be ordered to represent orientation accurately for large angles.) Using the data which led to Fig. 12, $\Delta f = 19.29^\circ$ and the pose correction computed is $[\Delta\alpha,\ \Delta\beta,\ \Delta\gamma]_{computed} = [0,\ 2.9^\circ,\ 19.4^\circ]$. After applying this correction the true pose is $[0,\ -2.9^\circ,\ 0.6^\circ]_{true}$ which has a feature variation $\Delta f = 0.584^\circ$. The computed pose correction is $[\Delta\alpha,\ \Delta\beta,\ \Delta\gamma]_{computed} = [0,\ 0.087^\circ,\ 0.587^\circ]$. Performing the correction, the true perturbation is now $[\Delta\alpha,\ \Delta\beta,\ \Delta\gamma]_{true} = [0,\ -3.0^\circ,\ 0.003^\circ]$.

The correction procedure would terminate at this point since, for this example, the data indicates that $\Delta f = 0$ would be obtained from the workpiece image.

This example points out the sensitivity of this feature to rotations in the plane and the lack of sensitivity about axes in the plane for this particular workpiece. Additionally, this example demonstrates the hazards associated with an underdetermined correction system.

RESEARCH ISSUES

Through our experiments, the influence of factors such as the number of picture elements subtended by the workpiece, lens selection, the image features used, and the characteristics of the workpiece will be studied.

Some additional research issues to be resolved are the systematic selection of the check pose at the visual check station and the placement of the two television cameras, the characterization of ^hT_w estimation errors and their effect on the amount of detailed knowledge which is required about the structures in the workspace and which is needed to accomplish collision avoidance.

Other research issues are concerned with improving the utility of the transportation algorithm by finding ways to minimize the number of iterations of the ^hT_w estimation procedure and of generalizing from discrete to continuous specification of the legal ways to grasp a workpiece.

SUMMARY

An algorithm was described which permits workpieces which are not precisely aligned with the robot hand to be transported without collision to the fixture. Computation of arm joint values which compensate for this misalignment was explained. Practical methods to avoid obstacles without requiring the complexity of geometric modeling have been presented. That is, a method to specify collision free (legal) ways to grasp a workpiece in the fixture and a method to specify safe trajectories from the check station to the fixture were enumerated.

ACKNOWLEDGEMENT

This material is based upon research partially supported by the National Science Foundation under Grant No. APR74-13935. Any publications generated as a result of the activities supported by this grant may not be copyrighted without the specific written approval of the Grants Officer and the authors. Any opinions, findings, and conclusions or recommendations in this publication are those of the authors and do not necessarily reflect the views of the National Science Foundation.

REFERENCES

1. Kelley, R., et al., "Workpiece Orientation Correction with a Robot Arm Using Visual Information", Proceedings of 5th International Joint Conference on Artificial Intelligence, Cambridge, Mass., August 1977, p. 758.

2. Birk, J., et al., "Orientation of Workpieces by Robots Using the Triangle Method", presented at ROBOTS I Conference, Chicago (Rosemont), Ill., November 1976, SME Technical Paper MR76-612.

3. Birk, J., et al., "Robot Computations for Orienting Workpieces", Second Report - Grant APR74-13935, Department of Electrical Engineering, University of Rhode Island, August 1976.

4. Birk, J., et al., "General Methods to Enable Robots with Vision to Acquire, Orient and Transport Workpieces", Third Report - Grant APR74-13935, Department of Electrical Engineering, University of Rhode Island, August 1977.

5. Seres, D., "Computer Controlled Robot with Visual Sensor for Advanced Industrial Automation", M.S. thesis, Department of Electrical Engineering, University of Rhode Island, 1975.

Reprinted from Robotics Today, Winter, 1980

Experiments in Part Acquisition Using Robot Vision

The vision system under development at NBS enables a robot to acquire

rectangular and curved parts randomly arrayed on a flat surface. Here's a

review of the system's hardware and software and an update on recent findings

ROGER N. NAGEL,
GORDON J. VANDERBRUG,
JAMES S. ALBUS,
and

ELLEN LOWENFELD
National Bureau of Standards

THE VISION SYSTEM being investigated at the National Bureau of Standards provides both depth and part orientation information to the robot control system. The principal components of the vision system are a solid state camera, a structured light source, and a camera interface system.

The solid state video camera produces a 128 × 128 raster image. The structured light source is a stroboscopic light which emits a plane of light through a cylindrical lens. The camera and light source are mounted on the wrist of the robot in the configuration shown in *Figure* 1.

The camera is oriented to position the columns of the image perpendicular to the flash plane, so that each column of the image has, at most, one intersection with the plane of light. By mounting the camera and light source at a fixed angle, it is possible to use triangulation to compute the distance from the robot to each point in the raster image. A calibration chart for the NBS vision system is shown in *Figure* 2. When an ob-

ject is in front of the robot, the plane of light forms a line segment image as shown in *Figure* 3. The height of the pixels in the line segment image represents the distance to the camera, while the shape of the line segment indicates the part orientation.

The final component of the hardware is a camera interface. The interface, which is attached to an 8-bit microprocessor, provides both control functions and data reduction. It furnishes software control of the flash

duration, camera clock rate, and iris adjustment. In addition, the interface will threshold the video signal and produce a run length representation of the resultant binary image. The threshold value is under software control and passed to the interface via the microprocessor. A block diagram of the interface is shown in *Figure* 4.

The interface produces two numbers per column in the raster image. The first number (called run length)

1. Solid state camera and light are mounted at a fixed angle to the robot's wrist.

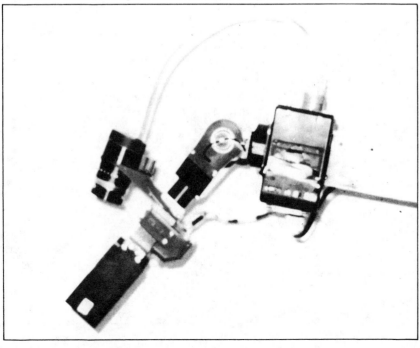

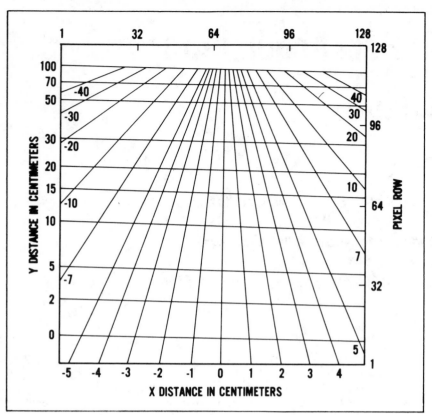

2. *The X and Y distances are measured in the coordinate system of the fingers. The X axis passes through the two fingertips and the Y axis is parallel to the wrist axis.*

is a count of pixels from the bottom of the column to the bottom of the flash line. The second number is the width of the flash line. Thus, the interface reduces the 128 × 128 raster array of gray scale values to 128 pairs of run lengths and line width values. These pairs contain all the information necessary to analyze the original image.

Software

The NBS vision system is under the control of the task decomposition hierarchy in the robot control system. When the vision system is invoked, it is given a maximum distance within which to look for the part. The vision system returns to the control system a status flag, the endpoints of the closest part edge (in the coordinate system of the wrist), and an edge characteristic.

The operation of the system is divided into two primary phases. In the first phase, the system acquires an image and evaluates it for image quality. In the second phase, the system analyzes the image to produce the data necessary for the control

system.

Emphasis is placed on acquiring an image which is easy to analyze, rather than on a complex analysis algorithm (such as one which can link disconnected portions of line) that may be required for poor quality images. Therefore, during phase one, a check is made for image quality.

Presently this phase can only adjust the value of the flash and threshold parameters. In the future, the vision system will also be able to adjust the iris setting.

Image quality is measured in terms of the attributes of the image and known characteristics of the system. For example, if the flash is too bright, the specific camera being used exhibits bleeding on the left of the image. Similarly, if the flash is too weak, the object breaks up into isolated line segments, or cannot be found at all. Image quality is also checked with respect to the width of the flash line. The plane of light should intercept the object and be seen in a small number of pixels (per column). Therefore, the width value is an indication of image quality. The detection of any of these image attributes results in appropriate changes in the previously mentioned image acquisition parameters.

The analysis phase of the vision system inspects the run length data and divides it into connected components called segments. Segments are isolated by gaps in the X direction (columns) or Y direction (rows). The algorithm proceeds from left to right, analyzing each segment as it is acquired.

There are three types of segments: straight lines, angles, and curves. The images to be analyzed represent

3. *The plane of light forms a line segment image of an object in the camera's field of view.*

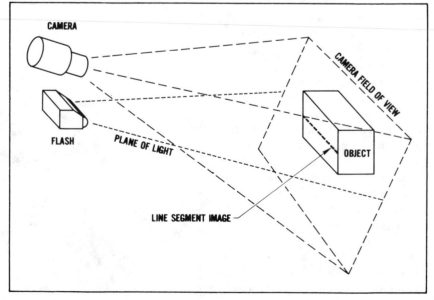

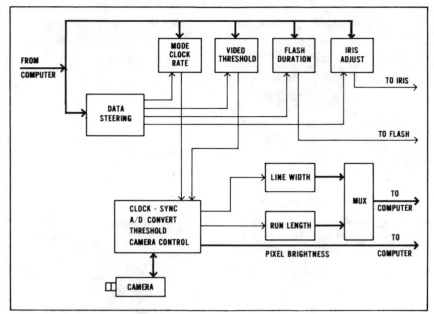

4. *This block diagram illustrates the NBS vision system interface hardware.*

the distance from the robot to the part. Thus, if one is examining convex parts, the above segments are a complete description of the possible images.

There are four types of border points defined for segments: left and right endpoints, minimum points, and vertices. All segments have left and right endpoints corresponding to the start and end of the segment. A curve has a minimum point which is defined to be the lowest row value in the curve. An angle segment has a vertex which is defined to be the corner point of the angle. Thus, line segments have two border points, while curves and angle segments have three, *Figure* 5.

To determine whether a segment is a line, angle, or curve, the slope at each point must be found. The right slope at a point is defined as the difference in Y-value between that point and the point dn values to the right, where dn is an integer parameter for the X offset. Similarly, the left slope at a point is defined as the difference in Y values between the point dn values to the left and that point. Note that slope computations are based solely on the Y values, because the X distances are constant. This permits the use of integer rather than real arithmetic in the slope computations with a resultant savings in computer

time. The computation of left and right slope is illustrated in *Figure* 6.

The algorithm which labels a segment as either a line, curve, or angle begins when 2 (dn) + 1 points to the right of the left endpoint have been found. This is due to the fact that slope calculations require at least that many points in the segment. In addition to the left and right slope computations, the algorithm computes the segment points with the minimum Y value (lowest row value) and those with the maximum and mini-

mum slopes in the entire segment. Proceeding from left to right as each new point in the segment is found, the algorithm computes the left and right slope. It then updates the segment minimum and maximum parameters discussed earlier.

In addition, the difference in left and right slopes for the point is computed. If this difference is greater than a specified parameter, or the left and right slopes differ in sign, then the point could be a vertex point (corner point) of an angle. Such a point is called a candidate point and the value of the difference in slopes for it is recorded. At the first point in the sequence after a candidate point which is itself not a condidate point, the algorithm chooses a vertex from the set of candidate points. The vertex is selected as the candidate point with the maximum slope difference and is entered in the border point table. The other candidate points are discarded. This procedure is repeated until the right hand endpoint of the segment is found.

When the right hand endpoint is

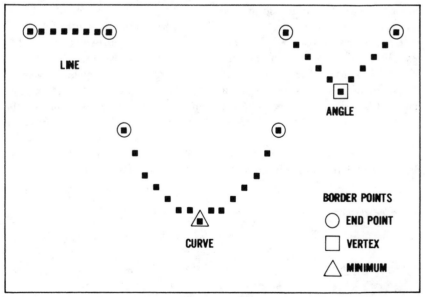

5. *Types of segment border points, including left and right endpoints, minimum points, and vertices. Line segments have two border points, curves and angle segments have three.*

slope calculations require at least that many points in the segment. In addition to the left and right slope computations, the algorithm computes the segment points with the minimum Y value (lowest row value) and those with the maximum and mini-

located, the algorithm has computed both endpoints, the minimum point, and the points with maximum and minimum slope. It also has found and labeled zero or more points as vertices. At this time the labeling procedure begins.

The first test is to determine if the segment represents a curve. There are three parts to the curve test:

▶ Enough points must have been found to compute slope [at least 2 (dn) + 1]

▶ The range between maximum and minimum slopes must be greater than a specified parameter

▶ The percentage of the curve between the last occurrence of the maximum and the first occurrence of the minimum must be greater than a specified parameter.

The first test insures that there is enough data to be significant. The second prevents a line that is noisy at the ends from being considered a curve. The third test separates curves and angles. Both curves and angles have their maximum and minimum slopes at the ends and a value near zero in the middle. The difference is that curves begin gradually changing slope at the endpoints, while angles retain their maximum slope until very near the middle. Therefore, a curve will have the majority of its length between the inner occurrences of the extrema, while an angle will have the majority of its length outside the inner occurrences of the extrema.

If the segment passes the curve test, any possible vertices found are discarded, and it is labeled a curve. If the segment is not found to be a curve, it is considered to be an angle or a line. If no vertices are found, the segment is classified as a line. If one or more vertices have been found, the segment is further analyzed to as-

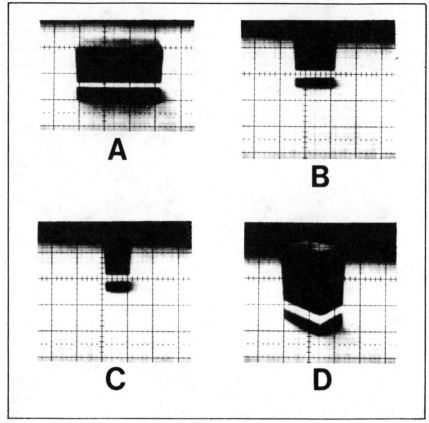

7. Gray scale pictures taken of a block reveal the expected lines and angle.

sign one vertex point (an angle can have only one), or to remove all vertices and call the segment a line.

This is done in two steps. First, a single vertex is assigned by choosing the vertex with the lowest row number. This is based on the vertex of an angle being the lowest point of the segment. Second, a test is made to see if the chosen vertex corresponds to a corner point. It could be the result of digitizing a line in a noisy en-

vironment. This test is based on the difference in Y (row) values between the possible corner point and the left and right endpoints. If neither difference is greater than a specified parameter, the vertex is removed and the segment is labeled as a line. Segments which have a vertex after these tests are labeled as angles.

When all the segments in the image have been analyzed and labeled, the segment with the closest border point to the robot is selected as the segment of interest to the control system. For a line, the endpoints are computed in robot coordinates, and the vision analysis is complete. For a curve, the endpoints and minimum point are computed. For an angle, further processing occurs.

In the case of an angle, the vision system will choose the left or right half based on which side of the angle is more nearly parallel to the robot's fingertips. This is done because the control system will then move the robot to face the part head on, and the chosen side will minimize the required motion.

6. The computation of left and right slope, used to determine whether a segment is a line, angle, or curve, is illustrated.

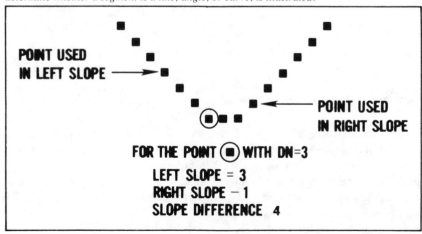

POINT USED IN LEFT SLOPE →

POINT USED IN RIGHT SLOPE ←

FOR THE POINT ⊙ WITH DN=3

LEFT SLOPE = 3
RIGHT SLOPE − 1
SLOPE DIFFERENCE 4

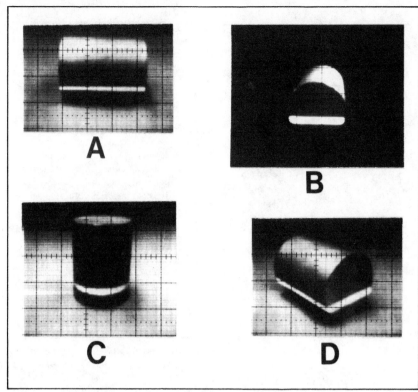

8. *Images of a cylinder show curve, lines, and angle. Only the bright light lines will remain after the images are thresholded.*

In all cases, a status flag indicating whether or not a part was found, the type of line segment, and the endpoints in robot coordinates are returned to the control system.

Experimental Results

The NBS vision system is operational and has been used in experiments in the acquisition and sorting of simple parts. In the experiments conducted, a pile of randomly oriented rectangular and cylindrical parts is placed at an arbitrary location on a table near the robot. The control system uses the vision system at three stages of the part acquisition sequence. The first is a faraway view used to locate and move up to the pile. The second view is close to the pile and is performed to isolate a single part, determine its orientation, and identify the side of the part facing the robot. The part is then pulled out of the pile and examined by a third, overhead view just prior to the part pickup.

Although the vision system does not know what type of part it is, the line, angle, and curve data from the vision system are used by the control system (which *does* know the part type and its dimensions) to identify the part side it is facing. For example, in a close view of a block, *Figure* 7, one expects to see a line or an angle, while in a close view of a cylinder, *Figure* 8, one expects a curve, line, or angle. The images in these figures are the gray scale pictures as seen by the vision system hardware.

The bright line, which is caused by the plane of light, will be the only thing left in the image after it is thresholded by the hardware. The data as reduced by the camera interface for the images in *Figures* 7 and 8 are plotted in *Figures* 9 and 10.

When the robot is facing a corner such as is shown in *Figure* 9d or 10d, the vision system will select the closest side—the one most parallel to the robot fingertips—by the algorithm described earlier. The control system then moves the robot to face the selected side head on. For the case of a block, it then measures the width of the side (from coordinates supplied by the vision system) and moves up for an overhead view. During this movement the robot hand is rotated so that the next side of the part presented is new. Thus, after the overhead view, the control system has visual measurements of two sides and knows the orientation of the block. (A block has three distinct width values. If two are known, the third can be inferred.) In the case of a cylinder, a similar set of logic in the control system is used.

Using the logic presented here, the NBS robot and vision system were programmed to pick up blocks or cylinders randomly located on a table and place them in a fixed orientation in a vise.

As an example of the image qual-

9. *Data for images in Figure 8, reduced by the camera interface, are plotted on this graph.*

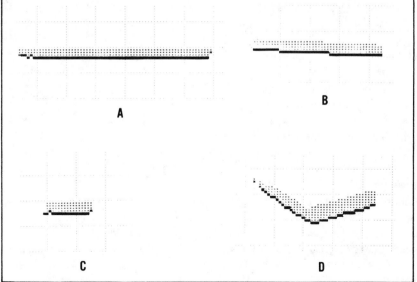

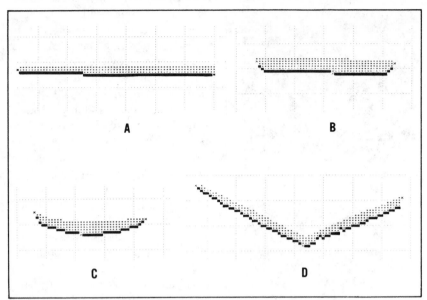

ings, 9th International Symposium on Industrial Robots. Washington, DC. March 1979.

10. *Plottings of images in* Figure *9. Black points at the bottom of the graphs in* Figures *10 and 11 are used in computing the run length values, while lighter x's represent the flash width in each of the columns.*

ity correction, *Figure* 11 shows an image which was rejected based on the bleeding at the left hand edge. The accepted version of this image appears in *Figure* 7b and the reduced data in *Figure* 9b. The vision system software keeps track of the parameters used on each view (far, close, and overhead). Therefore, in practice, the parameter changes required for image quality are limited to the first iteration of the part. Although the image quality check is always performed, subsequent views of the same object with the same parameters have produced quality images in the experiments conducted thus far.

To date, the vision system has performed well. It uses parameters to control the flash duration, threshold, and the decisions which find and discriminate among lines, angles, and curves. While the current parameter values are adequate for the experiments performed, it is expected that, with knowledge of the part size, color, reflectivity, etc., the parameters can be tailored to the part. Additional experiments in part acquisition and in creating and using depth maps for inspection are planned. A depth map can be acquired by creating many images across the face of a part as the robot moves the camera system incrementally. Further efforts lie in the area of characterizing curva-

ture and combining the primitive line segments into string descriptions of more complex parts. ■

Mr. Nagel is presently manager of robotics at International Harvester Science and Technology Laboratory, Hinsdale, IL. Dr. VanderBrug is product line manager at Automatix Inc., Burlington, MA.

REFERENCES

1. Agin, G. J. "Real Time Control of a Robot with a Mobile Camera." *Proceedings*, 9th International Symposium on Industrial Robots. Washington, DC. March 1979.

2. Barbera, A. J., Albus, J. S., Fitzgerald, M. L. "Hierarchical Control of Robots Using Microcomputers." *Proceedings*, 9th International Symposium on Industrial Robots. Washington, DC. March 1979.

3. Karg, R. and Lanz, O. E. "Experimental Results with a Versatile Optoelectronic Sensor in Industrial Applications." *Proceedings*, 9th International Symposium on Industrial Robots. Washington, DC. March 1979.

4. Martini, P. and Nahr, G. "Recognition of Angular Orientation with the Help of Optical Sensors." *The Industrial Robot*, Vol. 6, No. 2. June 1979.

5. VanderBrug, G. J., Albus, J. S., and Barkmeyer, E. "A Vision System for Real Time Control of Robots." *Proceedings*, 9th International Symposium on Industrial Robots. Washington, DC. March 1979.

6. Ward, M. R. et al. "Consight: A Practical Vision-Based Guidance System." *Proceedings*, 9th International Symposium on Industrial Robots. Washington, DC. March 1979.

—Adapted from "Experiments in Part Acquisition Using Robot Vision" by Roger N. Nagel, Gordon J. VanderBrug, James S. Albus, and Ellen Lowenfeld, National Bureau of Standards. MS79-784.

11. *Width of the flash line is large, a common characteristic of a poor image.*

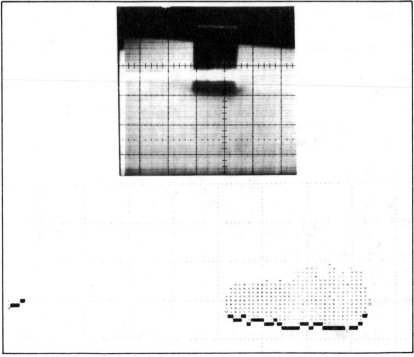

Presented at the Robots V Conference, October, 1980

Touch-Sensing Technology: A Review

By Leon D. Harmon
Case Western Reserve University

This is a survey and assessment of tactile sensing and feedback devices and systems for robots. Special emphasis is placed on touch sensing as it relates to industrial manipulators. The state-of-the-art in parameters (e.g., force, torque, compliance, slip), in transducers (e.g., conductive and semi-conductive materials and arrays, non-contact sensing), and in tactile pattern recognition is surveyed. Present application areas in manipulation and in prosthetics are outlined. The review concludes with consideration of outstanding problems, new opportunities, and emergent technology. A selected bibliography of 160 references is included.

I. INTRODUCTION

The purpose of the study, upon which this report is based, was to survey potential worldwide commercial application of touch-sensing technology. The aim was to determine the technical state-of-the-art, outstanding problems, and future opportunities in tactile sensing for automata. Primary concentration was directed to applications centered on industrial robotry and remote manipulation.

Emphasis throughout is on tactile sensing, defined as continuously variable touch sensing over an area within which there is spatial resolution. This patterned response is therefore distinct from force or torque sensing which usually is a simple vector-resultant measurement at a single point.

The primary input for this study was the identification and acquisition of a collection of technical articles, patents, and reports. Most of this material came from an extensive literature search which is detailed in the following section.

II. LITERATURE SEARCH

Five large literature databases were searched in early 1979 via the well-known Lockheed facility. These were:

1. NTIS   (National Technical Information Service)
2. INSPEC   (Physics Abstracts, Electrical and Electronics Abstracts, and Computer and Control Abstracts)
3. ISMEC   (Institution of Mechanical Engineers)
4. COMPENDEX   (Engineering Index)
5. SCI SEARCH   (Scientific Citation Index)

Seventeen key words and phrases were used to search these databases. Since some, like automated assembly and pressure transducers, if used alone, would produce a very large and largely irrelevant output, combinations of terms were carefully chosen. The 17 categories which were used in various appropriate combinations were:

1. Pressure transducers, sensors
2. Force transducers, sensors
3. Touch, contact sensors
4. Tactile sensing
5. Sensor displacement
6. Slip sensing
7. Grip, grasping
8. Stable grasping
9. Prehension
10. Multi jointed fingers
11. Mechanical hands
12. Artificial limbs
13. Industrial robots
14. Size, object, shape recognition
15. Contact pattern
16. Materials handling
17. Automated assembly

The output of this search was a listing (usually with abstracts) of 857 reports, publications, and patents from academia, industrial, and government sources. Most were from the U.S.; Japan was second; a few were from Russia and Germany; and there was a small scattering from England, France and Yugoslavia.

A fair amount of the Japanese literature is relatively inaccessible owing to publication in Japanese; English abstracts are generally provided but are sketchy. The Russian literature is often obtuse and philosophical; they publish little concrete detail.* The French and Yugoslavian literature is usually published also in English.

Of these 857 reports, approximately 300 were selected for relevance. Most of these were located either through local library facilities or by way of interlibrary loans and personal contact requests. The latter source, plus bibliographic references picked up from the publications, added another hundred or so citations. After study, slightly more than half of the judged relevant reports were retained as being sufficiently worthwhile. These 160 references are given in the bibliography at the end of this report and are cited throughout the report.

III. SCOPE OF FIELD

Industrial automation and remote manipulation by robot-like devices (flexible and programmable) are currently enjoying considerable growth and support, though less than projected several years ago. Laboratory research and development is extremely active, publications abound, and the expansion of conference activity and specialty journals indicates considerable interest and investment as well as forward motion. The locations identified below represent major efforts as perceived from the literature search and personal contacts.

1. Laboratories

    a. Academic - In the U.S.A., principal centers of activity in the field are at M.I.T., Purdue and Stanford. The Japanese universities concentrating on robotics are Chuo (Tokyo), Kyoto, Waseda (Tokyo), Tokyo Institute of Technology, and Nagoya. British action seems presently to be most evident at the University of Warwick. In Germany the principal effort is at the Fraunhofer Institute for Production and Automation (Stuttgart). There is an active center in Yugoslavia at the Mihailo Pupin Institute (Belgrade).

    b. Government - The principal U.S.A. establishments concerned are the National Bureau of Standards, the Jet Propulsion Labs (at Caltech), and the U.S. Naval Research Labs. France is represented by the AEC (Commissariat a l'Energie Atomique), Laboratoire d'Automatique et d'Analyse des Systems (Toulouse), and IRIA (Institute for Research on Information, Science, and Automation,

---

* However, it has been claimed that there is breadth and depth in the USSR work, particularly in undersea manipulators and articulated-limb vehicles (100, 144).

Rocquencourt, Le Chesnay). The principal Japanese effort is represented by the Electrotechnical Laboratories, although the government also subsidizes industrial research.

c. Independent and Industry – In the U.S.A., major factors are the C.S. Draper Labs., Inc., Stanford Research International, IBM, G.E., Westinghouse, Unimation, and Cincinnati Milacron. The Japanese are represented by Hitachi, Kawasaki, Mitsubishi, and Fujitsu.

2. Industries

In the U.S., there are two major factors, Unimation and Cincinnati Milacron. In 1979, Unimation accounted for about 80 percent of the market, and Cincinnati Milacron commanded something under 10 percent. The balance of 10 percent or so was accounted for by less than two dozen small companies. Thomas's Register listed 19. The total U.S. annual sales volume was just under 35×10^6, apportioned somewhat differently among the manufacturers, but accurate data are not easily obtained.

The European and Japanese market seems to be much more active than that in the U.S. This is largely illusory for the reason that in Europe and Japan any automatic machine is called a robot. It is claimed that the Japanese and Western European countries employ about 30,000 "robots". However, many of these are not robots by the usual U.S. definition (machines that are programmable, flexible, and which can be adapted to many jobs). Consequently, it is difficult to assess the market volume for those countries.

3. Conferences

There are three basic conferences:

a) Symposium on Theory and Practice of Robots and Manipulators:
Sponsored by CISM (Centre Internationale des Sciences Mechaniques) and IFToMM (International Federation for the Theory of Machines and Mechanisms), it is primarily theoretical. About 100 attend these meetings. The Fifth Annual Symposium will be held probably in Yugoslavia or Poland or England in 1981.

b) International Symposium on Industrial Robots:
Sponsored by the Society of Manufacturing Engineers and the Robot Institute of America (plus about seven other participating sponsors like NBS and SRI), it is predominantly applications oriented and attracts hundreds of attendees. The 11th Annual Symposium was held October 7-9, 1980 in Tokyo.

c) National Robots Conference and Exposition:
Sponsored also by the Society of Manufacturing Engineers and the Robot Institute of America, it is very applied, highly commercial, and quite popular. The fifth Conference was held in Dearborn, Michigan, October 28-30, 1980.

In addition, other occasional conferences and symposia bear on these topics. JPL/CalTech has sponsored two conferences on Remotely Manned Systems; the Institute of Electrical and Electronic Engineers, the National Bureau of Standards, and the National Science Foundation have organized workshops from time to time. Many of these meetings are represented in the reference list for this report.

4. Publications

Besides the conference symposia and workshop proceedings that represent contributions at the meetings listed above, there are three publications of interest. One,

The Industrial Robot, a quarterly journal, is now in its eighth year. It is an important source of general and overview information as well as of technical detail. Two other quarterlies, Robots Today, and Robotics Age, began in 1979.

There are two major resources for additional materials, including books and reviews. They are International Fluidics Services, Ltd. (who publish the proceedings of the International Symposium on Industrial Robots) and the Robot Institute of America (who sponsor those symposia).

An impressive array of literature on robotics as cited in (31) has been produced by the Japanese Industrial Robot Association (JIRA). A compendium of 112 pages of bibliography covering only a three year period is one example. Another lists U.S., U.K. and Japanese patents on 425 pages! They also have available a 342-page report of robot research in Japan and long-term forecasts of demands for robot machinery.

5. Personnel

There are an estimated 150-200 persons around the world full-time and long-term committed to research in robotics, manipulation, locomotion, and limb prostheses. Most of the leaders in this group are cited in the reference list. A number do not appear in those citations because their emphasis is more on systems and specialties not the direct concern of this report. Of course, several times the estimated number of researchers are part-time, short-term personnel, mainly students. And another group is that of the commercial developers, who often have little commitment to research.

IV. STATE-OF-THE-ART IN TOUCH-SENSING TECHNOLOGY

1. Introduction

A wide assortment of devices and systems exist for robot sensing of the environment. They fall into two generic classes, contact and non-contact. Contact, of course, is the domain of tactile sensing with which this survey is primarily concerned. Non-contact sensing principally includes optical, sonic, and magnetic ranging; some examples in these categories will be given in passing, where appropriate.

The state-of-the-art in tactile sensing is surprisingly primitive. Up until the last few years, touch feedback systems for robots and manipulators were very simple and relatively crude. The industrial systems still employ primitive devices; most of the more sophisticated and complex tactile sensors are in laboratory development, largely in academic or government settings and seldom in industrial use.

Unimation, with about 80 percent of the U.S. market, has so far felt little need to incorporate more than simple touch-sensing in its products. Their market capture has been accomplished with a tactile technology of surprising simplicity. Touch-sensing is accomplished in two ways: on-off switching for contact and poke-probe travel for force feedback. In the latter case, major emphasis is on analog-to-digital encoders which work by a sensing probe which is mechanically linked to a rotating digital code wheel (usually optical). When the probe touches an object, its (continuous) axial displacement rotates the code wheel and thus a (discrete) position signal is obtained. Both absolute and incremental codes are used: the absolute type of encoder reads out a unique number at each quantized position whereas the incremental encoder simply emits the same signal (e.g., a pulse) at each position, and thus a counter must be employed.

A similar and perhaps even simpler approach appears in one of the most popular and representative of the Japanese robots, Hitachi's HI-T-HAND (41, A4). This seven-degree-of-freedom articulator can feel for scattered objects, recognize forms, and position articles for firm grip. This is done using only metal contact detectors and pressure sensors of conductive rubber where the force sensing is limited to thresholded binary signals.

Unimation presently perceives no need for complex visual or tactile feedback.

However, they acknowledge that there are many areas where such features obviously could be offered to advantage. But, for today's applications, it is usually unnecessary to get so fancy or expensive. In prognosticating future needs for robot manipulators, the president of Unimation laid heavy stress on tactile sensing, including orientation, recognition, and physical interaction (31). Emphasis was also placed on the need for systems with visual-tactile coordination -- much like human functions.

Cincinnati Milacron's sensors are predominantly of three types: inductive proximity switches, limit switches, and photocell sensing. All are extremely simple and need little description. The inductive proximity sensors are used to register contact or near contact with ferrous materials. The limit switches, long-used in the industry, are generally no more glamorous than microswitches. The photocell arrangements consist of beam-interrupt sensing and straightforward angle-of-reflectance proximity sensing.

A simple force-feedback system is used by Cincinnati Milacron in some of their systems. In this case pneumatically operated grippers are used, and gripper force is estimated by monitoring the air pressure.

Other types of touch sensors have been used in a few applications, both in industrial and research robots, but they are not commonly employed so far. These include sensing of displacement by strain-gauges, potentiometers, gyroscopes, variable-conductance materials, and optical devices.

The general needs for sensing in manipulator control are proximity, touch-slip, and force/torque. The following remarks are taken from a discussion on "smart sensors" (17):

> "...specific manipulation-related key events are not contained in visual data at all, or can only be obtained from visual data sources indirectly and imcompletely and at high cost. These key events are the contact or near-contact events including the dynamics of interaction between the mechanical hand and objects.
>
> The non-visual information is related to controlling the physical interaction, contact or near-contact of the mechanical hand with the environment. This information provides a combination of geometric and dynamic reference data for the control of terminal positioning/orientation and dynamic accommodation/compliance of the mechanical hand.
>
> The terminal/compliance phases of manipulator motion are essential and quite intricate elements of manipulation. Adaptive and soft grasp of objects, gentle load transfer in emplacing or lifting objects, performing geometrically and dynamically constrained motions like fitting two parts together are typical examples of manipulator control engineering."*

In this same paper (17), there appear observations relating to the crudity of current technology and the many opportunities for extensive, further development. Another useful and illuminating tutorial overview on these topics appears in (11).

Although the industrial robots to date manage to do with rather primitive techniques to sense position, proximity, contact, force, and slip, all of these variables plus shape recognition have received extensive attention in research and development laboratories. In some of them a new generation of sophistication is just now beginning to emerge. Some of the more promising approaches are described in Section VII, Emergent Technology.

2. Parameters

   a. Force/Torque

---

* Discussion of the control engineering problems appears below in Section VI, Outstanding Problems and New Opportunities.

It is axiomatic that in general six articulators are needed for a robot arm. Three will position a wrist assembly, and three more orient an attached hand. From here on, specifications are less well defined since the choice of fingers, grippers, etc. and their degrees of freedom are open variables.

Tactile sensing requirements are far less precisely known, either theoretically or empirically. Most arm, wrist, hand, and finger sensors have been simple position and force-feedback indicators. Finger sensors have barely emerged from the level of microswitch limit switches and push-rod axial travel measurement.

In order to move objects under external dynamic constraints, a robot manipulator can in some special cases operate in open-loop mode. But the constraints on the regularity of the operational environment are so artificial and rigid that such systems have very little utility. When the world to be manipulated is less constrained, then at bare minimum force-and-torque feedback along three orthogonal hand-references are mandatory.

As late as four years ago there was no documented evidence for real-world applications (excluding master-slave ("teleoperator") systems) of force/torque, proximity, or touch sensors as terminal devices for manipulator control (11). That survey also noted the unexpected fact that force-and-torque sensing development dated back only to about 1972, while touch-and-slip art dated back to about 1966. And proximity sensing was then only some three years old.

Representative literature on force-and-torque sensing for manipulator control can be found in (1, 19, 34, 84, 103, 139, 141, 142, 143, A8). There are four distinct approaches to force sensing: 1) resultant (reaction) force sensing at manipulator joints (e.g., 82), 2) reaction force sensing at a work table or pedestal (e.g., 139), 3) force sensing at a wrist (e.g., 19, 139), and 4) passive (open-loop) compliance, such as remote-center compliance (e.g., 84, 142). Owing to extremely complex problems in structuring suitable computer control algorithms, the use of close-loop force feedback has so far been restricted to very simple force detection and neutralization techniques. Still, a notable achievement using such techniques (actually a combination of binary wrist sensing and passive compliance) is seen in the Hitachi HI-T-HAND, now six years old (41).

The state-of-the-art in force sensing is represented by the work at the Draper Labs. Two similar kinds of force sensor systems have been developed, each having six degrees of freedom. One is a pedestal (work table) sensor system; the other is a wrist sensor system. The wrist sensor transducer, for example, consists of four cantilever spring bars configured as a Maltese Cross. Each bar has bonded to it four silicon strain gauges, one on each face (143, A9). The work encountered forces (coupled via sliding ball joints) are translated into assembly (effector) forces via computer control. This arrangement can insert 1/2 inch diameter pegs into close tolerance (0.0005 inch) holes (78).

b. Compliance

There are two principal considerations in manipulators which relate to compliance. One has to do with the unwelcome fact that force cannot be measured directly; one can measure only displacement resulting from force, and thus the measurement is derivative, subject to error, hysteretic effects, etc. The other consideration relates to gripper-applied pressure distributions; fingers and hands must be compliant in order to insure that handling does not insult objects handled.

Industrial robot designers are extremely sensitive to the issue that compliance, particularly in force sensing, is not an ideal fact of life. Ideally, zero compliance is desired -- for two reasons: 1) compliance means uncertainty of sensor position and hence position of sensed surface; position must be computed from the measured force, and 2) compliance means reduced frequency response and, often, hysteresis.

Compliance could be put into software, i.e., "stop on touch" with preassigned force variables. However, the cost is a concern; high-bandwidth sensors imply high-bandwidth hardware, which in turn calls for high-bandwidth software -- all expensive.

Unimation and others are beginning to investigate the possibilities of software compliance, but the developments to date are rudimentary.

It was suggested nearly six years ago (33) that force sensing compliance in pedestals, for instance, could be reduced by taking advantage of the well-known microelastic behavior of metal. This could reduce hysteresis effects below the transducer's sensitivity threshold. So far as has come to light in the present survey, this idea has not been picked up.

Some approaches to gripper compliance are simplistic but do the job adequately. An ingenious example is Cincinnati Milacron's solution to the raw-egg handling problem. Overall gripper force is sensed and (binary) limited by gripper spring-loaded limit switches, while local pressure distributions are controlled by compliant, shaped, sponge-rubber gripper pads in self-centering jaws. Pad texture and friction characteristics control slip. As with most industrial robot technology in use so far, this empirical solution is hardly elegant but works well. An even simpler but very functional solution to the now common mechanized handling of raw eggs is obtained by a vacuum-operated accordioned rubber snout (70).

Perhaps the simplest force-limiting gripper found in this report's search is a device for tongs which handle radioactive materials (2). A current sensing switch in series with the motor that closes the tongs simply opens at the preset force (current). Similarly, but needing the additional control complexity of a computer, the well-known Stanford Research International's robot arm senses joint force simply by monitoring driving motor currents (49).

A large variety of specialized gripper mechanisms has been developed for industrial robots. This variety often reflects the need to introduce compliance as well as to satisfy particular object-handling requirements. The principal types are finger or tong type grippers, vacuum grippers, and magnetic grippers. Open-loop compliance is introduced via flexible fingers, self-aligning stiff fingers, padding, and spring loading. A 1973 survey of gripper design (70) illustrates many techniques.

One ingenious mechanism to control applied force consists of a segmented structure which, snake-like, wraps itself around a grasped object (51, 52). Control is obtained by monitoring tension in the internal tensioning cable system. This gripper conforms to objects of almost any shape and over a wide range of sizes while applying nearly uniform pressure all over. The interesting claim is made that the design lends itself to harvesting fruit, handling eggs, capturing animals, and transporting people.

Compliance to limit forces in non-gripping operations (like pin insertion) generally takes the form of series-inserted elastic members (e.g., 84, 134, 135, 142). In another approach, servo interaction is used; this is in some limb prosthesis applications where a "soft touch" is desirable for a mechanical hand (62). Feedback loops are arranged, for example, where sensed upward forces on a prosthetic wrist reduce downward shoulder torque, thus resulting in a softer touch. (This can be viewed as a small step in the direction of developing active compliance.)

c. Slip

Force and pressure sensing are, of course, vital elements in touch, though to date, as we have seen, industrial robots employ only simple force feedback. However, unless considerable gripper overpressure can be tolerated, slip sensing is essential to proper performance in many manipulation tasks. Information about contact areas and pressure distributions and their changes in time are needed in order to achieve the most complete and useful tactile sensing.

In contacting, grasping, and manipulating objects, adjustments to gripping forces are required in order to avoid slip and to avoid possibly dangerous forces both to the hand and to the workpiece. Besides the need for slip sensing transducers, there is the requirement for the robot's being able to determine at each instant the necessary minimum new force adjustments to prevent slip.

One of the earliest attempts to detect and correct slip used a piezo-electric crystal which sensed object slide relative to it (5). In the same year (1968) an ingenious device was reported which used strain gauges to report shear (slip) forces for prosthetic limbs (102). During the period 1972-1975 there was extensive device research on slip sensing. Techniques included the use of piezo-electric, semiconductor, and electromagnetic transducers where the motion of a workpiece-contacting stylus is sensed, and photoelectric and magnetic rolling-ball devices where discrete markings on the balls (optical or magnetic) produce A→D converted signals (A6, A11, A12).

In both the early stylus-motion slip detectors and the early ball-rotating types (where the balls rolled on an axle), only unidirectional slip could be detected for each transducer. This difficulty was overcome by a later development using a freely mounted (cup capture) optically marked ball (127). In this device, photo sensing of ball rotation in any direction yields all slip information in one transducer. Further, only tangential motion is detected, and thus noise rejection is superior.

All of the slip sensors cited so far are qualitative devices. The first report of a quantitative transducer is found in a development by Hitachi (72). Single-direction slip is optically measured by a rolling cylinder coupled to a radially slitted disc.

Until just recently, there had been no specific effort reported to assess quantitatively the direction of slip. A modification of the rolling ball scheme of (127) is the use of a surface-dimpled ball in contact with a sensing needle. The needle is fastened to a circular plate in such a way that slip-generated vibrations in a particular direction cause the circular plate to contact two or three pairs of surrounding electrodes, thus signalling slip direction in "any" direction (15, 17). Since there are 16 pairs of ring-surrounding electrodes, omnidirectionality is quantized into 16 directions.

This directional slip sensor (at JPL) is presently undergoing developmental tests in parallel with a simpler non-directional transducer. The second type also uses a captured rolling ball, one which is non-magnetic but studded with steel pins. As the ball rotates in response to a slipping surface, a magnetic sensing head registers pulses when the pins pass it.

3. Transducers

Since tactile sensing implies contact, this section is devoted principally to the technology of contact transduction. However, for the purposes of industrial robotry and remote manipulation, object position, shape, and even surface detail -- properties for which touch sensors are configured -- are at times sensed by non-contact means. Some of the more interesting types of remote "touching" are discussed at the end of this section.

As of about 1971 the only devices available for tactile sensing were microswitches (ABS-2), pneumatic jets (ABS-1), and (binary) pressure-sensitive pads (65, 71). These devices served principally as limit switches and provided little or no means for detecting shape, texture, or compliance. Still, such crude devices are used currently (e.g., 61, 124).

In the early seventies the search was already underway for shape detection (64, 122, ABS-2, ABS-3, ABS-8) and for "artificial skin" which could yield tactile information of complexity comparable to the human sense of touch. Two fields has primary interest: robotics and prosthetics.

An obvious methodology for obtaining a continuous measurement of force is potentiometer response to a linear (e.g., spring-loaded rod) displacement. Early sensors in many laboratories used such devices, and they still are in use today (e.g., 89, 129). (And one of Unimation's most widely used force feedback sensors develops signals from push-rod displacement; however, in this case optical code wheels are employed.) Specialized types of potentiometer arrangements have been developed, such as that in the "Belgrade Hand"* where resistive paint/electrode arrays compress compliant material to achieve continuous conductance change (98).

---

* This is one of the oldest continuous projects concerned with artificial grasping and touch sensing (for prosthesis). It was initiated in 1964.

One of the earliest attempts to develop transducers that were analog and which had more spatial extent than a single contact point used arrays of conductive-sponge (Motorola) elements; progressive cell-collapse caused progressive resistance change (65, 66). Another approach by the same workers used graphite filaments in fiberglass mats. These were deliberate attempts to circumvent the well-known hysteretic difficulties with conductive rubbers. (Note: Ref. 65 contains an interesting set of simple sketches for stress or pressure patterns for 13 classes of shapes.) Other early (and primitive) efforts to achieve spatial and, to some extent, quantized sensing used spring-adjusted switch arrays (36, 52, 66).

Another early attempt to obtain proportional force sensing used changing contact area between conductive rubber and copper (58). An informal but quantitative comparative overview of a number of such devices and techniques can be found in (20).

One of the most widely used force-transduction elements is the strain gauge. Many configurations have been used, both in laboratory exploration and in commercial application (e.g., 15, 17, 41, 105, 132, 139, 145).

The following four Subsections deal more explicitly with transducer types. The first three relate to contact sensing: conductive materials, semiconductors, and a small miscellany of others. The last Subsection treats non-contact or remote "touching".

a. Conductive materials and arrays

Proportional pressure or force sensing by conductive rubbers and polymers have been used in robots and manipulators in France (123), Germany (137), USSR (4), Japan (60), Great Britain (66), Yugoslavia (123), and the US (15, 121).

The choice of materials to be used for tactile sensing via conductive elements is sharply limited. Developmental models using graphite embedded in anelastic medium (66), conductive rubber (58, 123), and conductive foam (66, 121) have not been very satisfactory. Noise, nonlinearity, hysteresis, fatigue, long time-constants, low sensitivity, and drift, in various combinations have called for improved materials technology. (However, the performance encountered in (123) was claimed to be adequate for the function needed.)

Most of these materials fatigue after a few hundred operations. An independent laboratory study of them contains additional useful details (121).

A variety of other types of conductive and semiconductor material design for elastic carriers are described in (76). The most developed one of these materials is a piezo-resistive semiconductor compound whose description appears in the next Subsection.

The use of electrode arrays in conductive elastomers received strong impetus from the French (123) who are responsible for one of the basic patents in the art (63).

Tongs which could recognize cubes, spheres, cylinders, cones and pyramids used a pair of 3 x 3 arrays of conductive polymer buttons (46). The buttons were said to yield proportional signals, but no details were given.

In another 3 x 3 array using conductive plastic, small spike electrodes were pressed directly onto the plastic surface (137). Simple force-distribution pattern analysis was employed to orient workpieces automatically.

A prototype experimental 4 x 8 array using crossed electrode strips sandwiching a thin sheet of conductive elastomer is being tested at JPL (13, 15). Pressure distributions are studied on a visual display after microprocessing.

A 4 x 4 electrode conductive elastomer array was used to study detection of a small ensemble of surface discontinuities -- touch, point, pit, spur, ridge, crack, edge (121).

A long-term and widely-known program to develop artificial hands for amputees has been undertaken in Yugoslavia. The "Belgrade Hand" project started experiments with so-called "artificial skin" in collaboration with workers in France in 1975 (123). The aim was to achieve perception of roughness, hardness, pressure, and slip with a material which, like skin, was soft, elastic, and tough. A metallized foil surface on a conductive rubber sheet is used; sensing electrode arrays on the other side of the sheet provide distributions of force signals. Evidently, continuing experimental development is proceeding satisfactorily (23, 122).

b. Semiconductor sensors

Force transduction by solid-state devices has evolved rapidly during the past six or seven years, though many of the underlying developments have been available for more than a decade. For example, a 1967 patent (filed in 1964) disclosed important new advances in combatting low output and hysteresis (38). Piezo-diodes are described which are essentially hysteresis free and extremely stable. The material in this patent signals the basic start of our present bonded-straingauge technology.

In the following year the same inventor filed a patent for an electro-mechanical hand having tactile sensing. Issued in 1970, this patent (39) describes how piezo-diodes could be used as straingauges at prosthetic hand manipulator joints to provide linear force-feedback. The principal technology for force sensing in mechanical hands had been carbon-powder pressure transduction, a rather non-linear process.

A decade-old review of basic semiconductor piezo-junction effects and techniques appears in (147). Both needle and mesa are surveyed. Small, diaphragm-operated bridge-configured piezo-resistor pressure monitors subsequently were developed (e.g., 108, 109).

Recent technology has brought about semiconductor straingauges of great sensitivity, reliability, linearity, and versatility. It is these types of devices which typically have been used in robot manipulator force sensing. However, their costs are not trivial, usually several hundred dollars for a fully signal-conditioned unit. Consequently, their use has been rather limited. See (11) for a representative review of experimental approaches.

An interesting example of high-cost force-feedback semiconductor transduction is seen in the six-component wrist-mounted force sensor developed by Unimation. Sixteen strain gauges are used (arrayed in eight bridges). The selling price is $3,500. (Active marketing has not yet occurred because good control algorithms in software are not yet available to complement the device.)

The discrete-element semi-conductor force sensors are now being seriously challenged by two simpler and far less expensive devices, bonded doped-silicon bars and silicon-doped resistive elements formed by IC techniques. These piezo-resistive transducers bring the costs down by an order of magnitude, signal-conditioned units being available for just a few tens of dollars.

Piezo-resistive sensors such as those mentioned have considerable advantage over the elastomeric sensors described earlier with respect to sensitivity, linearity, noise characteristics, stability, speed, and hysteresis (being more crystalline). However, they do not compete well with respect to physical flexibility, spatial distribution, and low cost. The latter characteristics are represented well by elastomers in the search for "artificial skin", the ultimate anthropomorphic aspect of robot manipulators that appears to have great potential.

An intermediate approach holds promise for combining the advantages of semiconductor and elastomeric devices. This is the "Pressistor". This device consists of a force-variable resistance material which is applied in liquid form and then cured. Thus, it may be painted onto electrode arrays or impregnated into porous materials in arbitrary configurations. Sensitivity is controlled by thickness. Stability is said to be excellent.

Pressistor compounds are piezo-resistive semiconductor powders in organic polymers. They consist of microscopic bi-directional contact diodes suspended in a matrix. These diodes leak in proportion to applied force. In some configurations, avalanching occurs at a critical force, and solid-state switching occurs in about a nanosecond. For .005" thick films, compression repeatability of 10^{-6} inches is claimed (personal communication).

In other configurations, there is a reasonably good range of force/resistance characteristic, but the response is not very repeatable. Hysteresis is a considerable problem.

While these transducers do not yet lend themselves to precision applications, they may be converging to a very competitive technological advance. Costs may be more like 30 cents than 300 dollars, a factor of 10^3.

c. Other types of sensors

The types of contact transduction discussed so far principally were conductive and piezo-resistive elements. Other applicable techniques include piezo-electric, electromagnetic, hydraulic, optical, and capacitive sensors, though these are relatively rarely seen in robotics touch sensing.

An overview of sensor requirements and types appears in (132). Several exotic types of sensors are described in the review, including a pneumatically retractable delicate whisker which is a simple but effective non-intrusive poke-probe switch. A bi-stable metal dome that snaps to contact under pressure and which can be assembled into matrix arrays is another unusual device mentioned there.

A submersible claw gripper with hydraulic sensing is described in (32). The device is ultra simple, providing a differential pressure reading when the claw elements exert forces on hydraulic pistons.

A load-profile analyzer, consisting of an array of pressure sensitive transducers, was designed to indicate distributions of force over large areas (6). Sensing elements used in the design include variable inductance and piezo-electric elements. Variable-inductance proximity switches are used in industrial robots (e.g., Cincinnati Milacron) and are rugged and reliable.

Arrays of piezo-electric transducers have been used to (simplistically) model the tactile sense (61).

A novel transduction scheme for wrist sensors was employed in an SRI manipulator (49, 104). Each of the three component wrist motions are transduced by LED/photo-transistor pairs which sense the relative shadow area cast by a pin in the light path. The proportional pin intrusion is a function of the position of the wrist element to which it is attached. A similar use of proportional light sensing is used in the jaw touch-sensors of this manipulator; force applied to a compliant pad progressively shutters a light path. Both systems are simple, elegant, and linear.

Other touch sensing devices have been developed to sense touch by a human finger. Thus, in contrast to the devices discussed above where a machine is given means to feel when it touches something, the devices briefly mentioned below are meant to "feel" when something (a human finger) touches it. Thus this is passive touch rather than actual touch.

Methods for detecting the touch of a human finger include capacitance change, infrared beam interrupt, and echo ranging (73). Electrets provide a simple and reliable means for sensing capacitative changes (113). More complex but now inexpensive technology can sense finger electrical resistance by means of integrated circuit methodologies. Touch screens are useful devices, sensing the position of a finger over a specified area. One simple system uses strain gauges placed at the corners of the screen; thus, differential signals relating to finger position are produced (54).

d. Non-contact (remote) "touching"

Two types of sensing at a distance are useful to robots. One, of course, is vision in the sense of scene portrayal. The other is ranging, usually accomplished optically or by ultrasound, though R.F. and magnetic techniques also have been employed (26, 103).

Scene analysis lies outside the purview of this report, but ranging, particularly proximity ranging, is closely related to touch. Proximity sensing is useful for avoiding nearby obstacles, finding objects, and adjusting the positioning and attitude of the hand prior to grasping.

Surveys of proximity and touch sensing technology are found in (103) and (132).

Concentrated effort to develop proximity sensing for manipulators dates back only about six years. Emergent electronics technology has made possible a number of sophisticated, miniaturized systems that would not have been feasible much earlier.

Early optical triangulation ranging systems (ca. 1969) worked over ranges not less than several feet. The much closer distances required for robot manipulator sensing are accommodated by systems such as the one described in (15, 18, 57) where near-IR LED light sources work with silicon detectors. The arrangement is straight-forward; a fixed optical geometry defines a volume in space in which a reflecting object is located by triangulation. Typical sensing distances are a few centimeters; accuracy is on the order of a few tenths of a centimeter.* More complex arrays were developed to provide two-dimensional sensing, and thus bounded object regions became detectable. Experiments with similar technology are described in (22, 90, 132, A12).

Ultrasonic (time-of-flight) sensors are less commonly encountered but are useful (132). One unusual approach uses a matrix of transducers/sensors for primitive shape recognition, and another employs resonance phenomena in a compliant blanket to assess surface details of an object (23). This kind of system may be used both for range information and object shape/texture determination.

One laboratory experimental model of robot proximity sensing using pulse-ranging ultrasound uses up to 16 transducers operated by a microprocessor (66). Complex sonar-mapping strategies are being explored.

Magnetic sensors work with reluctance, Hall-effect, and eddy-currents. A cursory review is provided in (140). Outside of the Cincinnati Milacron use of inductive proximity switches, there seems so far to be little commercial robot manipulator use of magnetic techniques.

4. Pattern recognition

Automatic recognition of patterns, whether mechanical shape, handwriting, music, or stock market trends, is no easy matter. Most of our accomplishments to date are rudimentary and ad hoc. This is very obviously true for robot tactile sensing. However, as research in artificial intelligence proceeds briskly and interest in applied problems remains high, continuing progress may be anticipated -- but slowly, because the problems are extremely nontrivial.

One of the first forays into tactile shape and force-distribution recognition consisted of an interactive system; processed signals were displayed to a human operator for recognition (117). A flexible mirror is pressed against a surface, reflecting to the operator (via fiber optics, videon, and T.V. monitor) the image of a regular checkerboard grid. Surface irregularities and outline shape distort the grid field. Distortions representing a small number of geometric shapes were easily learned. The optical grid design was adopted after trying photoelastic stress pattern generators and moire pattern generators.

It is interesting to note that one of the most active current projects on auto-mated tactile sensing (JPL), a dozen years later, is studying sampled-field object representation by showing humans T.V. displays (13, 15).

A two-fingered 66-switch sensing system was reported to be able to discriminate six or seven different solid shapes. Experiments using propositional calculus were hand simulated. No details of any interest are given (56).

Similar studies have been conducted with 5 x 10 and 10 x 10 arrays (21). Conductive polymer sensing was used, but in binary mode only owing to poor transduction characteristics. Objects such as circles, rectangles, and "L" shapes were successfully recognized, but as the authors discovered, the great magnitudes of the downstream problems are just coming into focus.

In an ingenious study using a five-fingered robot hand, discrimination by groping was studied (60, 61).** Twenty-one pressure-sensitive on-off switches arrayed on fingers and palm provided signals as grasped objects were manipulated to determine shape. Discrimination between a cylinder and a square pillar was high.

---

* Precise ranging information in such systems is not in general achieved easily since received light depends on target reflectance.
** A similar approach for a three-fingered hand appears in (91).

This last study began with a consideration of human characteristics. While such anthropomorphic reflections are intriguing, the artificial constructs generally continue to be ad hoc, chiefly because we really know so little about living system design. Still, at the transducer level (for touch) there may be some payoff by considering skin and finger characteristics. This approach is seen in several of the references cited here (e.g., 50, 60, 61, 65, 102, 107, 127). At the other end of the system, consideration of the brain itself has stimulated interesting developments. In particular, a model of cerebellar action for robot control, though still primitive, is being developed vigorously (3).

A speculative essay on tactile shape recognition by parallel (two dimensional array) poke-probing is rather novel (94, 96). The probe array obtains a number of three-dimensional object samples simultaneously. Contour-following algorithms represent the object by a series of cross-section slices. However, the scene is barely set before the authors exit. A collection of descriptions of these techniques may be found in (A2).

V. PRESENT APPLICATIONS

1. Manipulation by industrial robots includes acts of recognition, acquisition, and handling. The real-world uses to which these systems have been put so far are many. Just a partial listing of major examples would include: assembling (from bicycles to automobiles), casting and molding (including sand handling), forging, grinding and polishing, heat treating, locomotion, master-slave manipulation (usually for hazardous environments), machining, painting, pouring, sorting, stacking, transporting, and welding.

2. Prosthetic and orthotic needs for the handicapped comprise the other important area for manipulators. This includes replacement artificial limbs for the amputee and sensing and assist devices for the paralyzed. There are an estimated 7,500,000 disabled persons in the U.S. (excluding rheumatoid arthritis) of which about half could be helped with presently available technology (4). However, even as in industry, the economics do not permit as much application as one might hope. Many of the prosthetic and orthotic devices and systems having the robot-type configurations of concern to this report are still in prototype development, though many of these are "in the field" for realistic testing and evaluation.

There has been strong and continual technological interplay between industrial robotic and prosthetic/orthotic developers. Problems of force feedback, slip, stable grasping, position sensing, light touch, etc. are common to both.

Typical developments in artificial hands are seen in (35, 62, 98, 122, 123) and representative work in teleoperator assists can be found in (9, 45, 129).

VI. OUTSTANDING PROBLEMS AND NEW OPPORTUNITIES

There are two main areas which are most in need of development. They are: 1) Improved tactile sensors and 2) Improved integration of touch feedback signals with the effector control system in response to the task-command structure. Sensory feedback problems underlie both areas, of course. What is needed is more effective comprehensive sensors (device R & D) and the sophisticated interpretation of the sense signals by control structures (system R & D).

Many people in the field continue to believe that visual sensing is most important -- at least in the sense of having greater ultimate potential and having the largest number of unsolved problems, some of very great difficulty. However, a fair number of people actively engaged in research and development of touch sensing feel that since this field has been so little plowed, a considerable number of important and useful developments can be achieved in the very near future. Further, touch sensing for robots is believed by robot manufacturers to be commercially feasible at a cost of around $2,000 (compared to vision at $7,000) per robot (A3).

Sensitive, dextrous hands head the problem list for manipulators just as sensitive, adaptable feet head the list for legged locomotion vehicles. Each application area has its own detailed special problems to solve; for example, the design approach for muddy-water object recovery and for delicate handling of unspecified objects in an unstructured environment differ vastly.

The problem of how to process and use information for control looms largest. Although the development of sensor devices and manipulators poses many challenging problems, these are relatively easy problems compared to the systems design achievements needed to use the feedback signals to greatest advantage. This includes the need to develop much more complex hierarchical control structures than are presently available. Algorithmic control procedures are needed in robot environments which may have a large number of variable couplings, shapes, inertias, surfaces, etc. Optimal control in real time (at the servo level) is the big problem. Operating systems for process control where either visual or tactile inputs of great complexity serve as inputs are literally in the stone age so far. Computer control, of course, is the critical core of these matters, and despite some excellent advances in programming languages, multi-sensory feedback, and combined sensing/manipulating, the level reached so far, even in well-controlled laboratory conditions, remains rather low.

As Bejczy puts it (16):

> As of now, no general theories exist which are aimed to analyze stability and optimality problems in pattern-referenced control, or to synthesize stable 'feedback logic'. Each case is based on ad hoc considerations and requires extensive experimental or simulation work for verification. A most challenging task would be the creation of a theoretical framework for the analysis and synthesis of pattern-referenced control systems.

Another statement of this problem is given by Vertut et al. in (129):

> The computer must have a complete mathematical model of the manipulator, must be capable of monitoring manipulator state at any moment, and must know the task to be executed, and provide adaptive control from knowing the environment.

An appreciation for the complexity and magnitude of this task will be obtained by reading (7, 8, 10, 19, 79, 87, 97, 118, 125).

It may well be that our present serial digital techniques and relatively simple, linear mathematics may not be the way to go. Complete formal analysis (e.g., equations of motion) for just a six-degree-of-freedom manipulator appears to be intractable (A10).

Distributed processing may hold promise and is just now receiving concentrated attention. We need parallel and perhaps analog techniques to interface sensor information to effector operation via job description. We need to find ways to make these systems flexible, adaptive, and heuristic. Researchers are increasingly gaining perspective on what looms as an acute problem area. An excerpt from a recent research proposal (W. R. Ferrell, University of Arizona) reflects the awareness.

> The general objective of the research is to obtain a working understanding of how to coordinate sensory input and manipulative activity in semi-automated contact (touch and force) sensing, and is based on the recognition that a "smart hand" is both a mechanical tool and a sensor network. Within the context of the tool/sensor duality principle for a "smart hand", the research encompasses electromechanical hand design, touch sensor design, manipulative-tactile information processing, reprogrammability for different objects, and local computation for tactile arrays. Developing the technology needed for high-volume production and general purpose computer controlled mechanical hands suitable for industrial advanced automation is the long range goal of the research effort.

Memory is needed in fingers; as in humans, there may be virtue in partial peripheral processing which puts less demand on central processor design.

A collection of observations on these and related considerations appears in (53, A7, A10). Recent assessment of robotic aids for the handicapped appears in (67, 68). A thoughtful and useful set of comments and projections concerning the future of robotics and teleoperators is given in (4, 29, 126, A1).

Despite the large variety of tasks already handled by these systems, many applications areas remain largely untouched so far. Emerging technology in improved touch, vision, and system competence will undoubtedly open many new market opportunities. Near-future new systems will be developed for undersea salvage and research prospecting (see Ref. 118 for a complete survey), space station operations, mining, microsurgery, and hazardous factory and rescue operations. And even agricultural harvesting by robots may not be too far off.

Remote palpation and manipulation in telediagnosis had a brief exploration in 1967 (116), but little if anything appears to have been done since. Boston's Logan Airport Remote Clinic (T.V. link to physician at Massachussetts General Hospital) is the most widely known of the experimental medical-examination-at-a-distance enterprises. It provides for vital-sign inspection, transmission, and evaluation by the remote physician. However, palpation or manipulation has not been introduced. The study cited above was intended to work into the Logan Airport facility, but the state-of-the-art was too immature in 1967.

There may be a potential market for such systems, especially as physician shortages continue. There are presently many paramedic and lay personnel remote medical service experimental trials all over the world, but there appears to be a total lack of remote touch, feel, and move facility. The most sanguine projections of present master-slave manipulator (teleoperator) systems even envisions remote surgery (4, 126).

Throughout all of the above lies the implicit need for improved tactile sensors. A really effective "artificial skin" is needed and would have wide application. Material developments that provided pressure, force, and shear signals of high linearity and fine spatial resolution are not yet available. Hysteresis problems remain to be resolved, and fatigue is a concern. This arena is not yet mature, but it may be on the verge of becoming so. Note should be taken of challenging emergent technology in semiconductor arrays (next Section).

VII. EMERGENT TECHNOLOGY

One of the newest developments in touch-sensing technology is that of reticular (Cartesian) arrays using solid-state transduction and attached micro-computer elements that compute three-dimensional shape. The approach is typified by the research of Marc Raibert at Cal Tech's JPL (unpublished, presently in feasibility study phase). Raibert plans to use a flexible IC wafer having a metal-electrode substrate. The device will be compact and the resolution high. Matrix analysis to compute touched shape will be done by a sophisticated combined processor; hence, the fingertip is a self-contained "smart finger". This is a quantum jump ahead of prior art, typified by (21, 123), for example, where small arrays of touch sensors use passive substrates and materials such as conductive elastomers. Resolution in such devices has been quite low, and hysteresis a problem.

An intriguing concommitant of this new approach is the information processing in the reticulated neural nets of some living systems, notably jellyfish and similar creatures. Both physiological studies and modeling research has disclosed that in coelenterates, for example, regular network lattices are capable of rather complicated spatial-to-temporal pattern transformation and classification. For example, similar (but not identical) temporal stimulus patterns applied to the net will produce similar distributions of propogated signal shape in the net, and vice versa. In related neurophysiological and modeling studies, regular (and rather simple) one- and two-dimensional arrays of interconnected sensors are capable of extracting a considerable variety of shape information. Conceivably such findings (and analysis) may have important bearing on new tactile sensory systems for automata.

The only other relevant bionic approach to manipulator technology is that of the cerebellar-like controller system being developed by James S. Albus at the National Bureau of Standards (3).

Another relatively new concentrated effort in sensing technology is that of ultrasound ranging in three dimensions. Although the actual techniques and concepts are not novel, some of the currently developing systems have new and important twists. A good example of this is found in the work at Cecil Equipment Company, Medina, Ohio. The ranging technique, intended primarily for industrial welding robots, uses time-of-flight technology with recognition of first arriving reflected patterns; an accuracy of 0.001" over a range of 1/8" to 6" is claimed. Still in development, the planned product line will operate at a working distance of about 2" and have a ranging accuracy of .005" to .010". These devices are intended to fit a considerable market need for non-contact sensing (e.g., welding, which is one of the largest applications areas).

A potentially promising new approach to "smart hands" is underway at Stanford (Mechanical Engineering Department) by Larry J. Leifer and his students. Intended primarily for rehabilitation applications (prostheses), two-finger, 11-linkage devices are being developed. These fingers will have optical sensors for pseudo touch; their range is short (1-4 mm), and resolution is high. They will use HP devices, recently available, which are employed both for emission and detection. "Fly's eye" arrays are under development using strobe illumination for long-range perception.

VIII.   BIBLIOGRAPHY

The accompanying references are in two sections. The first one, consisting of 147 entries, was assembled prior to the deadline data required to initiate report writing. The second list, containing 13 entries, includes those materials which were received later. These latter references are cited with an A preceding the reference number.

Not all of these references are cited in the text. This owes to the fact that some were not immediately relevant, some were redundant, and some were not of first-order value in general. Yet these 49 items are included for additional background information and may be of peripheral interest. They are: (12, 13, 14, 25, 28, 30, 37, 40, 42, 43, 44, 47, 48, 50, 55, 59, 69, 74, 75, 77, 80, 81, 82, 83, 85, 86, 88, 92, 93, 95, 99, 101, 104, 105, 106, 107, 110, 111, 114, 115, 119, 120, 122, 128, 130, 133, 138, 146, A5).

IX.   ACKNOWLEDGEMENT

I gratefully acknowledge the support for this survey which was provided by the LORD CORPORATION, Erie, Pennsylvania.

REFERENCES

1. Abraham, R. G. et al., State-of-the-art in adaptable-programmable assembly systems. SME Technical Paper MS77-757, 1977, 12 pp.

2. Adam, M. F., Grip accessory for remote-control manipulator tongs. U.S. Patent 3,815,761, June 11, 1974.

3. Albus, J. S. and J. M. Evans, Jr., Robot systems. Sci. Am., 234 (2), Feb., 1976, 77-87.

4. Alexander, A. D., III, A survey study of teleoperators, robotics, and remote systems technology. In: Proc. of the First Nat. Conf. on Remotely Manned Systems, California Institute of Technology, 1973, 449-458.

5. Baits, J. C. et al., The feasibility of an adaptive control scheme for artificial prehension. In: Proc. Instn. Mech. Engrs., Vol. 183, Pt. 3F, 1968-69, 54-59.

6. Barron, E. R. et al., Load profile analyzer in the attached specification. U.S. Patent 3,818,756, June 25, 1974.

7. Bejczy, A. K., Machine intelligence for autonomous manipulation. In: Proc. of the First Nat. Conf. on Remotely Manned Systems, California Institute of Technology, 1973, 337-396.

8. Bejczy, A. K., Algorithmic formulation of control problems in manipulation. In: Proc. Int. Conf. on Cybernetics and Society, IEEE, Sept. 23-25, 1975, 134-142.

9. Bejczy, A. K., Performance evaluation studies at JPL for space manipulator systems. NBS Special Pub. #459, Performance Evaluation of Programmable Robots and Manipulators. Workshop held at Annapolis, MD, Oct. 23-25, 1975, 159-174.

10. Bejczy, A. K., Issues in advanced automation for manipulator control. In: Proc. 1976 Joint Automatic Control Conf., Purdue Univ., Lafayette, IN, July 27-30, 1976, 700-711.

11. Bejczy, A. K., Effect of hand-based sensors on manipulator control performance. Mechanism and Machine Theory, Vol. 12, 1977, 547-567. Pergamon Press.

12. Bejczy, A. K., Voice command system for motion control. In: Int. Conf. on Telemanipulators for the Physically Handicapped, Rocquencourt, France, Sept. 4-6, 1978.

13. Bejczy, A. K., Sensor systems for automatic grasping and object handling. In: Int. Conf. on Telemanipulators for the Physically Handicapped, Rocquencourt, France, Sept. 4-6, 1978.

14. Bejczy, A. K., Manipulation of large objects. In: Third CISM-IFToMM Int. Symp. on Theory and Practice of Robots and Manipulators, Udine, Italy, Sept. 12-15, 1978.

15. Bejczy, A. K., Smart sensors for smart hands. In: AIAA/NASA Conf. on "Smart" Sensors, Hampton, VA., Nov. 14-16, 1978, 17 pp.

16. Bejczy, A. K., Manipulator control technology for space. In: Proc. NSF Workshop on the Impact of Required Research Activity for Generalized Robotic Manipulators, U. Fla., Feb. 8-9, 1978, 266-289.

17. Bejczy, A. K., Manipulator control automation using smart sensors. In: Electro/ 79 Convention, New York, NY, April 24-26, 1979, 16 pp.

18. Bejczy, A. K. et al., Evaluation of proximity sensor aided grasp control for shuttle arms. In: Proc. 15th Annual Conf. on Manual Control, Wright State Univ., Dayton, OH, March 20-22, 1979, 26 pp.

19. Bejczy, A. K. and R. L. Zawacki, Computer-aided manipulator control. In: Proc. of the First Int. Symp on Mini- and Microcomputers in Control, San Diego, CA, Jan. 8-9, 1979.

20. Binford, T. O., Sensor systems for manipulation. In: Proc. of the First Nat. Conf. on Remotely Manned Systems, California Institute of Technology, 1973, 283-291.

21. Briot, M., The utilization of an "artificial skin" sensor for the identification of solid objects. In: 9th Int. Symp. on Industrial Robots, Wash., D.C., March 13-15, 1979, 529-547.

22. Catros, J. Y. et al., Automatic grasping using infrared sensors. In: 8th Int. Symp. on Industrial Robots, Stuttgart, W. Germany, May 30-June 1, 1978, 132-142.

23. Clot, J. et al., Project pilote - Spartacus. In: Convention de Recherche, Toulouse, France, July, 1977, 21 pp.

24. Collins, C. C., Tactile image perception. In: Proc. IEEE Intocon. Tech. Papers, 37 (3), 1973, 1-4.

25. Collins, C. C. and J. M. J. Madey, Tactile Sensory Replacement. In: Proc. San Diego Biomed. Symp., 13, 1974.

26. Dixon, J. K., Robot Sensors. Internal Report, U.S. Naval Research Lab., 1978.

27. Dixon, J. K. et al., Research on tactile sensors for an intelligent naval robot. In: 9th Int. Symp. on Industrial Robots, Wash., D.C., March 13-15, 1979, 507-517.

28. Dobrotin, B. M., Robots in space exploration. In: Proc. IEEE Symp. on Systems, Man, and Cybernetics, Dallas, TX, Oct. 1974, 40-43.

29. Driscoll, L. C., Second- and third-generation industrial robots. In: Proc. of the First Nat. Conf. on Remotely Manned Systems, California Institute of Technology, 1973, 469-477.

30. Ehrreich, J. E., Manufacture of reinforced conductive plastic gaskets. U.S. Patent 3,194,860, July 13, 1965.

31. Engelberger, J. F., A robotics prognostication. In: Proc. 1977 Joint Automatic Control Conf., 1977, 197-204.

32. Fishel, K. R., Differential pressure transducer and readout for sensing claw grip force. U.S. Patent 3,759,092, Sept, 18, 1973.

33. Flatau, C. R., Arrangement of motions in a robot assembly machine. Working Paper 91, Mass. Inst. of Tech., Artificial Intel. Lab., Nov. '74.

34. Flatau, C. R., Force sensing in robots and manipulators. In: <u>Second CISM-IFToMM Symp. on Theory and Practice of Robots and Manipulators</u>, Warsaw, Sept. 14-17, 1976, 294-306.

35. Fletcher, J. C. et al., Tactile sensing means for prosthetic limbs. U.S. Patent 3,751,733, Aug. 14, 1973.

36. Folchi, G. A. et al., Tactile sensors operated by a warping plate and buckling beams. <u>IBM Technical Disclosure Bulletin</u>, <u>Vol. 18</u>, No. 2, July, 1975, 575-577.

37. Folchi, G. et al., Gimbal system for natural wrist motion for a manipulator. <u>IBM Technical Disclosure Bulletin</u>, <u>Vol. 17</u>, <u>No. 10</u>, March, 1975.

38. Fraioli, A. V., Solid state pressure transducer. U.S. Patent 3,323,358, June 6, 1967.

39. Fraioli, A. V., Electro-mechanical hand having tactile sensing means. U.S. Patent 3,509,583, May 5, 1970.

40. Gilliland, J. R., Electrically conductive resinous compositions. U.S. Patent 3,412,043, Nov. 19, 1968.

41. Goto, T. et al., Precise insert operation by tactile controlled robot. <u>The Industrial Robot</u>, Sept., 1974, 225-228.

42. Gray, W. E. and D. L. Pieper, Implications of non-space applications. In: <u>Proc. of the First Nat. Conf. on Remotely Manned Systems</u>, California Institute of Technology, 1973, 459-467.

43. Grossman, D. D. and M. W. Blasgen, Orienting mechanical parts by computer controlled manipulator. IBM Research Report, Aug. 21, 1974, 14 pp.

44. Guittet, J., Application of a systems science approach to the development of tools to aid the physically handicapped. Submitted to <u>Conf. on Systems Science in Health Care</u>, Paris, France, July 5-9, 1976, 8 pp.

45. Guittet, J. et al., The Spartacus telethesis: manipulator control and experimentation. In: <u>Int. Conf. on Telemanipulators for the Physically Handicapped</u>, Rocquencourt, France, Sept. 4-6, 1978, 79-95.

46. Gurfinkel, V. S. et al., Tactile sensitizing of manipulators. <u>Engineering Cybernetics</u>, <u>12</u> (<u>6</u>), 1974, 47-56.

47. Hill, J. W., Two measures of performance in a peg-in-hole manipulation task with force feedback. Presented at <u>13th Annual Conf. on Manual Control</u>, Mass. Inst. Tech., June 15-17, 1977, 17 pp.

48. Hill, J. W., Force-controlled assembler. Presented at <u>Robots II Conf.</u>, Detroit, MI, Nov. 1-3, 1977, 10 pp.

49. Hill, J. W., and A. J. Sword, Manipulation based on sensor-directed control: an integrated end effector and touch sensing system. Presented at <u>17th Annual Human Factors Society Convention</u>, Wash., D.C., Oct. 16-18, 1973.

50. Hill, J. W., and A. J. Sword, Touch sensors and control. In: <u>Proc. of the First Nat. Conf. on Remotely Manned Systems</u>, California Institute of Technology, 1973, 351-368.

51. Hirose, S. and Y. Umetani, The development of soft gripper for the versatile robot hand. Mechanism and Machine Theory, Vol. 13, 1978, 351-359, Pergamon Press.

52. Hirose, S. and Y. Umetani, Kinematic control of active cord mechanism with tactile sensors. In: Second CISM-IFToMM Symp. on Theory and Practice of Robots and Manipulators, Warsaw, Poland, Sept. 14-17, 1976, 249-260.

53. Horn, B. K. P., What is delaying the manipulator revolution? Working Paper 161, Artificial Intelligence Laboratory, Mass. Inst. Techn., Feb., 1978, 6 pp.

54. Howe, J. et al., Programming a touch screen device. Management Informatics, Vol. 3, No. 2, 1974, 69-85.

55. Inoue, H., Force feedback in precise assembly tasks. Memo No. 308, Artificial Intelligence Laboratory, Mass. Inst. Tech., August., 1974, 31 pp.

56. Ivancevic, N. S., Stereometric pattern recognition by artificial touch. Pattern Recognition, Vol. 6, 1974, 77-83, Pergamon Press.

57. Johnston, A. R., Proximity sensor technology for manipulator end effectors. Mechanism and Machine Theory, Vol. 12, 1977, 95-109, Pergamon Press.

58. Johnson, J. B. and T. R. Huffman, A tactile prosthetic device for the denervated hand. In: 11th Annual Rocky Mountain Bioengineering Symp. and Int. ISA Bio-medical Science Instrum. Symp., April 11, 1974, 9-10.

59. Kato, I. et al., Artificial softness sensing - an automatic apparatus for mea-suring viscoelasticity. Mechanism and Machine Theory, Vol. 12, 1977, 11-26 Pergamon Press.

60. Kinoshita, G. -I., Classification of grasped object's shape by an artificial hand with multi-element tactile sensors. Information-Control Problems in Manufacturing Technology, 1977, 111-118, Pergamon Press.

61. Kinoshita, G. -I, et al., A pattern classification by dynamic tactile sense in-formation processing. Pattern Recognition, Vol. 7, 1975, 243-251, Pergamon Press.

62. Kwee, H. H., The Spartacus telethesis: "soft touch". In: Conf. on Telemanip-ulators for the Physically Handicapped, Rocquencourt, France, Sept. 4-6, 1978, 305-308.

63. Legasse, J. et al., Tactile pick-up. U.S. Patent 4,014,217, March 29, 1977.

64. Larcombe, M. H. E., Tactile perception for robot devices. In: 1st Conf. on Industrial Robot Technology, Univ. of Nottingham, UK, March 27-29, 1973, 191-196.

65. Larcombe, M. H. E., Tactile sensing using digital logic. In: Proc. Shop Floor Automation Conf., Paper #9, Birniekill Inst., National Engineering Lab. East Kilbride, Dec. 11-13, 1973, 7 pp.

66. Larcombe, M. H. E., Tactile sensors, sonar sensors and parallax sensors for robot applications. In: 3rd Conf. on Industrial Robot Tech. and 6th Int. Symp. on Industrial Robots, Univ. of Nottingham, UK, March 24-26, 1976, C3-25-32, 25-32.

67. Leifer, L. et al., Robotic aids for the severely disabled - needs assessment. Internal Report, Stanford Univ., 1978, 17 pp.

68. Leifer, L. et al., Robotic aids for the severely disabled, feasibility assessment. Internal Report, Stanford Univ., 1978, 11 pp.

69. Lewis R. A., Adaptive control of a robotic manipulator. In: Proc. 1977 IEEE Conf. on Decision and Control, New Orleans, LA, Dec., 7-9, 1977, 743-748.

70. Lunstrom, G., Industrial Robot Gripper, The Industrial Robot, Dec., 1973, 72-82.

71. Mansfield, J. W. B., Recording of alarm devices. U.S. Patent 3,836,900, Sept. 17, 1974.

72. Masuda, R. et al., Slip sensor of industrial robot and its application. Electrical Eng. in Japan, Vol. 96, No. 5, 1976, 129-136.

73. McEwing, R. W., Touch displays in industrial computer systems, 79-81.

74. McGhee, R. B., Robot locomotion. In: Proc. Int. Conf. on Neural Control of Locomotion, Philadelphia, PA, Sept., 1975, 37 pp.

75. McGhee, R. B., Future prospects for sensor-based robots. In: Proc. Symp. on Computer Vision and Sensor-Based Robots, General Motors Research Lab., Sept., 1978, 21 pp.

76. Mitchell, R. J., Pressure responsive resistive material. U.S. Patent 3,806,471, April 23, 1974.

77. Mosher, R. S., Industrial manipulators. Sci. Am., 211 (4), Oct., 1964, 88-96.

78. Nevins, J. L., Sensors for industrial automation, Yearbook in Science and Technology, 1975, McGraw-Hill.

79. Nevins, J. L. and D. E. Whitney, Information and control issues of adaptable, programmable assembly systems for manufacturing and teleoperator applications. Mechanism and Machine Theory, Vol. 12, 1977, 27-43, Pergamon Press.

80. Nevins, J. L. and D. E. Whitney, Assembly research and manipulation. In: Proc. 1977 IEEE Conf. on Decision and Control, New Orleans, LA, Dec. 7-9, 1977, 735-742.

81. Nevins, J. L. and D. E. Whitney, Computer-controlled assembly. Sci. Am., Vol. 238, No. 2, Feb., 1978, 62-74.

82. Nevins, J. L. et al., A scientific approach to the design of computer controlled manipulators. Internal Report R-837. The Charles Stark Draper Lab., Inc., Cambridge, MA, Aug., 1974, 179 pp.

83. Nevins, J. et al., Exploratory research in industrial modular assembly. Internal Report R-1111. The Charles Stark Draper Lab., Inc., Cambridge, MA, Sept. 1, 1976 to Aug. 31, 1977, 167 pp.

84. Nevins, J. L. et al., Exploratory research in industrial modular assembly. Internal Report R-1218. The Charles Stark Draper Lab., Inc., Cambridge, MA, Sept. 1, 1977 to Aug. 30, 1978, 136 pp.

85. Nevins, J. L. et al., Robot assembly research and its future directions. Internal Report P-741. The Charles Stark Draper Lab., Inc., Cambridge, MA, Sept., 1978, 58 pp.

86. Nitzan, D., Machine intelligence research applied to industrial automation. In: 6th NSF Grantees Conf. in Production Research and Technology, Purdue, Univ., West Lafayette, IN, Sept. 27-29, 1978, 7 pp.

87. Okada, T., Object-handling system for manual industry. IEEE Trans. on Systems, Man, and Cybernetics, Vol. SMC-9, No. 2, Feb., 1979, 79-89.

88. Okada, T., On a versatile finger system. In: Proc. 7th Int. Symp on Industrial Robots, Oct. 19-21, 1977, 345-352.

89. Okada, T., Stable grasping and hardness evaluating of an object by highly mounted pressure sensors. Technical Note No. 19, Electrotechnical Lab., Tokyo, Japan, Aug., 1977.

90. Okada, T., A short-range finding sensor for manipulators. Electrotechnical Lab., Tokyo, Japan, May 4, 1978, 28-40.

91. Okada, T. and S. Tsuchiya, Object recognition by grasping. Pattern Recognition, Vol. 9, 1977, 111-119, Pergamon Press.

92. Orin, D. E. et al., Kinematic and kinetic analysis of open-chain linkages utilizing Newton-Euler methods. Mathematical Biosciences, Vol. 43, 1979, 107-130, Elsevier North Holland, Inc.

93. Orin, D. E. et al., Design of a manipulator arm for an interactive computer-controlled legged locomotion system. In: Milwaukee Symp. on Automatic Computation and Control, Milwaukee, WI, April 22-24, 1976, 167-173.

94. Page, C. J. et al., New techniques for tactile imaging. The Radio and Electronic Engineer, Vol. 46, No. 11, Nov., 1976, 519-526.

95. Paine, G., Microprocessors for realtime displays and control of space teleoperators. In: Proc. First Int. Symp on Mini- and Micro-computers in Control, San Diego, CA, Jan. 8-9, 1979.

96. Pugh, A. et al., Novel techniques for tactile sensing in a three dimensional environment. The Industrial Robot, March, 1977, 18-26.

97. Raibert, M. H., Motor control and learning by the state space model. Ph.D. Dissertation, Artificial Intel. Lab., Mass. Inst. Tech., Sept. 8, 1977.

98. Rakic, M., The "Belgrade hand prosthesis". In: Proc. Instn. Mech. Engrs., Vol. 183, Pt. 3F, 1968-69, 60-67.

99. Rechnitzer, A. B. and W. Sutter, Naval applications of remote manipulation. In: Proc. of the First Nat. Conf. on Remotely Manned Systems, California Inst. Tech., 1973, 493-503.

100. Rechnitzer, A. B., An overview of non-U.S. underwater remotely manned manipulators. Mechanism and Machine Theory, Vol. 12, 1977, 51-56, Pergamon Press.

101. Reichardt, J., Artificial arms, hands, and other extensions. In: Robots: Fact, Fiction, and Predition, 1978, 122-130, Penguin Books.

102. Ring, N. D. and D. B. Welbourn, A self-adaptive gripping device: its design and performance. In: Proc. Instn. Mech. Engrs., Vol. 183, Pt. 3F, 1968-69, 45-49.

103. Rosen, C. A. and D. Nitzan, Use of sensors in programmable automation. IEEE Computer Society Magazine, Dec., 1977, 12-23.

104. Rosen, C. et al., Exploratory research in advanced automation. First Semi-Annual Report. Stanford Research Institute, Menlo Park, CA, April 1, 1973 to Sept. 30, 1973.

105. Rosen, C. et al., Exploratory research in advanced automation. Second Report. Stanford Research Institute, Menlo Park, CA, Oct. 1973 to June 30, 1974.

106. Roth, B., Robots. Applied Mechanics Reviews, Vol. 31, No. 11, Nov., 1978, 1511-1519.

107. Sakai, I. et al., Approach and plan: most suitable control of grasping in industrial robot. In: 5th Int. Symp. on Industrial Robots, IIT Research Institute, Chicago, IL, Sept. 22-24, 1975, 525-532.

108. Samaun, S. et al., An IC piezoresistive pressure sensor for biomedical instrumentation. In: IEEE Int. Solid-State Circuits Conf., U. of PA, 1971, 104-105.

109. Samaun, K. D. W. and J. B. Angell, An IC piezoresistive pressure system for biomedical instrumentation. IEEE Trans. on Biomedical Eng., Vol. BME-20, No. 2, March, 1973, 101-109.

110. Saraswat, K. C. and J. D. Meindl, HV silicon-gate MOS integrated circuit for driving piezoelectric tactile displays. In: IEEE Int. Solid-State Conf., 1974, 164-165, 246.

111. Schmid, H. P. and G. A. Bekey, Tactile information processing by human operators in control systems.

112. Schraft, R. D., Industrial robots in Europe - market, applications and developments. Mechanism and Machine Theory, Vol. 12, 1977, 45-50, Pergamon Press.

113. Sessler, G. M. et al., New touch actuator based on the foil-electret principle. IEEE Trans. on Communications, Vol. COM-21, No. 1, Jan., 1973, 61-65.

114. Shannon, G. F., Characteristics of a transducer for tactile displays. Biomedical Engineering, June, 1974, 247-249.

115. Shannon, G. F., The case for sensory feedback on artificial limbs. Elec. Eng. Trans., 1975, 36-38.

116. Sheridan, T. B. et al., Tactile sensing for remote palpation and manipulation in telediagnosis. In: 20th Annual Conf. on Eng. in Med. and Biol., Boston, MA, Nov., 13-16, 1967, 23.3.

117. Sheridan, T. B. and W. R. Ferrell, Measurement and display of control information. Progress Report, Engineering Projects Lab., Mass. Inst. Tech., Oct. 1, 1966 - March 31, 1967, 13-22.

118. Sheridan, T. B. and W. L. Verplank, Human and computer control of undersea teleoperators. Man-Machine Systems Lab., Mass. Inst. Tech., July 15, 1978.

119. Skinner, F., Designing a multiple prehension manipulator. Mechanical Eng., Sept., 1975, 30-37.

120. Skinner, F., Multiple prehension hands for assembly robots. In: 5th Int. Symp. on Industrial Robots, Chicago, IL, Sept. 22-24, 1975, 77-87.

121. Snyder, W. E. and J. St. Clair, Conductive elastomers as sensor for industrial parts handling equipment. IEEE Trans. on Instrum. and Measure, Vol. IM-27, No. 1, March, 1978, 94-99.

122. Stojiljkovic, Z. and D. Saletic, Learning to recognize patterns by Belgrade hand prosthesis. In: 5th Int. Symp. on Industrial Robots, Chicago, IL, 1975, 407-413.

123. Stojiljkovic, Z. and J. Clot, Integrated behavior of artificial skin. IEEE Trans. on Biomed. Eng., Vol. BME-24, No. 4, July, 1977, 396-399.

124. Sword, A. J. and W. T. Park, Location and acquisition of objects in unpredictable locations. Technical Note 102. Artificial Intelligence Center, Stanford Research Institute, Menlo Park, CA, June, 1975.

125. Takase, K., Task-oriented variable control of manipulator and its software servoing system. In: Proc. of the IFAC Int. Symp. on Information-Control Problems in Manufacturing Technology, Tokyo, Japan, Oct., 17-29, 1977, 139-145.

126. Thring, M. W., Telechiric Mining. In: Second CISM-IFToMM Symp. on Theory and Practice of Robots and Manipulators, Warsaw, Poland, Sept. 14-17, 1976, 493-506.

127. Tomovic, R. and Z. Stojiljkovic, Multifunctional terminal device with adaptive grasping force. Automatica, Vol. 11, 1975, 567-570, Pergamon Press.

128. Ueda, M. and T. Shimizu, Sensors and systems necessary for industrial robots in the near future. 4th Int. Symp. on Industrial Robots, Tokyo, Nov. 19-21, 1974, 79-88.

129. Vertut, J. et al., Advance of the new MA23 force reflecting manipulator system. In: Second CISM-IFToMM Symp. on Theory and Practice of Robots and Manipulators, Warsaw, Poland, Sept. 14-17, 1976, 307-322.

130. Vukobratovic, M., Contributions to forming criteria for the evaluation of robots and manipulators. In: Performance Evaluation of Programmable Robots and Manipulators. Wash., D.C.: Nat. Bur. Standards, S.P. 459, Oct., 1976, 197-203.

131. Wallace, J. J., Creating a robot market. In: 5th Int. Symp. on Industrial Robots, Chicago, IL, Sept. 22-24, 1975, 29-45.

132. Wang, S. S. M. and P. M. Will, Sensors for computer controlled mechanical assembly. The Industrial Robot, March, 1978, 9-18.

133. Ward, M. R. et al., Consight: a practical vision-based robot guidance system. In: 9th Int. Symp. on Industrial Robots, Wash., D.C., March 13-15, 1979, 195-211.

134. Warnecke, H. J. and D. Haaf, Automatisches montieren mit programmierbaran montagessystemen. WT Zeitschrift fur industrielle Fertigung, Vol. 69, 1979, 145-150.

135. Warnecke, H. J. et al., A programmable assembly system with software-integrated optical sensor for process control and fast product change-over. Fraunhofer-Institute fur Producktionstechnik und Automatisierung, Stuttgart, Internal Rept., 1979, 17 pp.

136. Warnecke, H. J. and G. Herrmann, Applications, social impacts, education and training. In: Second CISM-IFToMM Symp. on Theory and Practical of Robots and Manipulators, Warsaw, Poland, Sept. 14-17, 1976, 509-515.

137. Warnecke, H. J. et al., Pilot work site with industrial robots. In: 5th Int. Symp. on Industrial Robots, Chicago, IL, Sept. 22-24, 1975, 71-86.

138. Warnecke, H. J. and M. Schweizer, Taktile sensoren fur programmierbare handhabungsgerate. WT Zeitschrift fur industrielle Fertigung, Vol. 69, 1979, 159-163.

139. Watson, P. C. and S. H. Drake, Pedestal and wrist force sensors for automatic assembly. In: 5th Int. Symp. on Industrial Robots, Chicago, IL, Sept. 22-24, 1975, 501-511.

140. West, D., Touch-me-not. Elec. Products Mag., Jan., 1976, 39-43.

141. Whitney, D. E., Force feedback control of manipulator fine motions. In: Proc. 1976 Joint Automatic Control Conf., July 27-30, 1976, 687-693.

142. Whitney, D. E. et al., What is the remote center compliance (rcc) and what can it do? Internal Report P-728, The Charles Stark Draper Lab., Inc., Cambridge, MA, Nov., 1978.

143. Whitney, D. E. et al., Robot and manipulator control by exteroceptive sensors. In: Proc. 1977 Joint Automatic Control Conf., 1977, 155-163.

144. Will, P. M. and D. D. Grossman, An experimental system for computer controlled mechanical assembly. In: Proc. IEEE Intercon. Session 13, New York, 1975, 18 pp.

145. Will, P. M., Computer controlled mechanical assembly. Research Report. IBM T.J. Watson Research Center, Yorktown Hts., NY, 1975, 31 pp.

146. Woo, T. C., Progress in shape modeling. Computer, Dec., 1977, 40-46.

147. Wortman, J. J. and L. K. Monteith, Semiconductor mechanical sensors. IEEE Trans. on Elec. Devices, Vol. ED-16, No. 10, Oct., 1969, 855-860.

ADDITIONAL REFERENCES

A1. Alexander, III, A. D., Impacts of telemation on modern society. In: Proc. 1st CISM/IFToMM Symp., Vol. II, On Theory and Practice of Robots and Manipulators, Udine, Italy, 1974, 121-136.

A2. Briot, M., La stereognosie en robotique application au tri de solides, Thése, Universite Paul Sabatier de Toulouse (Sciences), Nov., 14, 1977, 183 pp.

A3. Evans, J. M. et al., eds, Proc. NBS/RIA Workshop on Robotic Research, Williamsburg, VA, July 12-13, 1977, 34 pp.

A4. Goto, T. et al., Compact packaging by robot with tactile sensors. In: Proc. 2nd Int. Symp. on Ind. Robots, Chicago, IL, May 16-18, 1977, 149-159.

A5. Hanafusa, H. and H. Asada, A robot hand with elastic fingers and its application to assembly process. In: Information-Control Problems in Manufacturing Technology. Oxford: Pergamon, 1978, 127-138.

A6. Matsushima, K. and K. Hasegawa, Study on the industrial robot with adaptability. Bull. Tokyo Inst. of Tech., No. 123, 1974, 115-129.

A7. McGhee, R., Conclusions for the NSF Workshop on Robotics. In: Proc. NSF Workshop on the Impact on the Academic Community of Required Activity for Generalized Robotic Manipulators, U. of Florida, Feb. 8-10, 1978, 13-17.

A8. Shimano B. and B. Roth, On force sensing information and its use in controlling manipulators. Report, Mechanical Engineering Department, Stanford University, May, 1973.

A9. (Anoynmous), Preliminary work on implementing a manipulator force sensing wrist. Report, A. I. Lab., Stanford Univ., Stanford, CA.

A10. Tesar, D., Conclusions for the NSF Robotics Workshop. In: Proc. NSF Workshop on the Impact on the Academic Community of Required Research Activity for Generalized Robotic Manipulators, U. of Florida, Feb. 8-10, 1978, 8-12.

A11. Ueda, M. et al., Tactile sensors for an industrial robot to detect a slip. In: Proc. 2nd Int. Symp. on Ind. Robots, Chicago, IL, May 16-18, 1972, 63-76.

A12. Ueda, M. et al., Sensors and systems of an industrial robot. Memoirs of the Faculty of Engineering, Nagoya Univ., Vol. 27, No. 2, 1975, 163-207.

A13. Yonemoto, K., The present status and the future outlook of industrial robot utilization in Japan. Report, Japan Industrial Robot Assn., Nov., 1978, 46 pp.

Reprinted from Robotics Today, Spring, 1980

Adaptive Control

Giving Robots the Power to Cope

Adaptive control

frees the robot

to change its path

and speed

and thus perform

its assigned task

with optimum

efficiency and accuracy

HANS COLLEEN
Manager of Engineering
Industrial Robot Systems
ASEA Inc. (U.S.A.)

Can a robot be made more flexible? More versatile? And more able to adapt to change? The answer to all three questions is Yes. Somewhat paradoxically, the basic concept of a robot is that of a machine capable of repetitive motions, and thus suitable for performing some production or material handling function over and over again. It does this without tiring and without getting bored, while re-maining oblivious to its working environment.

But industry wants more than this from its growing investment in automation. In response to this demand, ASEA now offers its robots with adaptive control, *Figure* 1. It's a completely new type of control system in which one or more sensors provide information to the robot to modify and refine its movements. In effect, it provides the robot with a sense of touch.

Adaptive control offers the following advantages: less critical accu-racy during initial programming; fewer points required for any particular task; reprogramming not required for minor position changes; and less demand on peripheral equipment, which can be both simpler and less expensive.

ASEA's adaptive control now achieves both point adaptivity and curve adaptivity. Point adaptivity provides various ways for the robot to search for and find the position of a point, or points, not precisely determined in advance—for example, the starting point from which the robot will perform a programmed operation. It also provides for automatically correcting the positions of program points. ASEA robots have had this capability for years.

Curve adaptivity, recently made available commercially, allows the robot to work along a curve, or curves, not exactly localized but whose shape is defined by information from a sensor. This capability was added to ASEA robots in mid-1979.

Adaptive control, in short, refines the programmed pattern and speed of movement in response to changes in dimensions and location of the work-

1. Holding a workpiece against a fixed deburring tool, this ASEA robot makes precision moves with the aid of adaptive control.

piece(s), or to changes in the workplace. Principal applications include arc welding, grinding, deburring, and various assembly operations.

The fundamental principle behind ASEA's adaptive control system is that the robot's position or motion is directly modified by sensor signals

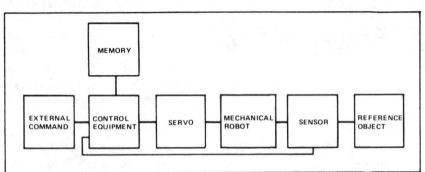

2 *Block diagram illustrates use of feedback signals from sensors to modify robot position and motion.*

which, in turn, are generated by the robot's movements during the work cycle. The particular manner in which the sensor signals influence the robot's movement pattern can be defined in the user program. A block diagram representing the adaptive control function is shown in *Figure* 2.

The sensor interface included in the adaptive control system is capable of receiving signals from up to eight three-level sensors with alarm limits. The control system, however, can only process signals from four sensors simultaneously.

The sensor signal is represented by two bits of digital information which can report four different states of the system, as indicated in *Figure* 3 and explained as follows:

00—signals no correction is needed

10—signals that a correction in the positive direction is needed

01—signals that a correction in the negative direction is needed

11—signals an "alarm" situation exists and needs attention

The capabilities associated with point adaptivity and curve adaptivity significantly enhance the robot's versatility. Point adaptivity provides four different types of searching for unknown locations and the capability of making automatic adjustments to the program based on feedback. The four types of searching include:

• Coarse searching. Used when the exact location of a position is not needed. The sensor(s) in the control program will interrupt the robot's movement in the programmed direction.

• Fine searching. Used when an exact position is required. Provides for stopping the robot's movement at the exact point where the change in the sensor signal occurs.

• Delayed searching. Interrupts the robot's programmed movement after the sensor(s) has been activated for a prescribed period. It can be used, for example, to find an outer corner using just one sensor perpendicular to the plane of movement. It is also useful for recognizing and distinguishing an outer corner or edge from variations in the shape of the curve or in the surface.

• Free searching. Used, for example, to find an inner corner or internal angle. It defines only the main direction of the search and stops the robot at the point where all of the several programmed sensors are satisfied.

Unlike standard systems, adaptive control permits searching along several spatial coordinate axes simultaneously. This makes it possible to carry out highly sophisticated searching operations and simplifies the design of fixtures and workpieces.

Curve adaptivity comprises both velocity control and contouring. Velocity control, primarily useful for deburring, permits selection of the optimal speed of robot movement for a particular operation. Even where the shape of the curve to be followed is well defined, there can still be a problem in choosing the best speed. In deburring a surface with burrs of different sizes and positions without adaptive control, the fixed speed selected must be low enough to handle the largest burrs encountered. This often means that the speed will be lower than it needs to be during most of the cycle. However, with adaptive

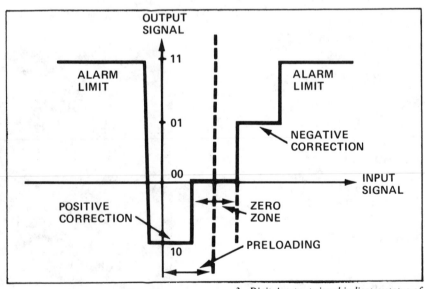

3. *Digital output signal indicates status of system and corrections required to achieve desired program.*

control providing speed correction, the sensors can furnish information by which the speed is modified according to the size and location of the burrs. Robot speed is varied for maximum efficiency. This eliminates unnecessarily slow robot movements,

4. *ASEA robot gripper with special deburring tool incorporating adaptive control sensor for deburring workpiece held in fixed position.*

reducing cycle time and increasing production.

Contouring provides for following a curve of unknown shape. It enables the robot to find and follow curves that are parallel to, or at an angle from, the programmed curve. The robot moves in the programmed direction and, when the sensors are activated by a deviation from this pattern, compensates so as to keep the sensors satisfied. Adjustment is performed with the help of a correction vector for each sensor direction. By selecting the appropriate sensors, the sensing direction can be made perpendicular to the programmed path.

Automatic program adjustment is an additional feature. By using point adaptivity and curve adaptivity together, it is possible to adjust the user program automatically. This can be done in real-time, i.e., while the robot is operating. Rough programs are prepared initially, using the conventional teaching method. When these are executed for the first time they are adapted automatically in the robot's memory, according to the actual conditions and positions encountered. This automatic program adjustment feature considerably reduces programming time and also allows the operator to use essentially the same program for pieces having similar, but different, shapes. This capability is especially valuable for arc welding small batches of parts, or parts having large tolerances. It is also useful where it is difficult or expensive to make several fixtures, or where it is extremely awkward to simultaneously execute a program and search for variations and make corrections.

Experience indicates it is almost impossible to develop a universal sensor; specials must be designed for individual applications. Special sensors have been developed and tested for arc welding and deburring.

Adaptive control has been tested extensively under laboratory and pilot scale conditions. It has also been applied to an actual production operation—the deburring of castings for the automotive industry. In this application the problem is to achieve a constant point of intersection between machined and cast surfaces. A robot without adaptive control is not able to handle such variations. For this application, a special deburring sensor was incorporated in the tool-holder mounted on the robot's hand, *Figure 4.*

Certain limitations must be accepted with adaptive control. It can only compensate for rather small variations—for example, those resulting from out-of-tolerance castings. Also, it is not suitable for weaving motions, such as in painting and certain kinds of welding.

The program for adaptive control requires somewhat more complex instructions than a standard program. This, in turn, reduces the space available in memory for the user program. The memory is limited to 4K words and cannot be doubled by connecting an extra read/write memory. However, with adaptive control, a simpler and shorter program can be developed which compensates for this limitation. And by using a tape recorder as a mass memory, the capacity can be increased up to thirty times. ∎

Presented at the Assemblex IV Conference, October, 1977

The Remote Center Compliance System And Its Application To High Speed Robot Assemblies

By Paul C. Watson
Consultant

A new class of devices has been devised which can be used in the wrist of a robot arm to convert a simple arm with limited degrees of freedom into a high speed assembler of precision parts. The assembly is accomplished simply and economically without any external energy source or sensory feedback, using the built-in mechanical preperties of the remote center device. A new theory of operation of this assembly system together with experimental results showing the exact forces and moments which occur during typical insertions is presented. The application of these devices to future assembly systems is also discussed.

INTRODUCTION

Remote Center devices such as the remote center compliance make it easy for machines to make assemblies which formerly were either difficult or impossible for machines to do, and which therefore required human operators. These devices are adaptive, in that they can accommodate variations in the work pieces and the assembler, and permit easy and reliable insertions and assemblies. In their present form, they can assemble chamfered parts with any clearance fit, no matter how high or low the precision of the particular fit is. The only requirement is that the presentation of the parts is within the limits of the chamfer size.

The present paper considers the basic assembly task of inserting a round peg into a round hole. This simple task can easily be generalized to include many others, such as driving threaded screws, placing objects with holes in them over pins, push fits, and many others.

The remote center device, which is usually in the form of a compliance, is designed to have an effective center or pivot point at the assembly interface, such that a force applied at the center does not produce rotation, and such that a moment applied to the piece does not produce displacement. These properties permit the device to adapt to misalignments of the hole or peg, and to offsets of either the hole or the peg or both. The contact and frictional forces produced at the contact points tend to move the peg into alignment and permit it to move into the hole. A detailed consideration of exactly how this happens forms the body of this paper.

The Stages of an Insertion

In order to show that the remote center device can meet the requirements for a low-force high speed insertion, we consider the detailed stages which can occur during the insertion. The forces which exist during each of these stages of the insertion can be calculated and measured. The following discussion is concerned with the problem of inserting a peg into a hole, using the remote center compliance. A coordinate system is used, located at the center of compliance and oriented along the undeflected direction of the peg. The insertion is then considered to be motion of the hole, onto the peg such that the deflections which occur are determined by the constraints introduced and by the forces and moments developed by these constraints.

As the hole moves onto the peg, it first strikes on the chamfer. The chamfer crossing can involve contact at the leading edge, the trailing edge, or on a line in the chamfer surface. In any case, we consider the cross section such that motion occurs to the right, in the plus x direction.

When the chamfer crossing is completed, the peg will be fully entered but not inserted any appreciable distance, as shown in Figure 3B. At this time, the page will have assumed an initial angle due to the action of the force at the contact point. As the insertion proceeds, branch occurs to either top or bottom point contact as shown in Figures 3C and 3E.

Top Point Contact

The top point contact occurs if the initial angle is greater than the tilt angle of the hole (measured counterclockwise). In the case of top point contact, the effect of deeper insertion is to rotate the peg more closely into alignment with the hole. Nonetheless, the right hand leading edge of the peg may strike the other side of the hole, producing the type of two-point contact shown in Figure 3D.

Top point contact can also lead to line contact, where the peg contacts the left hand side of the hole as shown in Figure 3G. In the case of top point contact, the tendency is to move toward line contact although a phase of positive type two point contact may intervene.

The tilt angle of the peg during top point contact is given by

$$\theta_{TP} = \frac{\dfrac{C_{22}D}{C_{11}}(G + \mu R)}{1 + \dfrac{C_{22}G}{C_{11}}(G + \mu R)} \; ,$$

and the insertion force is given by

$$F_{TP} = \frac{\mu}{C_{11}}(D + G\theta).$$

The quantities involved are defined in Figure 4A.

It will be noted from these equations that the initial tilt of the peg can be made to be zero by locating the center of the compliance at a distance μR above the end of the peg. If this is done, the force is only due to the displacement, and is given by $\mu D/C_{11}$.

Bottom Point Contact

In case the angle of the hole is greater than the initial angle of the peg, a condition of bottom point contact as shown in Figure 3E will be obtained. In the case of bottom point contact, the situation in respect to the peg changes very little as the insertion proceeds, and the forces and the peg angle stay nearly constant. However, if the right hand side of the peg strikes the top of the hole, a negative type of two point contact will be obtained as shown in Figures 3F and 4B.

The tilt angle of the peg in this phase is given by

$$\theta_{BP} = \frac{\dfrac{C_{22}}{C_{11}}(D + ZE)(A - \mu R)}{1 + \dfrac{C_{22}}{C_{11}} A \, (A - \mu R)} \; ,$$

and the insertion force is

$$F_{ZBP} = \frac{\mu}{C_{11}}(D + ZE - \theta A).$$

Once again, if $A = \mu R$, the force reduces to $\mu D/C_{11}$. The geometry for this condition is shown in Figure 4B.

Insertion Histories

As the peg is inserted, its possible wobble angle compared to the hole reduces. The limits of wobble are shown in the graph of Figure 6 as the positive two point contact, and the negative two point contact lines. These lines form a sort of funnel which will guide the insertion. Several possible initial angles of the peg are shown at A, B, and C. If the initial point is less than the hole tilt angle (below the center of the funnel), bottom point contact is obtained and the insertion proceeds along the heavy line starting from A.

If the initial angle is greater than the hole tilt angle, conditions such as those at B and C are obtained. In this case a top point line is followed. In case B, the peg tilts toward the left edge of the hole until a line contact is

achieved, and then stays in the hole without ever contacting
the other side. In case C, the two point contact line is
reached, and the rest of the insertion proceeds along the
"positive 2 point contact" line. If the hole is very deep, or
the clearance large, the history may leave the two point line
and move toward the hole tilt angle line as shown.

These histories represent the angular displacement in the
peg coordinates, and thus represent the moment applied to the
peg by the hole. Graphs of this type can be obtained experimentally by performing an instrumented insertion.

Another important characteristic of a particular insertion
is the insertion force history. The force history can be predicted and it also can be measured. The forces to be expected
in the various phases of the insertions are given in the discussions of those phases. We next consider the condition of
two point contact. In this condition, the hole is tilting the
peg compared to the natural angle it would assume if it were
not constrained by the hole. Depending on the direction of
rotation of the peg, there is either a positive type of two
point contact, with the peg tilted as far counterclockwise as
possible within the limits of the hole, or there is a negative
type with the peg being tilted clockwise until it contacts the
hole at two points. For the positive type of two point contact,
the situation shown in the next figure applies.

Positive Type Two Point Contact

We next analyze this type of contact to find the force
and moment histories.

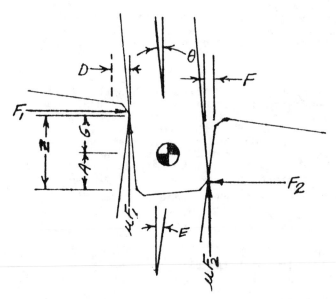

Figure 5A. Positive Type Two Point Contact.

From Geometry

$$X = D + G(E + \frac{F}{Z})$$

$$\theta = E + \frac{F}{Z}$$

406

The forces can be found from

$$\frac{\theta}{C_{22}} = (F_1 - F_2)(A - \mu R) - F_1 Z$$

$$\frac{X}{C_{11}} = (F_1 - F_2)$$

yielding

$$F_1 = \frac{1}{Z}\left[\frac{X}{C_{11}}(A - \mu R) - \frac{\theta}{C_{22}}\right]$$

$$F_2 = F_1 - \frac{X}{C_{11}}$$

If $F_1 > 0$, and $F_2 > 0$, necessary for this condition, the insertion force is

$$F_Z = \mu(F_1 + F_2), \text{ or}$$

$$F_Z = \mu\left[\frac{X}{C_{11}}\left(\frac{2}{Z}(A - \mu R) - 1\right) - \frac{2\theta}{Z C_{22}}\right]$$

Negative Type Two Point Contact

The force and moment histories for this contact can be obtained similarly:

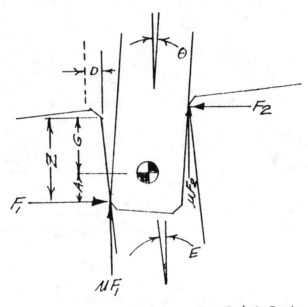

Figure 5B. Negative Type Two Point Contact.

From Geometry,

$$\theta = E - \frac{F}{Z}$$

$$X = D + EG + \frac{FA}{Z}$$

The forces are

$$F_2 = \frac{1}{Z}\left[\frac{\theta}{C_{22}} - \frac{X}{C_{11}}(A - \mu R)\right]$$

$$F_1 = \frac{X}{C_{11}} + F_2$$

If $F_1 > 0$, and $F_2 > 0$, necessary to obtain this condition, the insertion force is

$$F_Z = \mu\left[\frac{X}{C_{11}}\left(1 - \frac{2}{Z}(A - \mu R)\right) + \frac{2\theta}{Z C_{22}}\right]$$

Discussion

After the chamfer has been crossed, one point contact is obtained. If either bottom point or top point contact is obtained, it is clear that the insertion can proceed as long as the available insertion force is greater than that given in the expressions for each case. The important thing about these solutions is that the tilt angle of the peg is obtained thus allowing prediction of the exact depth at which two-point contact begins.

When two point contact exists, states of moving, jamming, and wedging are possible. These states have been and are being studied by Simunovic, and reference should be made to his work for a detailed discussion. It is sufficient here to mention that jamming is stoppage of motion due to insufficient or incorrect force, while wedging is associated with parts deformation and requires some active strategy to overcome.

Jamming would occur if the available insertion force is less than the amounts given in the two point analyses in this paper. It can be seen from the expressions that the critical phase of the insertion is at the smallest value of Z (insertion depth) at which two point contact exits. This is because the available moment arm to generate the required moment to tilt the peg into alignment with the hole has its smallest value, and the forces applied will necessarily become largest.

In particular, if two point contact is obtained at very small insertion depths, the insertion becomes critical and the exact depth determines the force required which may become very large. Also at these low depths, various nonlinearities and surface roughness effects become significant although they are not covered in this paper.

Conclusion

The results obtained in this paper allow the prediction of the actual insertion forces and moments during any desired insertion. The present approach also defines the phases of the insertion, and permits determination of the force and moment impulses necessary to accomplish the changes from one phase to another.

A digital simulation has been designed, allowing study of the various insertion histories under conditions of various clearances and initial errors in hole location and tilt. These constitute a useful design tool in the design of actual insertion systems for various purposes.

By adapting to parts and assembler variations, the remote center inserter can obtain the following advantages:

1) The possibility of damage to the assembly machine is minimized.

2) Parts damage is minimized, reducing the scrapping of parts due to failed assemblies, and also minimizing the possibility of later warranty and service parts due to completed assemblies with damaged parts being sold.

3) Prevents down time due to jams of assembly processes.

4) Avoids scrapping of marginal parts which are actually usable.

5) Makes possible automation which was previously impossible.

Acknowledgement

The author wishes to express his gratitude to all his former colleagues at the Charles Stark Draper Laboratory for discussions and interactions which have contributed to the present results. Special mention should be made of the work of S. Simunovic who has done extensive work in the force friction analysis field, and also of S.H. Drake who has done similarly extensive work in the field of compliant assembly.

References

1. Nevins, J., D. Whitney, M. Dunlavey, S. Drake, D. Killoran, A. Kondoleon, C. Mogged, D. Seltzer, S. Simunovic, S. Wang, and P. Watson, "Exploratory Research in Industrial Modular Assembly, Fifth Report," C.S. Draper Report R-1111, August 1977.

2. Drake, S.H., Paul C. Watson, and Sergio N. Simunovic, "High Speed Robot Assembly of Precision Parts Using Compliance Instead of Sensory Feedback," presented at the proceedings of the 7th International Robot Symposium, Toyko, 1977.

3. Simunovic, S., "Force Information in Assembly Processes," presented at the 5th International Symposium in Industrial Robots, Chicago, September 1975. Proceedings published by Society of Manufacturing Engineers.

4. Watson, P.C., "A Multidimensional System Analysis of the Assembly Process as Performed by a Manipulator," presented at the 1st North American Robot Conference, Chicago, October, 1976.

5. Watson, P.C., and S.H. Drake, "Pedestal and Wrist Sensors for Automatic Assembly," presented at 5th International Symposium on Industrial Robots, Chicago, 1975.

6. Lynch, P.M., "Economic-Technological Modeling and Design Criteria for Programmable Assembly Machines," Ph.D. Thesis, M.I.T. Mechanical Engineering Department, June 1977.

7. Drake, S.H., "Using Compliance in Lieu of Force Feedback for Automatic Assembly," Ph.D. Thesis, M.I.T. Mechanical Engineering Department, August 1977.

APPENDIX

Digital Simulation

A digital simulation of the insertion has been designed. As an illustration of some results, the following case is shown:

Radius of Peg	6 mm
Location of Center (A)	3 mm
Hole offset (D)	1 mm
Hole tilt (E)	.017 rad (1°)
Diametral Clearance (F)	.024 mm
Mu	.3
Delta Z	.5 mm per step
No. of points	10

Using these parameters, the results which follow were obtained. In this case the initial angle is halfway between the top point and bottom point values or .00927. Since this is less than the hole tilt (.017), bottom point contact is obtained and the insertion proceeds until the bottom point contact angle intersects the two point contact negative angle, at the tenth point, Z equal to 5 mm. By varying the parameters in the simulation, all of the various cases can be studied, and the effects of varying the center location and the effects of piece parts tolerances can be studied.

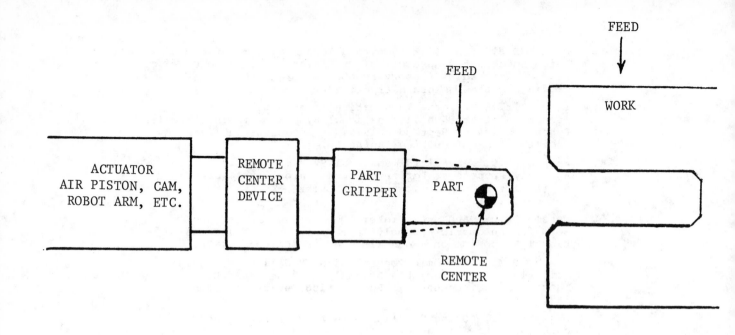

Figure 1 TYPICAL INSERTION CONFIGURATION

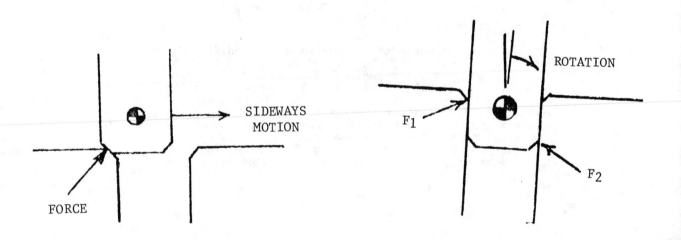

Figure 2A RESPONSE TO FORCE

Figure 2B RESPONSE TO MOMENT

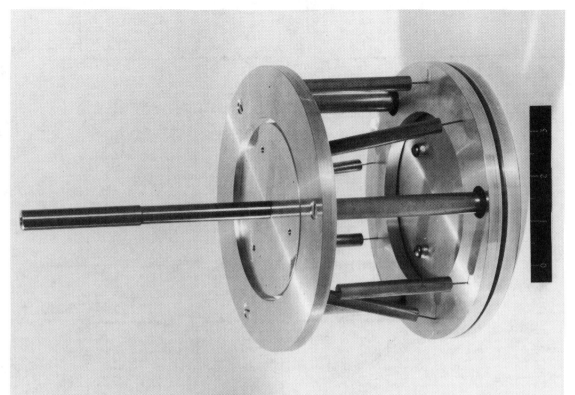

Fig. 1C Conceptual Model of A Remote Center Inserter.

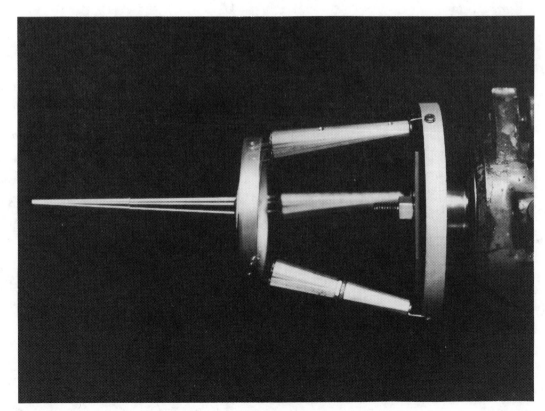

Fig. 1A Operating Principle of One Form of Remote Center Device

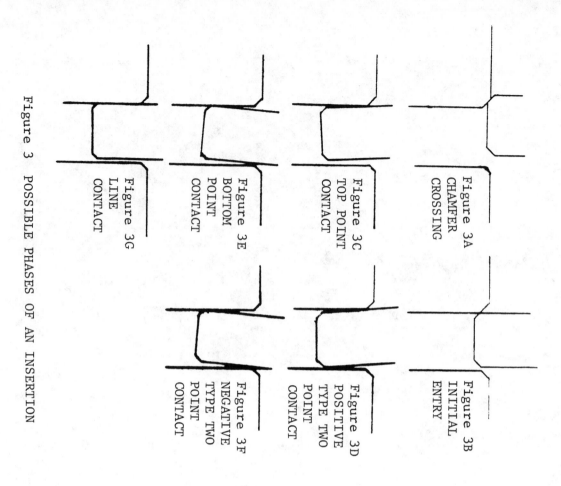

Figure 3A
CHAMFER
CROSSING

Figure 3B
INITIAL
ENTRY

Figure 3C
TOP POINT
CONTACT

Figure 3D
POSITIVE
TYPE TWO
POINT
CONTACT

Figure 3E
BOTTOM
POINT
CONTACT

Figure 3F
NEGATIVE
TYPE TWO
POINT
CONTACT

Figure 3G
LINE
CONTACT

Figure 3 POSSIBLE PHASES OF AN INSERTION

Figure 4A TOP POINT CONTACT GEOMETRY

Figure 4B BOTTOM POINT CONTACT GEOMETRY

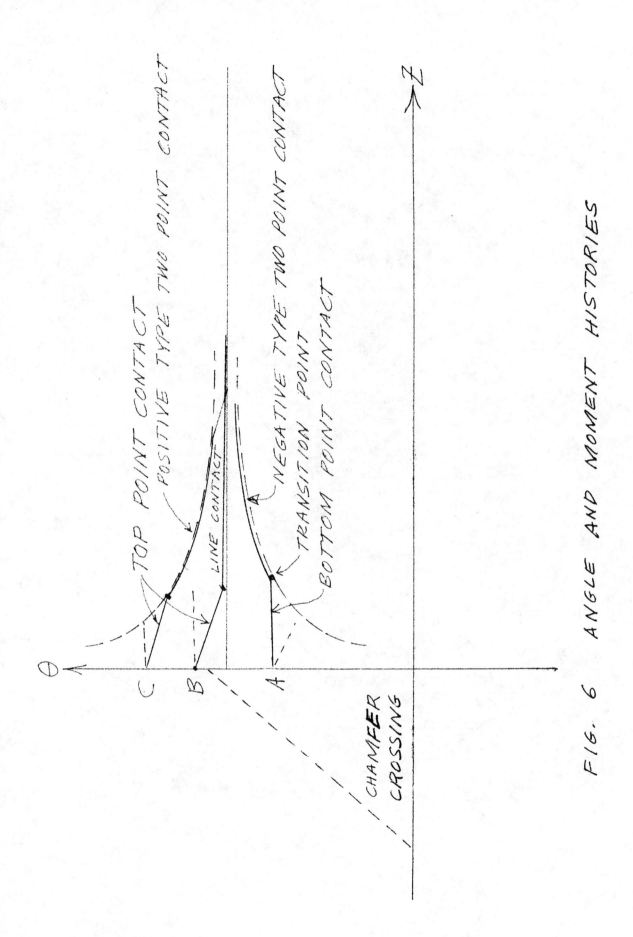

FIG. 6 ANGLE AND MOMENT HISTORIES

413

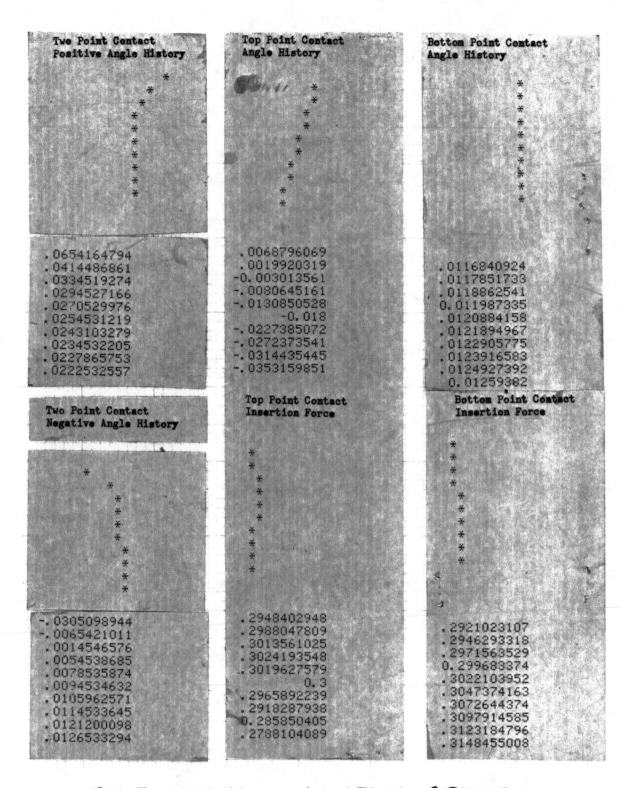

Two Point Contact Positive Angle History

```
                 *
              *
             *
             *
             *
             *
             *
             *
             *
             *
             *
```

```
.0654164794
.0414486861
.0334519274
.0294527166
.0270529976
.0254531219
.0243103279
.0234532205
.0227865753
.0222532557
```

Top Point Contact Angle History

```
               *
               *
              *
              *
             *
             *
            *
            *
           *
           *
```

```
 .0068796069
 .0019920319
-0.003013561
-.0080645161
-.0130850528
        -0.018
-.0227385072
-.0272373541
-.0314435445
-.0353159851
```

Bottom Point Contact Angle History

```
               *
               *
               *
               *
               *
               *
               *
               *
               *
               *
```

```
.0116840924
.0117851733
.0118862541
0.011987335
.0120884158
.0121894967
.0122905775
.0123916583
.0124927392
0.01259382
```

Two Point Contact Negative Angle History

```
           *
           *
           *
           *
           *
           *
           *
           *
           *
           *
```

```
-.0305098944
-.0065421011
.0014546576
.0054538685
.0078535874
.0094534632
.0105962571
.0114533645
.0121200098
.0126533294
```

Top Point Contact Insertion Force

```
         *
         *
         *
         *
         *
         *
         *
         *
         *
         *
```

```
.2948402948
.2988047809
.3013561025
.3024193548
.3019627579
        0.3
.2965892239
.2918287938
0.285850405
.2788104089
```

Bottom Point Contact Insertion Force

```
           *
           *
           *
           *
           *
           *
           *
           *
           *
           *
```

```
.2921023107
.2946293318
.2971563529
0.299683374
.3022103952
.3047374163
.3072644374
.3097914585
.3123184796
.3148455008
```

FIG. 7   DIGITAL SIMULATION RESULTS

414

Presented at the Robot IV Conference, October/November 1979

The Remote Axis Admittance—A Key to Robot Assembly

By Paul C. Watson
Assembly Associates

The Remote Axis Admittance (RAA)[1] is a new device to perform the fine motion of parts mating. It can be mounted on the end of a robot arm, and will perform the final phases of assembly when the parts are in close proximity and when they are in contact. The RAA can be adapted to a number of different types of tasks, including the simple insertion, such as the chamfered peg and hole, the insertion of edges in slots, the multiple insertion, and the chamferless insertion. The RAA has optional built-in sensing and actuation, depending on the task, and uses a microcomputer type of controller to monitor and switch its different modes of operation. The RAA also has built-in safety features, and inherent internal damping.

INTRODUCTION

The first adaptive mechanism which made the practical robot inserter possible was the author's remote center compliance (RCC) mechanism (Ref. 2). The RCC performs the simple insertion of chamfered round pieces by a simple passive mechanism. The present paper describes an improved assembler, the RAA which can be adapted to a wider variety of assembly tasks.

The RAA operates by projecting virtual axes of rotation to locations at or near the assembly interface. These axes permit the contact forces and torques induced by the mating parts to produce motions of rotation, translation and tilting in directions to perform the assembly. In cases where straight line motion is required, the corresponding axis is projected to a large distance (infinity). The motions of the various axes are sensed and used to operate mechanical mode switching according to instructions from an optional microcomputer. In the case of more complex types of assembly tasks, it is also possible to power the axes, to produce programmed motions. The RAA is also arranged to yield under overload conditions, preventing damage to the assembler and the parts.

The RAA can be used together with many different types of robot arms to perform parts fetching and assembly.

TASK DESCRIPTIONS AND PARTS MATING

The work considered in the remainder of this paper consists of assembly of rigid objects. Previously reported results described the solution of the basic assembly task of inserting a peg into a hole with the aid of chamfers. This task and other assembly tasks are separated into a gross motion portion and a fine motion portion. The gross motion consists of the large motions necessary to obtain the part to be assembled and bring it close to its destination. Gross motions are easily performed by almost all of the robot arms available today since the required accuracy is relatively low. Fine motion occurs when the parts are in contact or almost in contact, and is used to accomplish the actual assembly or mating of the parts. The fine motion part of the assembly, however, is more difficult since the fine motion consists of the actual mating or insertion of the parts to be assembled and thus requires motions which are comparable to the tolerances and clearances of the parts.

Parts mating, or the fine motion part of assembly actually consists of three separate things. The first is the definition of the parts themselves and how they are to be assembled. This includes the shape of the parts, the tolerances, and the possibilities of defective parts. All of these must be specified, to permit the design of an appropriate system.

When the parts have been defined, a strategy or method is required to determine a sequence of forces or actions which will accomplish their assembly. This strategy corresponds to software in a computer, and represents a series of instructions or procedures which will accomplish the desired assembly. Just as in a computer, there are phases in a typical assembly corresponding to the various states of the assembly. When the assembly is in these phases, transitions are necessary to get from each state to the following state. These transitions correspond to branch points or decision points in the computer program.

In some cases, the branch points can be accomplished in the hardware. This occurs in the RAA chamfered insertion, since the states involved were

1. Approach

2. Chamfer crossing

3. Insertion

and the transition from 2 to 3 occurs automatically as a result of the inherent decoupling of the motions.

When the strategy, consisting of a set of operations which will accomplish the required assembly has been determined, the third aspect of parts mating can be designed. This is the hardware which is arranged to accomplish the required strategy. In some cases, the hardware can be passive, and in other cases it may be necessary to either employ switching of states or even active feedback, depending on the nature of the task.

Recapitulating, parts mating consists of

1. Specifcations of the parts to be assembled.

2. Definition of a strategy or procedure for accomplishing the assembly.

3. Hardware (and software, if appropriate) to accomplish the strategy.

In most cases, the hardware can consist of various members of the RAA family.

THE REMOTE AXIS ADMITTANCE [1]

The RAA (Remote Axis Admittance) is a generic name for a new family of assembly devices to perform fine motion parts mating tasks. Various RAA's can be adapted depending on the type of task to be performed. The principle of the RAA is a projected Axis or center line which can serve as a center of tilting or rotation. The basic element which projects a single center line can be concatenated with other elements to provide other axes, such that the total motion supported by the RAA contains all of the components desired. If a number of axes are projected, a virtual mechanism or mechanical admittance can be caused to exist in space. The way in which such axes or center lines are obtained is shown in Fig. 1. This figure shows four connections, which can bend to permit the part and gripper to rotate around either of the two axes defined by the intersections of the projections of the connections. Each of these axes corresponds to one degree of freedom. If the spacing of one axis is allowed to become large (approaching infinity), the corresponding motion appoaches translation.

Most assembly tasks require two 2-axis RAA elements, providing four degrees of freedom.

The RAA element is provided with an elastomeric filling which provides initial tension to the connections, and which also provides damping for the motion. The principle of the 2-element RAA is shown in Fig. 2. The RAA elements are also provided with mechanical locks such that any of the axes can be disabled as required. This permits adjustments of the axis location according to the requirements of the pieces being assembled.

The RAA has the following features:

1. Design flexibility for different assembly tasks.

2. Overload protection. In case of jamming due to any cause such as defective parts, the RAA yields without harm, preventing damage to the assembler.

3. Damping.
   The RAA is provided with a means of damping unwanted oscillations. This avoids any awkward clamping and unclamping mechanism as was required with the RCC and enhances the speed of operation.

4. Sections of one RAA may be used to provide different motions. Alternately, sections may be concatenated to provide a sequence of operations.

THE INSTRUMENTED RAA

Since rotations and translations induced by the assembly

forces are inherently decoupled by the RAA, it is possible to attach pick-up transducers to measure the forces and torque components independently. When a branch point in the operation is required, the corresponding force component can be monitored, and used to establish the branch point at the proper time. If the gross motion device is an arm such as a multijointed robot arm, the branch points can be used as inputs to the arm controller, obtaining the desired assembly sequence.

If either controllable locking devices, or possible servo actuators are attached to desired axes of the RAA, a self contained fine motion system can be obtained which can operate independently of its gross motion device. This permits operation of the RAA assembly system using nearly fixed automation such as a pick and place machine.

Locking devices are used to introduce branch points in the assembly procedure. In some cases, the inherent decoupling of the RAA device is not sufficient to provide the necessary transitions between the phases of the insertion. In these cases, the sections of the RAA may be deactivated by clamping them mechanically.

Additionally, it may be necessary to add fine components of motion in some of the axes. This is accomplished by using either stepper meters or servos, driving the corresponding axes.

Computer Controlled Branch Points

If the instrumented RAA signals are fed into a computer, calculations and decisions can be made during the assembly. When changes are required from one mode to another, or commands are required to control the gross motion device, the computer output can be used. If a powered RAA is employed, the computer output can be used to control the actuators of the powered RAA. A microprocessor is ideal for these uses, and produces a fully flexible "smart" assembly system.

A system block diagram showing the arrangement of the instrumented and powered RAA is shown in Fig. 3.

THE SIMPLE CHAMFERED INSERTION

We next consider some other tasks which can be accomplished with the RAA. The basic assembly task is the chamfered insertion of a peg into a hole. This task has been analyzed by the author and by others (Ref.3 and 6) and the conditions for jam and wedge-free insertions have been obtained. In order to accomplish this task, two RAA elements are used, producing two crossed rotational axes to allow tilting of the peg in any plane, and two distant axes to allow lateral translation. The translational part of the RAA system is used to permit a chamfer to guide the piece to be inserted into the entrance of the hole. After the piece has entered the hole, the rotational part of the RAA takes

over, and permits the piece to align itself with the hole. In
this case, no sensing or power source is required,since the
two phases of the operation occur automatically. In the first
phase, the chamfer crossing, the chamfer is crossed without
causing tilt of the part, since the net force is lateral. In
the second phase, the object is past the chamfer, and the tilt
is induced by the corresponding contact force or forces (Ref.3
contains discussions and solutions for the details of this insertion).

THE CHAMFERLESS INSERTION

Once the hole has been entered, no matter how slightly,
the standard RAA conditions can guarantee the insertion without
jamming or wedging. The problem, then is to locate the hole and
achieve the initial entry. This is done by tilting the peg, as
shown in Fig. 4A. The degree of tilt is such that the positioning
errors of the robot, plus the parts tolerance of the parts,
permit the corner of the peg to be inside the hole as shown in
Fig. 4A. This is achievable, no matter how close the fit is.

The RAA is in a state such that axis A-A' permits tilting
of the peg around itself, and the axis B-B' permits translation
in the direction into and out of the paper. The robot is now
moved in the direction M1 until contact is made in the region
E-E'. This condition is shown in Fig. 4B. This contact is
sensed by instrumented axes. When the motion in direction M1
is complete, the part will be seated in a direction in and out
of the paper.

Depending on the part, and the clearance, contact may be
made in the region F-F'. Motion M1 is continued until points
G and G' come into contact. This is detected by the increased
reistance to motion M1. The effect is either to tilt the
part into the hole without contact in the region F-F', or else
to contact in the region F-F' and slide and rotate the part
until it is jammed but not wedged in the hole.

A branch is now made to the insertion mode, and the part
pulled into the hole until it bottoms. Depending on the clearance,
and the coefficient of friction, other strategies are possible.

THE EDGE IN A SLOT INSERTION

A strategy similiar to that used in the chamferless insertion
can insert an edge in a slot. In the round chamferless insertion,
the transverse motion required is translation to locate the part
in the Y-Y' direction. In the edge in a slot insertion,an additional degree of tilting may be required, to permit entry of
the entire edge. After one corner of the edge is entered, a
second degree of translation and tilting,similiar to that
used in the round chamferless insertion occurs, thus completing
the entry.

NON-SYMMETRICAL AND MULTIPLE INSERTIONS

In the case of the non-symmetrical insertion, the optimal locations of the tilting axes depend on the separation of the contact points and the effective clearance ratio. Since the RAA can have different axis locations in different azimuths, it can adapt to this situation and provide an optimal insertion.

Conclusion

The Remote Axis Assembler presented in this paper represents a simple and effective approach to a variety of fine motion assembly tasks. It can be used in conjunction with a robot to provide an extremely flexible and effective assembler.

Because the assembler can adapt to parts and assembler variations, and because it has built-in overload protection, the RAA system produces the following advantages:

1. The possibility of damage to the assembly machine is minimized.

2. Parts damage is minimized, reducing the scrapping of failed assemblies, and reducing the incidence of warranty claims.

3. Down time due to jams of the assembly process is reduced.

4. Can adapt to marginal parts, insuring utilization of all actually useable parts.

5. Makes possible automation which was previously impossible.

6. Has Programmability, permitting use with different styles of products.

Future work consists of application of the principles of the RAA to individual designs for numerous different types and sizes of real world parts.

References

1. Watson, P.C., and Drake, S.H., "Pedestal and Wrist Force Sensors for Automatic Assembly" Proceedings of the 5th International Symposium on Industrial Robots, Chicago, Sept. 1975, pp. 501-512

2. Watson, P.C., "A Multidimensional System Analysis of the Assembly Process as Performed by a Manipulator" S.M.E., Dearborn, Mich., 1976 (Technical Paper MR76-613)

3. Watson, P.C., "The Remote Center Compliance System and its Application to High Speed Robot Assemblies" S.M.E., Dearborn, Mich., 1977 (Technical Paper AD77-718)

4. Lynch, Paul M., Economic-Technological Modeling and Design Crteria for Programmable Assembly Machines

Charles Stark Draper Laboratory, Cambridge, MAss., June 1976 (Report T-625)

5. Abraham, R.C., Stewart R.J.S., and Shum, L.Y., "State-of-the-Art in Adaptable-Programmable Assembly Systems" S.M.E., Dearborn, Mich. 1977 (Technical Paper MS77-757)

6. Simunovic, Sergio N., An Information Approach to Parts Mating, Doctor of Science Thesis, M.I.T., Cambridge, Mass., 1979

7. Hill, John W., "Force Controlled Assembler", S.M.E., Dearborn, Mich., (Technical Paper MS77-749)

[1] Patent applied for. The various RAA devices covered in this paper are covered in the application.

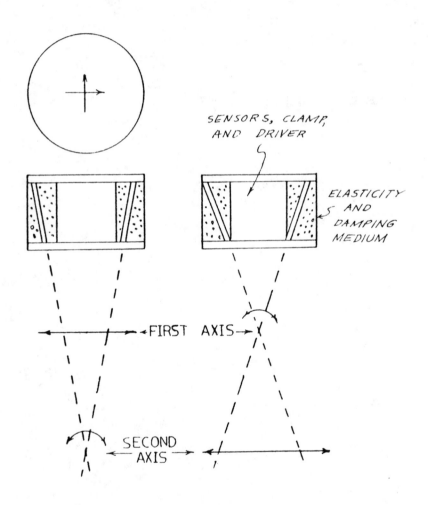

FIG 1 SINGLE ELEMENT R. A. A.

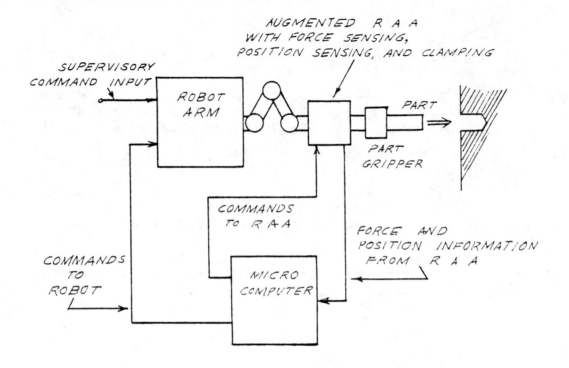

Fig. 3   Augmented RAA System Block Diagram

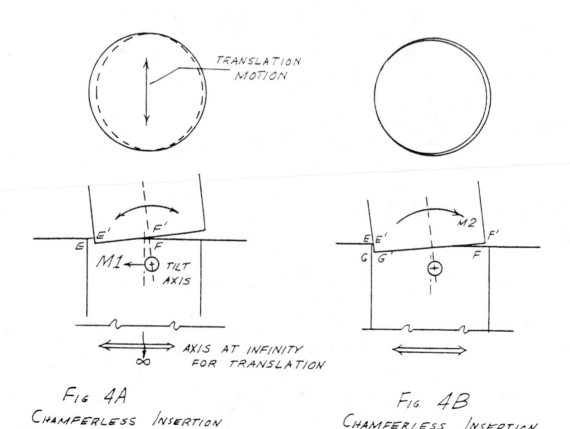

Fig 4A
Chamferless Insertion
Phase 0

Fig. 4B
Chamferless Insertion
Phase 1

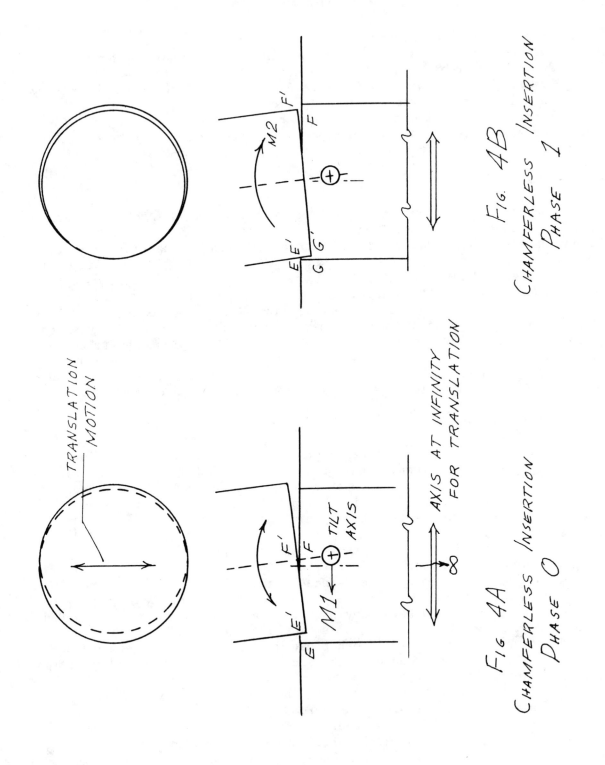

TRANSLATION MOTION

AXIS AT INFINITY FOR TRANSLATION

∞

M1

TILT AXIS

E'

F'

F

E

FIG 4A

CHAMFERLESS INSERTION

PHASE 0

E E'

G G'

M2

F'

F

FIG. 4B

CHAMFERLESS INSERTION

PHASE 1

Presented at the Robots IV Conference, October, 1979

Use of Sensory Information for Improved Robot Learning

By Donald S. Seltzer
Charles Stark Draper Laboratory, Inc.

Robots "learn" or acquire knowledge by three different methods. Currently, the most popular method is through being "taught" by an operator during an on-line preparation phase. A second method is the use of off-line geometric data bases and software aids. A third, and as of now rarely used method is learning from on-line experience.

The role of tactile sensory information in aiding learning is discussed for all three cases. A new type of tactile sensor developed in part for this purpose is described.

1.0 INTRODUCTION

Robots are by nature programmable devices. They are embodied with a number of useful abilities, such as generalized motion. These general purpose capabilities allow robots to be manufactured in quantity for end use in a wide range of applications. To make use of its general abilities in the performance of a specific task, the robot must first learn what to do, by acquiring application-specific information. This information includes both the logical procedure to be followed, and physical data, usually geometric, concerning the task. There are a number of methods by which robots can acquire task specific knowledge. On the simplest level, they can be taught by being literally led through the motions. A more sophisticated approach is to explain to the robot what it is to do. Finally, the robot can be allowed to learn some things by itself while on the job.

1.1 Teach By Showing

Teach by showing was the earliest method developed for teaching industrial robots, and it remains the predominant method used today. Teach by showing involves leading the robot through the desired motions, usually with the aid of a teach pendant or joy stick. Generally, only the endpoints of trajectory segments are recorded within the robot's memory, though in some cases, as with paint spraying robots, the entire path is recorded.

In the simplest systems, the arm must be manually driven, axis by axis. Straight line motion can be difficult to achieve. This is particularly true of rotary joint arms, since straight line motion requires the simultaneous movement of several axes at different rates. Consequently many robot arms on the market today can be driven in several user oriented coordinate frames. The most common is "world coordinates", a Cartesian frame fixed to the work area. Another useful Cartesian frame is hand coordinates, which are referenced to the end effector and continually shift as the arm moves.

The primary advantage of the teach by showing method is its simplicity. Relatively little training of personnel is required. Because no prior programming experience is needed, teaching and operating can be left to blue collar shop floor workers. Besides keeping support costs down, it probably lessens union resistance to the introduction of this new technology.

Another advantage is that minor program touchups can be done quickly, with minimal turnaround time.

An inherent weakness with teach by showing is that logic and data are too closely linked. Task logic is a specification of the sequence in which the different arm functions are to be carried out. For a simple robot, this might be a list of arm motions and tool functions to be performed. Distinct from this is the specification of real world data, such as actual position coordinates. With teach by showing, pushing the record button on the teach pendant both specifies the logic (a motion is to occur) and defines the data (the endpoints of the trajectory). By not keeping the two separate, it becomes difficult to modify an existing taught program. For many industrial robots in use today, an addition or deletion of a motion in the middle of a program necessitates the complete reteaching of all subsequent program steps.

An additional disadvantage is that all teaching is done on the production site. Thus, all associated production facilities are tied up during the teaching operation.

1.2. Off-line Programming

It is obviously advantageous if some of the logic and data specification can be done in advance without tying up production equipment. This off-line programming corresponds to the second method of robot learning, telling it what to do. The capability for this off-line preparation requires increased system computer resources. With the continual decrease of hardware costs, however, more robot systems are being offered with off-line teaching. Examples include Unimation's VAL[1] and Olivetti's SIGMA[2] systems. Cincinnati Milacron's T^3 robot[3] also has some limited off-line programming features.

Off-line programming can encompass just the logic, just the data, or both. Off-line specification of the task logic generally implies a high level, problem-oriented language. As a minimum, this language must include commands for arm motions and end effector operations. Additional commands can add to the system's versatility and ease of use. Examples include I/O statements, branching instructions, and arithmetic operations.

Use of an off-line programming language offers a number of advantages over teach by showing. Describing a task in a high level language serves to better define and document the process. Various software tools help to simplify the job and reduce errors. Subroutines and macros, for instance, can be used to define frequently repeated subtasks, shortening what would otherwise be a tedious and time-consuming teaching chore. Editing routines and symbolic reference to data allow easier modification of existing programs. Finally, the more powerful off-line programming languages allow the flexible use of sensor information for adaptive control.

These additional capabilities come at the expense of increased hardware and software costs. In particular, there is the danger of replacing blue collar production workers with higher paid white collar programmers. To a certain extent, this problem can be alleviated by the utilization of a programmer hierarchy, in which the programming chores are divided into several levels, permitting the use of production workers for the operation and for some of the programming of the robot[4].

Task associated data can also be generated in advance off-line. The data may come from an existing part geometry data base such as might be generated by a CAD (Computer Aided Design) program. Alternatively, the data may be defined by way of an algorithm which permits it to be computed at run-time, such as in a palletizing operation. Or, the data might be entered directly by the programmer, derived from his own knowledge of the task to be performed.

A problem with off-line generated data is that it is only an approximation of the real world. Real objects are not totally uniform, but vary in both their dimensions and their locations. Even robots vary significantly from unit to unit in terms of absolute accuracy and repeatability. Consequently, significant errors often exist, so that some on-line correction is necessary. Typically, this is done by "touch up" teach by showing.

1.3. On-line Experience

The final technique of robot learning is from on-line experience. As stated in the previous section, off-line generated data is generally an imperfect representation of the real world. These imperfections are revealed during attempted execution of the task. Some form of "touchup" teaching is required.

This does not mean that data generated on-line, through teach by showing, is immune from the problem. On the contrary, arm motions that are taught under quasi-static conditions will probably not repeat exactly during high speed playback. If the playback errors are large enough, some touchup may be required.

Once a program has been correctly taught, it is unlikely to run forever without further corrections. Over a long term, tools, fixtures, and jigs wear or shift in position. Even robot arms will experience wear, affecting their repeatability. New batches of parts may exhibit dimensional variations. Consequently, it is often necessary to modify the robot's initial data, based upon on-line performance. Currently, this is done by an operator who first determines that a correction is required, and then performs the necessary reteaching. In principle, it is possible to equip the robot with sensory devices that will allow it to learn from its own experience.

2.0. ROLE OF TACTILE INFORMATION

Perhaps the most difficult tasks to teach robots are those involving contact between rigid parts. Examples can be found in assembly, fabrication, and in some material handling operations. The difficulty arises because small position and orientation errors generate very large forces, causing jamming of parts, and possibly damage to the parts, tools, fixtures, or the robot.

During on-line teaching, the operator is guided only by his sense of sight. This is often unsatisfactory for a number of reasons.

a. The operator must be close to the work interface. In crowded work areas, this can be difficult and hazardous to the operator.

b. The position and orientation errors which give rise to these errors can be small, and not easily detected by eye.

c. Transforming the visual information into corrective arm motions is not an easily learned operator skill.

d. The region of contact, which provides the error information, may be obscured by the arm, fixturing, or the parts themselves.

Sensing devices, to supplement the operator's information, would be of obvious use. Vision, to a certain extent, only duplicates the operator's own sight. There are some advantages in respect to safety — a camera can be located in an area that is unsafe for a person. Cameras can also be mounted in places inaccessible to a person, such as the end of a robot arm. However, camera vision is as useless as human vision when the region of contact is obscured.

Force sensing devices, on the other hand, can provide the tactile information which the operator totally lacks. For common tasks such as insertion and edge-following, large predictable forces and moments can be measured and used in a corrective feedback loop, in a strategy called active accommodation[5,6]. A successful commercial example of active accommodation is Hitachi's Hi-T-Hand, a device for performing close tolerance insertions [7]. Besides insertion-type tasks, active accommodation can be used for orientation of one surface with respect to another, and edge following, useful for grinding, routing, and welding operations.

Tactile information can also be utilized in passive accommodation techniques, in which the error forces themselves cause the proper realignment of the parts in contact. Passive accommodation has been successfully demonstrated in the form of the RCC, or Remote Center Compliance[8,9,10,11, 12]. The RCC is a compliant mechanical structure, capable of movement in five degrees of freedom. During insertion-type tasks, the forces and moments arising from position and orientation errors cause structural deflections in the RCC which correct for these errors. In a sense, the RCC is a complete mechanical force feedback control loop. Its advantage over active accommodation methods is that it is cheaper and faster. The RCC structure is of low mass and has a high mechanical bandwidth, allowing it to perform tight insertions in less than .2 seconds. Active feedback schemes, on the other hand, are limited by the much lower servo bandwidth of the entire robot arm, and typically require three to five seconds to accomplish the same task.

Unlike the active force sensing approach, however, the RCC can not measure the errors it encounters, an ability useful for robot learning purposes. This suggests the possibility of a marriage of the two techniques, a combination of an RCC and a force sensor feedback loop. The RCC would act as a fast error absorber. The force sensor would measure the error and correct the data base.

The feasibility of such an approach was investigated at Draper Labs in the latter part of 1978[13]. Several assembly tasks were taught on-line for an automobile alternator, using a four axis electrically driven arm, with a compliant wrist. A six d.o.f. pedestal sensor measured the forces applied to the workpieces. A force display unit allowed the operator to watch and correct for the error forces as he taught the tasks.

A conclusion of this study was that force sensing does provide useful information for teaching purposes. It can be used as feedback to the operator during teach by showing, or direct feedback to the control computer for automatic steering and correction.

It was also noted during this study that the movements of the compliant structure were an excellent source of visual feedback for the operator. If, for instance, position misalignment caused the compliant wrist to deflect to the left, the proper corrective action would be to steer the arm in that direction. These visual clues provided information essentially similar to the force displays, and actually more convenient for the operator to use.

3.0. THE INSTRUMENTED REMOTE CENTER COMPLIANCE

The results of the above mentioned study suggest the usefulness of being able to measure the state of the RCC. The lateral deflections of the RCC represent lateral position errors, while angular deflections are a measure of angular error.

Use of a force sensor is one means of measurement. Because the RCC consists of a number of linear spring elements, there is a linear relationship between applied forces and the RCC's state. There are certain practical problems, though, that arise from this method of implementation. The normal working range of the RCC is only a few pounds, yet typical robotic applications will subject it to several hundred pounds of force. Rigid mechanical stops protect the RCC from damage form these excessive loads. It is more difficult to design protective stops for a force sensor. If the sensor is instead built to sustain these peak forces, it is unlikely to have sufficient resolution to measure the forces in the RCC's normal operating range. Furthermore, the compliance of the force sensor may adversely affect the operation of the RCC.

An alternative is to directly instrument the RCC. An instrumented RCC, or IRCC, will act as a high bandwidth error absorber, and will sustain large force overloads as does a simple RCC. It will also provide sensory information similar to that of a force sensor.

An RCC consists of a rigid mounting strucutre which attaches to the robot, a second rigid structure to which the end effector attaches, and a compliant structure which joins the two. Instrumenting the RCC requires the measurement of the relative position and orientation of the second structure with respect to the first. This can be done with position sensing transducers, preferably of a non-contacting type. The number of transducers must be at least equal to the number of degrees of freedom being measured. A greater number may also be used, to improve the quality of the data or to lessen non-linear effects.

RCC's typically have five degrees of freedom. There are two lateral degrees of freedom in the plane perpendicular to the main axis of the RCC, which can be denoted by X and Y. There are also two rotational degrees of freedom, θx and θy. The final degree of freedom is θz, rotation about the axial, or Z direction. In many applications, it is unimportant to monitor θz, thus reducing the number of required transducers to four.

There are numerous ways to position these transducers, but a preferred arrangement is shown in Figure 1. The four transducers shown are fixed to the mounting structure (not shown) and measure distance to the center shaft of the end effector structure. Alternatively, the transducers could be mounted on the end effector structure.

In the preferred arrangement, two of the transducers, X_1 and Y_1, are mounted in an XY plane, aligned with the X and Y axes. A second set of transducers, X_2 and Y_2 are similarly aligned in a parallel plane located a distance S from the first. This second plane is also a distance L from the center of compliance.

These four transducers are sufficient to determine X, Y, θx, and θy. A fifth transducer must be added if it were necessary to measure θz.

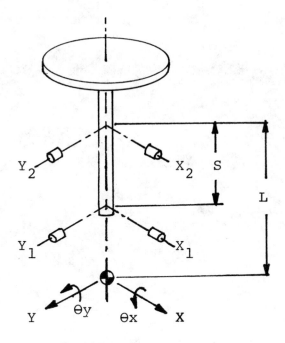

FIGURE 1.

Using standard linear approximations, it can be shown that the relationship between the state of the RCC $(X, Y, \theta x, \theta y)$ and the transducer outputs (X_1, X_2, Y_1, Y_2) is:

$$
\begin{bmatrix} X \\ Y \\ \theta x \\ \theta y \end{bmatrix} = \begin{bmatrix} L/S & 1-(L/S) & 0 & 0 \\ 0 & 0 & L/S & 1-(L/S) \\ 0 & 0 & -1/S & 1/S \\ 1/S & -1/S & 0 & 0 \end{bmatrix} \begin{bmatrix} X_1 \\ X_2 \\ Y_1 \\ Y_2 \end{bmatrix}
$$

Two implementations of the four d.o.f. IRCC have been built to date. One utilizes analog transducers, while the other uses digital sensor elements. Both are based upon the Draper Lab standard Model 4B RCC.

3.1.   Analog IRCC

The analog unit contains four proximity detectors manufactured by Kaman Scientific. These transducers work on the principle of inductive coupling with nearby metal surfaces. Output from the electronic drive and detection circuitry is a 0-2 volt DC signal proportional to displacement. The four transducers are mounted on the end effector member of the IRCC,

Figure 2. Analog Instrumented Remote Center Compliance. Two of the proximity sensors show as dark dots within their plastic support structure. Proximity to the housing is measured.

measuring its position and orientation with respect to the mounting case (Figure 2). The associated electronics are remotely located. The units selected have a normal operating range of 0-80 mils (2 mm), but are being used over an extended range of 0-200 mils (5 mm) with somewhat reduced linearity. Useful resolution appears to be better than .5 mils.

3.2. Digital IRCC

The second IRCC that was built utilizes optical devices for measuring displacement. Each of the four displacement sensors consists of a LED light source, a photosensitive element, and a thin metal shutter. The light source and photosensor are mounted on the fixed case of the IRCC with the LED illuminating the surface of the photosensor. The shutter, attached to the end effector mounting structure, is located between the LED and photosensor, partially shadowing the photosensor. Deflections of the compliant structure cause the shutter to move, varying the amount of light which reaches the photosensor (Figure 3).

A simple silicon photovoltaic cell could have been used as the photosensitive element. Instead, a Reticon linear photodetector was chosen. The Reticon device is a linear array of 256 photoelements located on .984 mil (25 microns). Self-contained timing and addressing circuitry allow the individual photo elements to be accessed as if they comprised a shift register. The output of each cell is a pulse whose amplitude is proportional to the incident light. Those that are illuminated by the LED have a high amplitude, while those cells which are covered by the shutter produce essentially zero output. The output of each array is a

Figure 3. Digital Instrumented Remote Center Compliance. The LED's and photosensor circuits are located on the outer structure, while the shutter is attached to the center shaft.

train of 256 pulses, the number of which are of full amplitude being proportional to the displacement of the shutter. A comparator converts these pulses to binary logic levels. A complete scan of the arrays is currently done 50 times a second. The upper limit on the scanning rate, governed by the photoarrays, is 4000 times a second.

The pulse trains from the four arrays are connected to a series of sum and difference counters. Four outputs are produced, $(X_1 + X_2)$ and $(Y_1 + Y_2)$ which are representative of the X and Y displacements; and $(X_1 - X_2)$ and $(Y_1 - Y_2)$ which are proportional to θx and θy rotations. The data output interface for each is a parallel three digital BCD number.

3.3   Other Implementations

The first two prototypes of the IRCC were designed specifically for the Model 4B unit, which has a translational range of .2 inches (.5 cm). Much larger ranges are possible with similar transducers. The Kaman proximity detectors are readily available in ranges up to 1.2 inches (3 cm). Reticon sensor arrays are manufactured in sizes up to 2.048 inches (5.2 cm). There are numerous design alternatives, including analog photo position detectors, linear potentiomenters, and LVDT's, which would permit measurement ranges of up to 2 feet (.6 m).

4.0.   THE IRCC AS AN AID TO ROBOT LEARNING

The IRCC is a unique device offering the advantages of both passive and active accommodation techniques. As a passive device, it is a high speed error absorber. As an active device, it is capable of measuring relative position errors with resolution on the order of .1 mm, and angular errors to 10^{-3} radians. It can be used advantageously in conjunction with

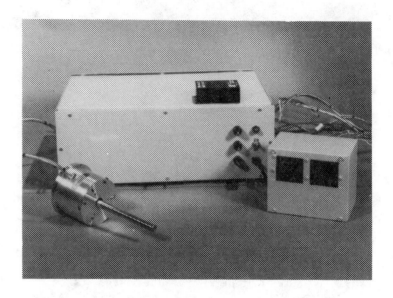

Figure 4. Position and angle errors can be read by the operator, using the IRCC, power and logic module, and visual display unit shown above.

all three methods of robot learning.

When used as an aid to on-line teach by showing, the compliant nature of the IRCC will allow the operator to quickly accomplish difficult tasks involving part contact. The instrumented portion of the device will then measure remaining position and angle errors, allowing further refinement of the arm endpoint location. This could be accomplished in either of two ways. A visual display unit such as the one pictured in Figure 4 will give direct feedback to the operator. Or, the error signals can be directly read by the control computer for automatic error nulling. In either case, teaching should progress more rapidly. Initial teaching will be more accurate, lessening the likelihood that extensive touchup teaching will be required.

Use of the IRCC during teaching should also improve user safety. The additional feedback information may allow the operator to accomplish teaching from a safe distance, avoiding the all too common practice of standing next to the work interface.

The IRCC can also correct for errors in data bases created by off-line programming methods. One type of error is characterized by a position and/or angular offset in the reference frame, due to misalignment of the fixturing with respect to the robot, or inaccuracies in the robot itself. This error can be compensated for by an initial calibration procedure in which the arm is required to drive to a number of reference positions located on the work area. The IRCC can measure the error at each reference location, providing sufficient data for the computation of the coordinate frame misalignment.

Individual locations may also need correction due to inaccurate models used to represent real parts. If the error is within the range of the IRCC, it can be measured and corrected.

In a similar manner, the IRCC can assist in robot learning through on-line experience. Used as a performance monitor, it will be able to detect errors during operation. Problems such as tool and fixture wear, robot drift, or part batch variations will be evident due to error readings consistently in the same direction. This information can be used to automatically update the data base without halting production.

5.0. SUMMARY

A new type of tactile sensing device, the Instrumented Remote Center Compliance, has been developed. The IRCC combines the best features of active and passive accommodation techniques. It can be used advantageously in conjunction with the three methods of robot learning: on-line preparatory, off-line programming, and on-line experience.

ACKNOWLEDGEMENT

The author wishes to acknowledge the contributions of T. L. De Fazio, J. Nevins, and P. Watson in developing the concept of the IRCC, and of E. Consales, J. Ford, and R. Roderick in the design and construction of working models.

This work was funded by the National Science Foundation under grant no. ATA74-18173 A03.

REFERENCES

1. "User's Guide to VAL, A Robot Programming and Control System," Unimation, Inc., Danbury, Conn., February 1979.

2. Salmon, M., "SIGLA: the Olivetti Sigma Robot Programming Language," 8th International Symposium on Industrial Robots, Stuttgart, Germany, 1978.

3. Cunningham, C., "Robot Flexibility Through Software," 9th International Symposium on Industrial Robots, Washington, D.C., 1979.

4. Seltzer, D., "A Hierarchical Programming Approach to Robot Assembly," PROLAMAT '79, Ann Arbor, Mich., May 1979. SME paper MS79-180.

5. Whitney, D., "Force Feedback Control of Manipulator Fine Motions," Journal of Dynamic Systems, Measurement, and Control, Trans. ASME, June 1977, pp. 91-97.

6. Whitney, D., and J. Nevins, "Servo-Controlled Mobility Device," U.S. Patent 4,156,835, filed May 29, 1974, issued May 29, 1979.

7. Goto, T., T. Inoyama, and K. Takeyasu, "Precise Insert Operation by Tactile Controlled Robot 'HI-T-HAND' Expert 2," 4th International Symposium on Industrial Robots, Nov. 1974.

8. Watson, P.C., "A Multidimensional System Analysis of the Assembly Process as Performed by a Manipulator," Proceedings of the 1st North American Robot Conference, Chicago, Oct. 1976.

9. Drake, S., "Using Compliance in Lieu of Sensory Feedback for Automatic Assembly," Doctor of Science thesis, MIT Mech. Eng. Dept., Sept. 1977.

10. Whitney, D., and J. Nevins, "What is the Remote Center Compliance and What Can It Do?," 9th International Symposium on Industrial Robots, Washington, D.C., 1979.

11. Watson, P., "Remote Center Compliance System," U.S. Patent 4,098,001, filed Oct. 13, 1976, issued July 4, 1978.

12. Drake, S., and S. Simunovic, "Compliant Assembly System Device," U.S. Patent 4,155,169, filed March 16, 1978, issued May 23, 1979.

13. Nevins, J., et al., "Exploratory Research in Industrial Modular Assembly," C.S. Draper Laboratory Report No. R-1218, Sept. 1978.

Manufacturers and Distributors

Programmable, Servo-Controlled Point-to-Point Robots

The American Robot Corporation
P.O. Box 10767
Winston-Salem, North Carolina 27108

(919) 748-8761

Armax Robotics, Inc.
27888 Orchard Lake Road
Farmington Hills, Michigan 48018

(313) 964-3311

ASEA Inc.
4 New King Street
White Plains, New York 10604

(914) 428-6000

Bra-Con Industries
12001 Globe Road
Livonia, MI 48150

(313) 591-0300

Cincinnati Milacron
4701 Marburg Avenue
Cincinnati, Ohio 45209

(513) 841-8189

Cybotech
P.O. Box 88514
Indianapolis, Indiana 46208

(317) 298-5893

General Numeric Corporation
390 Kent Avenue
Oak Grove Village, Illinois 60007

(312) 640-1595

PRAB Robots, Inc.
5944 East Kilgore Road
Kalamazoo, Michigan 49003

(616) 349-8761

Reis Machines
1426 Davis Road
Elgin, Illinois 60120

(312) 741-9500

Thermwood Corporation
P.O. Box 436
Dale, Indiana 47523

(812) 937-4476

Unimation Inc.
Shelter Rock Lane
Danbury, Connecticut 06810

(203) 744-1800

United States Robots
1000 Conshohocken Road
Conshohocken, Pennsylvania 19428

(215) 825-8550

Programmable, Servo-Controlled Continuous Path Robots

Advanced Robotics Corporation
Newark Ohio Industrial Park, Bldg. 8
Hebron, Ohio 43025

(614) 929-1065

Automatix Inc.
217 Middlesex Turnpike
Burlington, Massachusetts 01803

(617) 273-4340

Binks Manufacturing Company
9201 West Belmont Avenue
Franklin Park, Illinois 60131

(312) 671-3000

Cybotech
P.O. Box 88514
Indianapolis, Indiana 46208

(317) 298-5893

DeVilbiss
300 Phillips Avenue
Toledo, Ohio 43692

(419) 470-2169

Hobart Brothers Co.
600 West Main Street
Troy, Ohio 45373

(513) 339-6011

Nordson Corporation
555 Jackson Street
Amherst, Ohio 44001

(216) 988-9411

Shin Meiwa
5-25 Kosone-cho 1 chome
Nishinomiya-shi
Hyogo-ken
JAPAN

(0798) 47-0331

Thermwood Corporation
P.O. Box 436
Dale, Indiana 47523

(812) 937-4476

Programmable, Non-Servo Robots for General Purpose

Armax Robotics, Inc.
27888 Orchard Lake Road
Farmington Hills, Michigan 48018
(313) 964-3311

Automated Systems Engineering
49763 Leona Drive
Mt. Clemens, Michigan 48045
(313) 949-6960

Auto Place/Copperweld Robotics
1401 East 14 Mile Road
Troy, Michigan 48084
(313) 585-5972

Gametics, Inc.
15645 Sturgeon
Roseville, Michigan 48066
(313) 778-7220

Industrial Automates, Inc.
6123 West Mitchell Street
West Allis, Wisconsin 53214
(414) 327-5656

Manca/Leitz Inc.
Leitz Building
Rockleigh, New Jersey 07647
(201) 767-7227

Mobot Corp.
980 Buenos Ave.
San Diego, California 92110
(714) 275-4300

PRAB Robots, Inc.
5944 East Kilgore Road
Kalamazoo, Michigan 49003
(616) 349-8761

Seiko Instruments Inc.
2990 West Lomita Boulevard
Torrance, California 90505
(213) 530-3400

Wear Control Technology Inc.
189-15 Station Road
Flushing, New York 11358
(212) 762-4040

Programmable Non-Servo Robots for Die Casting and Molding Machines

The Rimrock Corporation
1700 Rimrock Road
Columbis, Ohio 43219
(614) 471-5926

Sterling Detroit Company
261 East Goldengate
Detroit, Michigan 48203
(313) 366-3500

Mechanical Transfer (Pick-and-Place) Devices

Automation Devices, Inc.   (814) 474-5561
Automation Park, P.O. Box AD
Fairview, Pennsylvania 16415

Richard D. Dane Corporation   (617) 387-1775
45 Appleton Street
Everett, Massachusetts 02149

Fraser Automation   (313) 979-4500
37900 Mound Road
Sterling Heights, Michigan 48077

I.S.I. Manufacturing Inc.   (313) 294-9500
31915 Groesbeck Highway
Fraser, Michigan 48026

The Kent Machine Company   (216) 928-2125
4445 Allen Road
Stow, Ohio 44224

Livernois Automation Company   (313) 278-0200
25200 Trowbridge
Dearborn, Michigan 48123

R&I Manufacturing Company Inc.   (203) 283-0127
91 Prospect Street
Thomaston, Connecticut 06787

Robomation Corporation   (313) 547-5130
P.O. Box 652
Madison Heights, Michigan 48071

The Van Epps Design & Development Company Inc.   (216) 661-1337
4760 Van Epps Road
Cleveland, Ohio 44131

Robotic Organizations

British Robot Association
Mr. T.E. Brock
International Fluidics Services Ltd.
35-39 High Street
Kempston
Bedford MK42 7 BT
UNITED KINGDOM

Association Francaise de Robotique Industrieele
   (AFRI)
Mr. J. Chabrol
Secretary General
60 Allee de la Foret
92360 Meudon de la Foret
FRANCE

Societa Italiana perla Robotics Industriale
   (S.I.R.I.)
c/o Professor M. Samalvico
Instituto di Elettrotechnica ed Elettronics
Politechnico di Milano
Piazza Leonardo da Vinci 32
20133 Milano
ITALY

Japan Industrial Robot Association
Mr. K. Yonemoto
Executive Director
Kikai Shinko Kaikan Building
3-5-8 Shiba-koen
Minato-ku
Tokyo 105
JAPAN

Robot Institute of America
Bernard Sallot
Executive Director
One SME Drive
P.O. Box 930
Dearborn, Michigan 48128

Robotics International of SME
One SME Drive
P.O. Box 930
Dearborn, Michigan 48128

Belgian Institute for Regulation & Automation (BIRA)
ROBOTICA
Mr. Frank Denis
FN Industry
Rue de Page 69/75
B-1050 Broxelles
BELGIUM

Swedish Industrial Robot Association (SWIRA)
c/o Mekanforbundet
Storgt. 19
S-114 85 Stockholm
SWEDEN

INDEX

M

Machine descriptions, **268-271**
Machine loading, 215-216, 240
Machine tool operations, 138
Machining, **152-157**, 202
Maintenance, 49, **114-124**, 164, 218
Management
  back-up, 84
  commitment, 181-182
  direction, 129
  indifference, 183
  middle, 76-77
  monitoring, 184
  operations, 173
  plant, 77-78
  public relations, 190-197
  reassessment of measures, 187
  resistance, **169-175, 181-189**
  shop support, 60
  support, 182-183
  undermining, 183
  upper, 76
Manipulator system, 258
Material savings calculation, 143
Material transfer, 138
Mechanical transfer devices, 27
Medium technology, 210
Memory, 16, 17, **21-26**
Metal inert gas seam welding, 241
Microprocessor, 9, 16
MIG, See: Metal inert gas seam welding, 241
Milling center, 92
Milling machine, 92
Minicomputer, 9, 243
Mirror image, 234-235
Molding, 138
Monitor, **347-350**
Motorcycle frames, 209
Moving base line tracking, 299
Moving line applications, **294-308**
Munitions, 164

N

National Bureau of Standards, 369
NBS, See: National Bureau of Standards
Negative type two-point contact, 407
Non-contact (remote) touching, 385-386
Non-economic factors, 127-128
Non-path related functions, 238
Non-servo robots, **6-7**
Non-symmetrical and multiple insertions, 420
Number installed, 21-26

O

Object detection, 341
Objectives, 62
Occupational Safety and Health Act, 30, 36, 165
Off-line programming, 425
Off-the-shelf automation, **36-40**
Off-the-shelf purchases, 72
On-line experience, 426
Operation phase, 349-350
OSHA, See: Occupational Safety and Health Act
Output contacts, 86-87
Output signal, 402
Overspray, 143

P

Packing corrugated shipping containers, 70
Painting, 151, 267
Palletizing, 140, 283
Palletizing conveyor, 283
Parent equipment modifications, 64
Part acquisition experiments, **369-374**
Part programming, 348
Pattern recognition, **386-387**
Payback time, 146
Pedestal force sensor, 328
Perspective depth measurement, 354-355
Pick and place, 19, 29
Pickup problem, 348
Pinn mill, 290
Pitch, 15, 319
Plant layout, 75, 81-82
Plant management, 77-78
Plastic molding, 140, 163
Plastics, **214-215**
Pneumatic control systems, 280
Pneumatic maintenance program, 118-119
Pneumatic nut-runners, drills and impact wrenches, 222
Point-to-point robots
  classified, 5
  common characteristics, 9
  control system, 225
  significant features, 9
Pose correction, 366
Post perturbations, 364
Positioning accuracy, 17, **21-26**
Positive type two point contact, **406-417**
Potential, 55
Potential application, **21-26**
Power, 17, **21-26**